The Semantics and Proof Theory of the Logic of Bunched Implications

APPLIED LOGIC SERIES

VOLUME 26

SCOPE OF THE SERIES
Logic is applied in an increasingly wide variety of disciplines, from the traditional subjects of philosophy and mathematics to the more recent disciplines of cognitive science, computer science, artificial intelligence, and linguistics, leading to new vigor in this ancient subject. Kluwer, through its Applied Logic Series, seeks to provide a home for outstanding books and research monographs in applied logic, and in doing so demonstrates the underlying unity and applicability of logic.

The titles published in this series are listed at the end of this volume.

The Semantics and Proof Theory of the Logic of Bunched Implications

by

DAVID J. PYM
University of Bath, U.K.

KLUWER ACADEMIC PUBLISHERS
DORDRECHT / BOSTON / LONDON

A C.I.P. Catalogue record for this book is available from the Library of Congress.

ISBN 1-4020-0745-0

Published by Kluwer Academic Publishers,
P.O. Box 17, 3300 AA Dordrecht, The Netherlands.

Sold and distributed in North, Central and South America
by Kluwer Academic Publishers,
101 Philip Drive, Norwell, MA 02061, U.S.A.

In all other countries, sold and distributed
by Kluwer Academic Publishers,
P.O. Box 322, 3300 AH Dordrecht, The Netherlands.

Printed on acid-free paper

Printed in the Netherlands.

Contents

List of Figures

List of Tables

Preface

This is a monograph about logic. Specifically, it presents the mathematical theory of the logic of bunched implications, **BI**: I consider **BI**'s proof theory, model theory and computation theory. However, the monograph is also about informatics in a sense which I explain. Specifically, it is about mathematical models of resources and logics for reasoning about resources.

I begin with an introduction which presents my (background) view of logic from the point of view of informatics, paying particular attention to three logical topics which have arisen from the development of logic within informatics:

- Resources as a basis for semantics;
- Proof-search as a basis for reasoning; and
- The theory of representation of object-logics in a meta-logic.

The ensuing development represents a logical theory which draws upon the mathematical, philosophical and computational aspects of logic. Part I presents the logical theory of propositional **BI**, together with a computational interpretation. Part II presents a corresponding development for predicate **BI**. In both parts, I develop proof-, model- and type-theoretic analyses. I also provide semantically-motivated computational perspectives, so beginning a mathematical theory of resources.

I have not included any analysis, beyond conjecture, of properties such as decidability, finite models, games or complexity. I prefer to leave these matters to other occasions, perhaps in broader contexts. However, I should remark that progress has already been made on some of these topics and I provide the appropriate references. Indeed, in all respects, the work presented herein should be considered merely a first step towards an understanding of bunched logics and a mathematical theory of resources.

Since this work is a research monograph, not a textbook, I have taken the liberty of assuming, without taking very much care to be uniform in my assumptions, that readers will have some background in logic, topology, algebra, category theory, the semantics and implementation of programming languages, and a general knowledge of computer science.

Acknowledgments

I am most particularly grateful to my students, Pablo Armelín, Jules Bean and Samin Ishtiaq, and to Peter O'Hearn and Hongseok Yang for many very helpful discussions about this work and for pointing out numerous errors and omissions. The following papers have all contributed greatly to this monograph, although it contains much more besides:

- [O'Hearn and Pym, 1999] and, later, [Pym, 1999], which originated the logic **BI**;
- [Ishtiaq and Pym, 1998, Ishtiaq and Pym, 1999];
- [Pym et al., 2000] by Pym, O'Hearn and Yang; and
- [Armelín and Pym, 2001].

Moreover, many conversations with Pablo, Samin, Peter and Hongseok have helped to resolve many technical and conceptual difficulties.

I am also grateful to many other colleagues for many very helpful discussions about this work and for pointing out various faults. They include Marcelo Fiore, Dov Gabbay, Didier Galmiche, Chris Hankin, James Harland, Martin Hyland, Guy McCusker, Daniel Méry, Dale Miller, Gordon Plotkin, John Power, Uday Reddy and Edmund Robinson. The comments of several anonymous referees have also been helpful. I also thank Dov Gabbay for his encouraging of this work. The typesetting has been done in LaTeX [Goossens et al., 1994, Taylor, 2002].

Much (though not all) of the work presented herein was carried whilst I was at Queen Mary, University of London, as Reader in Logic and then Professor of Logic, during which time I was also an EPSRC Advanced Fellow. I am grateful to both Queen Mary and the EPSRC for their support.

In a long and both technically and conceptually complex original work such as this, it would be naïve to expect an absence of errors or omissions. Any remaining errors or omissions are my responsibility alone.

Foreword

I very warmly welcome David Pym's book on Bunched Implications. In the past three decades, logic has been deeply influenced by its extensive applications in computer science. Not only do we have new logical systems and logical methodology at play but also the relative balance of old logic subjects has shifted.

Of special importance is the prominence which intuitionistic logic and linear and substructural/resource logics have achieved in computer science, a well as their algebraic–topological and their categorical semantics. On the methodological front the combination, fibring and products of logics as well as their fibred semantics have emerged as a new methodology in the landscape of logics.

This book studies in detail the combination of intuitionistic and substructural implications, given the name "bunched implications". It examines a variety of proof-theoretic formulations for the logic as well as the major types of semantics for them; the possible world Kripke semantics, the algebraic–topological semantics and the categorical semantics.

The book is among the first examples of an in-depth study of one of the new kinds of logics.

It is very natural to "hit" upon the logic of bunched implications. Let $\mathbb{L}_1$ and $\mathbb{L}_2$ be two logics with implication $\Rightarrow_1$ and $\Rightarrow_2$. Assume these logics are characterised by semantics and models of the form $\mathbf{M}_1 = (S_1, \mathbb{A}_1, a_1, h_1)$ and $\mathbf{M}_2 = (S_2, \mathbb{A}_2, a_2, h_2)$, where S_i is a set of possible worlds, $a_i \in S_i, h_i$ is the assignment to the atoms and $\mathbb{A}_i$ is a family of relations/functions used to define the recursive truth table for the connectives of $\mathbb{L}_i$.

Combining the two languages allows us to form the language $[\mathbb{L}_1, \mathbb{L}_2]$ where wffs can be formed by freely using connectives both from $\mathbb{L}_1$ and from $\mathbb{L}_2$. Thus we can form for example the mixed formula $A = (p \Rightarrow_1 (q \Rightarrow_2 p))$.

There are various ways of providing semantics for the combined language, ranging from the most general fibred semantics (where minimal interaction exists between the languages) to products (where essentially the languages are required to commute). A very common combination is *dovetailing*. The semantics for dovetailing has the form $(S, \mathbb{A}_1, \mathbb{A}_2, a, h)$, obtained by putting both semantical conditions $\mathbb{A}_1$ and $\mathbb{A}_2$ side-by-side and joining the requirements on h of both logics. Fibring, or dovetailing, or forming products of logics is an automatic methodological recipe and is done in the same way to any two logics.[1]

If we perform dovetailing on intuitionistic $\to$ with the Kripke semantics $(S, \sqsubseteq, h)$ and on substructural $-\!\!*$ with the semigroup semantics $(S, \cdot, e, h)$ we automatically get the semantics of the form $(S, \sqsubseteq, \cdot, e, h)$

[1]It may be illuminating for the reader to see how this works. The idea is very simple. Consider $A = (p \Rightarrow_1 (q \Rightarrow_2 p))$. From the point of view of language $\mathbb{L}_1, A$ has the form $p \Rightarrow_1 X$, where X is atomic. $\mathbb{L}_1$ does not recognise $X = (q \Rightarrow_2 p)$, because $\Rightarrow_2$ is not in the language. Let $\mathbf{M}_1 = (S_1, \mathbb{A}_1, a_1, h_1)$ be a model of $\mathbb{L}_1$ and start evaluating $t \vDash_1 A$, for $t \in S_1$. In the inductive course of evaluation of $\Rightarrow_1$, we will have occasion to evaluate $s \vDash X$ for some points $s \in S_1$ appropriately related to t via the relations and functions of $\mathbb{A}_1$. If X were a real atom of $\mathbb{L}_1$, then the assignment h_1 would have given us the value. But $X = (q \Rightarrow_2 p)$ is not a real atom. How do we get a value for $s \vDash_1 X$? The answer is that we *fibre* a (possibly set of) model(s) of the language $\mathbb{L}_2$, with each point $s \in S_1$. Let $\mathbb{F}_{1,2}$ be the fibring function and write $\mathbb{F}_{1,2}(s) = \mathbf{M}_2^s = (S_2^s, \mathbb{A}_2^s, a_2^s, h_2^s)$ and let

$$s \vDash_1 X \text{ iff } a_2^s \vDash_2 X \text{ (in } \mathbf{M}_2^s).$$

The model $\mathbf{M}_2^s$ knows how to give a value to X.

The above is *fibred* semantics for the combined language. The function $\mathbb{F}_{1,2}$ assigning to each s a model $\mathbf{M}_2^s$ is a *fibring function*. Of course we also need an $\mathbb{F}_{2,1}$ for passage from $\mathbb{L}_2$ models to $\mathbb{L}_1$ models.

Dovetailing is obtained by insisting that $s = a_2^s$. A little calculation shows that we can take models of the form $(S, \mathbb{A}_1, \mathbb{A}_2, a, h)$ and evaluate $\mathbb{L}_i$ connectives using $\mathbb{A}_i$ respectively.

Note that if $\mathbb{L}_1$ and $\mathbb{L}_2$ come with proof procedures, a similar way of fibring or dovetailing their proof procedures can be given. We deal, in the proof theory, with databases Δ in the combined language and we have two proof modes, the $\mathbb{L}_1$ and the $\mathbb{L}_2$ modes. The fibring functions $\mathbb{F}_1$ and $\mathbb{F}_2$ act on databases Δ and yield databases $\mathbb{F}_1(\Delta)$ and $\mathbb{F}_2(\Delta)$. If we want to prove from Δ a formula B whose main connective is in $\mathbb{L}_2$, then $\mathbb{F}_2(\Delta)$ tells us which parts of Δ is available to use in the proof. These ideas are systematically developed in my book: D. M. Gabbay, *Fibring Logics*, Oxford University Press, 1998.

The combination of logics is done methodologically and not on a particular logic to logic basis. Thus we have no choice in what we get, once the components to be combined are given!

As an example of what can be done, one of the chapters of my book shows that dovetailing any logic with Łukasiewicz infinite-valued logic is an automatic way of making it fuzzy (in the sense commonly adopted by the fuzzy logic community).

satisfying the condition below (arising from the persistence of intuitionistic connectives):

$$x \sqsubseteq x' \text{ and } y \sqsubseteq y' \text{ imply } x \cdot y \sqsubseteq x' \cdot y'.$$

This condition is put forward in Section 2.3 of Chapter 1 of the book.

Thus the logic of bunched implications is not just a favourite logic with a wide range of applicability, but is a very natural dovetailing of two very well known implications, and its semantical interpretation and proof theory should be methodologically obtainable from their respective semantics and proof theory.

This book is an in-depth study of some properties of this combination, and is therefore also, in addition to its other qualities, a serious contribution to the area of fibring logics!

It is most welcome to our series.

Dov M. Gabbay, FRSC
Augustus De Morgan Professor of Logic
King's College London

Introduction

David J. Pym

Informatics may be defined as *the science of the structure, complexity and communication of information.* As such, informatics is concerned with the study of the structure, behaviour, interactions and construction of natural, artificial and abstract systems. It has philosophical, mathematical, computational and social aspects. It has emerged in the wake of the (electronic) computer and the central function of the computer, the transformation of information, is its unifying notion.

Logic is usually defined as the *science of reasoning.* We suggest that a better definition is that it is the study of the *structure of information.* It has its roots in *grammar* and the *semantics of natural languages* but has been of central importance in *mathematics* and its *foundations.* With the growth of the computing-driven sciences, *i.e.*, of informatics, logic has developed in new and challenging ways. Building on the established model theory and recursion theory, informatics has driven new emphases on proof theory and constructivity.

Entirely new concerns have arisen, however, with the following being leading examples:

- Program logic: We want to reason about the behaviour of programs. The leading example of program logic is Hoare's logic [Apt, 1989]. The basic propositional assertions are

  PRE-CONDITION **Program** POST-CONDITION.

  The PRE-CONDITION and POST-CONDITION are assertions in a predicate logic and **Program** is a procedure written in a (typically imperative) programming language. The pre- and post-conditions may, for example, be assertions about the computer memory used by the procedure.

- Processes and Nets: Computational systems consist in networks of communicating devices. Each device, such as a processor, a computer, a printer, a scanner or a user, must interact with its peers. For this interaction to occur, the devices must both exchange and process information. To facilitate reasoning about the behaviour of such networks, two leading logical formalisms have been proposed, namely Petri Nets [Reisig, 1998] and Process Calculi [Milner, 1975, Hoare, 1985, Milner, 1989, Milner, 1999]. Nets provide a graphical description of networks and their connectivity of devices whereas process calculi use algebraic methods to model the transmission of information.

- Resources: Whenever a procedure executes, resources are consumed. Resources may, for example, be spatial, such as a computer's memory, temporal (such as CPU cycles) or monetary (such as the coins required to obtain goods from a vending machine). More delicately, resources may also be dynamic, such as processes [Hansen, 1973]. The challenge, the addressing of which is begun herein by the development of **BI**, *the logic of bunched implications*, is to provide a mathematical model which adequately describes such apparently diverse phenomena and to obtain from it a logic which may be used to reason about resources.

- Logical frameworks: When a programmer writes a program he describes to a computer a model of an external phenomenon, *i.e.*, the application. Logically, we may think of the corresponding situation in which one logic (the object-logic) is represented by another (the meta-logic):

Logic	Programming
Object-logic	Application model
Meta-logic	Programming Language

  This idea extends the range of mathematical logic from the study of given systems to the study of the representation of families of systems within a given system. Work on this topic began in earnest with the LF logical framework [Harper et al., 1987, Harper et al., 1993, Pym, 1990, Avron et al., 1992, Pym and Wallen, 1991, Pym and Wallen, 1992, Pym, 1995b, Pym, 1996, Pym, 1995a] and continues to be a major topic. However, LF's basis in intuitionistic logic (via the $\lambda\Pi$-calculus) leads to difficulties in representing program logics, such as Hoare's logic, of the kind described above [Mason, 1986]. The reason for this is the failure of the semantics of intuitionistic logic to account for the spatial properties of the resources to which

Hoare's logic inherently refers. A logical framework, RLF, based not on intuitionistic logic but rather on a substructural logic which has a semantics based on resources, provides a better analysis of program logics.

We begin with this introductory chapter in which we survey the background to our work on **BI**. Starting from a semantic view of logical consequence in terms of truth we go on to consider a range of formal calculi for constructing proofs. We then consider how the semantic view of consequence is affected if "possible worlds" are interpreted as (and constructed as) "resources". We move on to consider the relationship between the construction of proofs and the use of resources. We conclude with a sketch of the idea of logical frameworks, the study of the representation of systems of logic in a formal meta-logic, and consider the possibilities for a semantics based on resources.

Consequences, Truth and Proof

Logic may be seen as the study of *consequences*, *i.e.*, assertions that the truth of a given proposition follows from the truth of a given *collection* of *propositions*. Propositions are *declarative statements*. We can give a simple definition, as described in Hodges [Hodges, 1993], as follows:

> A proposition is that situation which is described by an English phrase which may be substituted for X in
>
> *It is the case that* X.
>
> so as to give a grammatically correct English sentence.

Examples are phrases like "the earth is flat", "the sun orbits the earth" or "I have enough coins to buy a chocolate bar from the vending machine." Barwise [Barwise and Perry, 1983, Barwise, 1989], and others, have developed *situation theory* in order to further analyse this linguistically-derived perspective on the notion of proposition in terms of *situations* and *infons*. Devlin [Devlin, 1990] provides a thorough description of these ideas. Mathematically, propositions are denoted by the *formulæ* of a formal language.

Logic is about more than propositions, however; it is also about reasoning. In classical logic (**CL**), intuitionistic logic (**IL**) and linear logic (**LL**), the basic notion of reasoning is captured by the idea of a *consequence relation* [Tarski, 1956, Scott, 1974, Avron, 1991]

$$\phi_1; \ldots ; \phi_m \vdash \psi_1; \ldots ; \psi_n$$

between finite sequences of propositions. It should be read as follows:

If we have all of the ϕs, then we have at least one of the ψs.

A common restriction is to the case in which $n = 1$. Formally, a consequence relation on set of formulæ is a binary relation $\vdash$ between finite sequences of formulæ such that:

1 Reflexivity: for every formula ϕ, $\phi \vdash \phi$;

2 Transitivity (or Cut): if $\Gamma \vdash \Delta; \phi$ and $\phi; \Gamma' \vdash \Delta'$, then $\Gamma; \Gamma' \vdash \Delta; \Delta'$.

Additional axioms which may be taken include:

3 Exchange: if $\Gamma \vdash \Delta$, then $\rho(\Gamma) \vdash \sigma(\Delta)$, for permutations ρ and σ;

4 Weakening: if $\Gamma \vdash \Delta$, then $\Gamma; \Gamma' \vdash \Delta; \Delta'$

5 Contraction: if $\Gamma; \Gamma \vdash \Delta$ or $\Gamma \vdash \Delta; \Delta$, then $\Gamma \vdash \Delta$.

Consequence relations are typically realized in two ways, model-theoretically and proof-theoretically. In **CL**, the key semantic notion is *truth*. Explanations of truth in mathematical logic usually begin with the idea of a *truth table* in which a proposition is assigned a *truth value*, 0 (false) or 1 (true). The assignment of truth values is performed by induction on the structure of propositions, connective by connective. For example. the truth table for classical implication is the following:

ϕ	ψ	$\phi \supset \psi$
0	0	1
0	1	1
1	0	0
1	1	1

Here the idea is that we assume, inductively, that we have assignments of truth values for ϕ and ψ — there are four possible combinations — and proceed to assign a value to $\phi \supset \psi$ in each case. We can write similar tables for conjunction, $\wedge$, disjunction, $\vee$, and negation, $\neg$, as follows:

ϕ	ψ	$\phi \wedge \psi$
0	0	0
0	1	0
1	0	0
1	1	1

ϕ	ψ	$\phi \vee \psi$
0	0	0
0	1	1
1	0	1
1	1	1

ϕ	$\neg\phi$
0	1
1	0

Mathematically, we think of such an assignment of truth values, or *model*, as a function

$$\mathcal{I} : \mathbf{Prop} \to \{0, 1\}$$

from the set of propositions to the two-element set. It is then convenient to define $\mathcal{I} \models \phi$, read as "$\mathcal{I}$ satisfies ϕ", by

$$\mathcal{I} \models \phi \quad \text{iff} \quad \mathcal{I}(\phi) = 1.$$

Starting from this point, we can define the notion of *semantic consequence for truth* in a given model $\mathcal{I}$:

$$\phi_1; \ldots ; \phi_m \models_{\mathcal{I}} \phi \quad \text{iff} \quad \mathcal{I} \models \phi_i, \text{ for each } 1 \leq i \leq m, \text{ implies } \mathcal{I} \models \phi.$$

A stronger notion is *semantic consequence for validity*, defined as follows:

$$\phi_1; \ldots ; \phi_m \models \phi \quad \text{iff} \quad \text{for all } \mathcal{I}, \mathcal{I} \models \phi_i, \text{ for each } 1 \leq i \leq m, \text{ implies } \mathcal{I} \models \phi.$$

These ideas are the very beginning of *classical model theory*, the area of logic which is perhaps mostly deeply integrated with mainstream pure mathematics. By adding *quantifiers*, such as $\forall$, or "for all", and $\exists$, or "there exists", and *theories*, or collections of special symbols and axioms, to the analysis described above, model theory is able to provide a logical study of important mathematical structures. For example, the *model theory of fields* is a major area in its own right. Its axioms include propositions such as

$$\forall x.(x + 0 = x), \quad \forall x.\forall y.\forall z.(x \times (y + z) = x \times y + x \times z)$$

and

$$\forall x.((x \neq 0) \supset \exists y.(x \times y = 1)),$$

where $+$, 0, $\times$ and 1 are *function* symbols used to build the *terms* of the logic, and $=$ is a special *predicate* symbol, taken in addition to the logical connectives and quantifiers. The equality symbol, $=$, is used to build the atomic propositions by *predicating* terms: if s and t are terms we can form the proposition that they are equal by writing $= (s, t)$ or, more simply, $s = t$. Similarly, we write $s \neq t$ as a shorthand for $\neg(s = t)$. From this point of view, a field is a model which satisfies these (and some other) axioms.

Moving on to **IL**, we must adopt a more sophisticated semantics (see [van Dalen, 1983] for an extended discussion). Heyting's formalization of intuitionistic predicate logic, arithmetic and set theory (see also Glivenko and Kolmogorov) [Heyting, 1989, Girard et al., 1989] adumbrated the now-familiar proof-interpretation or BHK semantics but did not provide a framework for a theory of models supporting a notion of truth and a corresponding completeness theorem. The central idea is that the facts which hold in the world may be discovered by exploring the world. As more of the world is explored, so more facts are discovered.

The analogue of classical truth for intuitionistic logic is found in topology [Tarski, 1956, van Dalen, 1983, Lambek and Scott, 1986] (and other references therein) but may be very clearly understood from the work of Beth [van Dalen, 1986] and Kripke [Kripke, 1965], in which the exploratory character of the semantics is clearly captured. The central issue concerns *implication* (from which follows a treatment of negation). In **CL**, it may be defined in terms of disjunction and negation,

$$\phi \supset \psi \quad \text{iff} \quad (\neg\phi) \vee \psi,$$

but in **IL**implication is primitive. Kripke's solution [Kripke, 1965] is to require that a model be structured as an ordered set of *possible worlds*:

$$v \sqsubseteq w \quad \text{means} \quad \text{“}v \text{ is accessible from } w\text{”}.$$

An explorer, the "creative subject" [van Dalen, 1983], can move from his current world to any world which is accessible from it. At each world, we have a collection of known *basic facts* which hold there. We write

$$w \models^{\mathcal{M}} \phi$$

to denote that the proposition ϕ holds, *i.e.*, is true, at world w in the model $\mathcal{M}$. We can now define the truth of implicational propositions as follows:

$$w \models^{\mathcal{M}} \phi \supset \psi \quad \text{iff} \quad \text{for every } v \sqsubseteq w, \text{ if } v \models \phi, \text{ then } v \models \psi,$$

i.e., $\phi \supset \psi$ holds at w just in case at every v which is accessible from w, if ϕ holds at v then ψ holds at v. Thus the meaning of implication is dependent on the (order) structure of the model.

Both **CL** and **IL** also have well-developed proof-theoretic descriptions, *i.e.*, formal systems which characterize their consequences and which may be independently mathematically analysed. There are three main types of system.

1 **Hilbert-type** systems. In a presentation of a logic as a Hilbert-type (or just Hilbert) system consists of a collection of axioms together with a collection of inference rules. Hilbert systems for **CL** give a good sense of the idea:

Axioms:

I: $\phi \supset \phi$;
K: $\phi \supset (\psi \supset \phi)$;
S: $(\phi \supset (\psi \supset \chi)) \supset ((\phi \supset \psi) \supset (\phi \supset \chi))$;

Rules:

MP: If ϕ and $\phi \supset \psi$, then ψ.

A proof in a Hilbert system consists of a sequence of propositions in which the last proposition follows from the preceeding propositions by one of the axiom schemata or one of the rules.

For predicate logic, we must add the rule of *Generalization*,

G: If $\phi(x)$, then $\forall x.\phi(x)$.

Hilbert systems for **IL** may be obtained by varying the axioms [Mendelson, 1987].

2 **Natural deduction** (ND) systems. Introduced in Gerhard Gentzen's paper from 1934, "Untersuchungen über das logische Schliessen" ("Investigations into logical deduction") [Gentzen, 1934], natural deduction systems for **CL** and **IL** are described by pairs of rules which manipulate proofs by either *introducing* a connective into a proof or *eliminating* it from a proof. Proofs are constructed by starting with assumptions and deriving conclusions. As such, that process may be represented as trees.

A good example is provided by the formulation in natural deduction of *reasoning by cases*, which may be summarized as follows:

- Let ϕ_1, ϕ_2 and ψ be propositions;
- Suppose (i) that we have a proof of ψ assuming ϕ_1 and (ii) that we have a proof of ψ assuming ϕ_2;
- Suppose (iii) we have a proof that $\phi_1 \vee \phi_2$ holds;
- From (i), (ii) and (iii), we can construct a proof of ψ.

In a natural deduction presentation of **CL**, this argument is described by the rule of $\vee$-*elimination*, $\vee$E for short, in contrast to its corresponding $\vee$-*introduction* rules:

$$\dfrac{\phi_1 \vee \phi_2 \quad \begin{matrix}[\phi_1] \\ \vdots \\ \psi\end{matrix} \quad \begin{matrix}[\phi_2] \\ \vdots \\ \psi\end{matrix}}{\psi}\,\vee E \qquad \dfrac{\phi_1}{\phi_1 \vee \phi_2}\,\vee I_1 \qquad \dfrac{\phi_2}{\phi_1 \vee \phi_2}\,\vee I_2$$

Notice that we have *discharged* our assumptions ϕ_1 and ϕ_2: given that we have a proof of $\phi_1 \vee \phi_2$, we need not retain the assumptions in order to get a proof of the conclusion.

The rules for implication provide another example:

$$\frac{\phi \quad \phi \supset \psi}{\psi} \supset \mathrm{E} \qquad \frac{\begin{array}{c}[\phi] \\ \vdots \\ \psi\end{array}}{\phi \supset \psi} \supset I$$

So suppose that we have proofs of ψ from either ϕ_1 or ϕ_2 and that we have a proof of χ assuming ψ. Then the following is an example of a proof of χ assuming $\phi_1 \vee \phi_2$:

$$\frac{\dfrac{\phi_1 \vee \phi_2 \quad \begin{array}{c}[\phi_1] \\ \vdots \\ \psi\end{array} \quad \begin{array}{c}[\phi_2] \\ \vdots \\ \psi\end{array}}{\psi} \vee E \qquad \dfrac{\begin{array}{c}[\psi] \\ \vdots \\ \chi\end{array}}{\psi \supset \chi} \supset I}{\chi} \supset E. \tag{I.1}$$

So we can see that the assumptions made in a proof are represented as the *undischarged* leaves of the tree — in this case, just $\phi_1 \vee \phi_2$.

The pairing of rules for introducing and eliminating connectives is the key characteristic of natural deduction. The MP rule of a Hilbert system corresponds to $\supset$-elimination and the G rule corresponds to $\forall$-introduction.

The key property which a natural deduction system **N** may have is *normalization*: in any proof in **N**, all occurrences of an introduction rule immediately[1] followed by the elimination rule for the same occurrence of the same connective may be eliminated from the proof so as to yield a proof in **N** of the same conclusion from the same assumptions. For example,

$$\frac{\phi \qquad \dfrac{\begin{array}{c}[\phi] \\ \vdots \\ \psi\end{array}}{\phi \supset \psi} \supset I}{\psi} \supset E. \tag{I.2}$$

[1] Up to some permutabilities of rules [Kleene, 1968].

is a proof of ψ assuming ϕ. However, such a proof is what we started out with in the right-hand branch of the proof tree. The $\supset I$ rule is immediately followed by the $\supset E$ rule which gets us back to a proof of ψ assuming ϕ. The introduction followed by the corresponding elimination is a "pointless detour": we could have just used our original proof. Eliminating all such pointless detours leads us to the *normal form* of a proof.

The consequences established using natural deduction rules may be represented as *sequents* (from the German "Sequenzen") and natural deduction rules may be represented in *sequential* or *linearized* form, in which the assumptions made globally, at the leaves of the proof tree, are represented *locally* within the rules. For example, the $\vee$-elimination and $\vee$-introduction rules go as follows:

$$\frac{\Gamma \vdash \phi_1 \vee \phi_2 \quad \Gamma, \phi_1 \vdash \psi \quad \Gamma, \phi_2 \vdash \psi}{\Gamma \vdash \psi} \; \vee E$$

and

$$\frac{\Gamma \vdash \phi_1}{\Gamma \vdash \phi_1 \vee \phi_2} \; \vee I_1 \qquad \frac{\Gamma \vdash \phi_2}{\Gamma \vdash \phi_1 \vee \phi_2} \; \vee I_2.$$

For example, the proof in (I.1) is represented as

$$\frac{\dfrac{\Gamma \vdash \phi_1 \vee \phi_2 \quad \Gamma, \phi_1 \vdash \psi \quad \Gamma, \phi_2 \vdash \psi}{\Gamma \vdash \psi} \vee E \qquad \dfrac{\Gamma, \psi \vdash \chi}{\Gamma \vdash \psi \supset \chi} \supset I}{\Gamma \vdash \chi} \supset E. \tag{I.3}$$

Two important things may be seen from this example.

- Firstly, that discharge corresponds to removing formulæ from the left-hand side of $\vdash$: Note that we can see two versions of this, one in which the discharged formula simply moves to form part of the right-hand side ($\supset I$) and one in which the discharged formulæ are witness by a formula on the right-hand-side $\vee E$.
- Secondly, that the rôle of the Γ is somewhat arbitrary. In particular, we could replace Γ with $\Gamma; \Gamma'$, *i.e.*, do a Weakening operation,

  $$\frac{\Gamma \vdash \phi}{\Gamma; \Gamma' \vdash \phi} \; \textit{Weakening},$$

  and still have a perfectly good proof of χ, with more (unused) assumptions. A related structural is that of Contraction,

  $$\frac{\Gamma; \Gamma \vdash \phi}{\Gamma \vdash \phi} \; \textit{Contraction},$$

in which duplications of assumptions are removed.

The notion of the *usage*, associated with Weakening Contraction, of formulæ in a proof is a central concern of *substructural logics*. In particular, linear logic [Girard, 1987] takes neither Weakening nor Contraction for arbitrary formulæ.[2]

Linear logic's denial of arbitrary Weakening and Contraction has the consequence that conjunction (also disjunction) splits into additive and multiplicative,

$$\frac{\Gamma \vdash \phi \quad \Gamma \vdash \psi}{\Gamma \vdash \phi \wedge \psi} \; \wedge I \quad \text{and} \quad \frac{\Gamma \vdash \phi \quad \Delta \vdash \psi}{\Gamma ; \Delta \vdash \phi \otimes \psi} \; \otimes I,$$

together with corresponding elimination rules, the two forms being interderivable only in the presence of both Weakening and Contraction.

Whilst linear logic may seen rather weak, it recovers the full strength of intuitionistic logic via an exponential, !: formulæ prefixed with !, and only such formulæ, may be Weakened and Contracted in the usual way. Whilst an occurrence of ϕ is taken to denote exactly one copy of ϕ, an occurrence of $!\phi$ may be viewed as denoting infinitely many occurrences of ϕ. Moreover, assumptions without the prefix may be converted to have it, using linear logic's rule of Dereliction,

$$\frac{\Gamma ; \phi ; \Delta \vdash \psi}{\Gamma ; !\phi ; \Delta \vdash \psi} \; \mathbf{LL}'s \quad Dereliction,$$

but not *vice versa*.[3]

The Cut rule, a form of substitution, is formulated as:

$$\frac{\Gamma, \phi \vdash \psi \quad \Gamma \vdash \phi}{\Gamma \vdash \psi} \; Cut.$$

[2]In systems which have Weakening, Contraction and Exchange,

$$\frac{\Gamma ; \phi ; \psi ; \Delta \vdash \chi}{\Gamma ; \psi ; \phi ; \Delta \vdash \chi} \; Exchange,$$

the collection of assumptions can be viewed as a set: In the absense of Contraction, it can be viewed as a multiset.

[3]Linear logic comes in single-conclusioned and multiple-conclusioned forms, including classical linear linear logic (**CLL**) . The multiple-conclusioned forms admit a multiplicative disjunction, together with an additional exponential, ?, dual to !. In neither form is the structure of the sequent rich enough to admit a decomposition of implication except via !.

The step into sequential form suggests a further development, the sequent calculus, again introduced by Gentzen [Gentzen, 1934].

3 **Sequent calculi** (SC). Sequent calculi, too, were introduced in Gentzen's 1934 paper [Gentzen, 1934]. In sequent calculi, the idea of pairs of introduction and elimination rules is replaced by that of pairs of introduction rules, one rule for introducing a connective on the *left* of a sequent and one rule for introducing it on the *right*. For example, the rules for $\vee$ are rendered as follows:

$$\frac{\Gamma, \phi_1 \vdash \psi \quad \Gamma, \phi_2 \vdash \psi}{\Gamma, \phi_1 \vee \phi_2 \vdash \psi} \vee L$$

and

$$\frac{\Gamma \vdash \phi_1}{\Gamma \vdash \phi_1 \vee \phi_2} \vee R_1 \qquad \frac{\Gamma \vdash \phi_2}{\Gamma \vdash \phi_1 \vee \phi_2} \vee R_2.$$

So sequent calculi may be seen as manipulating the consequence relation of the logic directly.

The key property which a sequent calculus **S** may have is *Cut-elimination*: if $\Gamma \vdash \Delta$ is provable in **S** using Cut, then it is provable in **S** without using Cut. The standard sequent calculi for both **CL** and **IL** both admit Cut-elimination. The standard sequent calculus for the modal logic S4 also admits Cut-elimination but the situation for S5 is more complex [Fitting, 1983].

We shall develop our discussion of the sequent calculus more in the sequel.

The relationship between the semantic notion of truth ($\models$) and the syntactic notion of proof ($\vdash$) is very important in logic in general and in computational settings in particular. In logic in general, it is important to establish *soundness* ($\phi_1; \ldots ; \phi_m \vdash \psi$ implies $\phi_1; \ldots ; \phi_m \models \psi$) and *completeness* ($\phi_1; \ldots ; \phi_m \models \psi$ implies $\phi_1; \ldots ; \phi_m \vdash \psi$) theorems. In more computational settings, the equivalent theorems are concerned to establish that the *execution dynamics* of program correponds to its *intended meaning*. This is the topic of *operational* and *denotational* semantics. A programming language for which these two ideas coincide is said to be *fully abstract*. See [Winskel, 1993] for a discussion of these topics. The key unifying concept here is that of *representation*, a concept to which we shall return in the sequel.

We have seen that both semantic entailment (or satisfaction), $\models$, and syntactic entailment, $\vdash$, may be used to describe consequence relations. Can we take a more general, unifying stance ? One possible answer is provided by the notion of *truthmakers* [Read, 2000, Restall, 1999]. Truthmakers are an appropriate notion for us for the following reasons:

- They are an attempt, which is prior to a particular choice of semantics, to account for the conditions under which propositions hold. Our view of truthmakers is that they are *situated evidence* [Barwise and Perry, 1983, Barwise, 1989];

- A situated view of evidence facilitates a comparative analysis of truth-functional semantics, proof-theoretic semantics and *resource semantics*.

We consider the Truthmaker Axiom which may be found, for example, in [Read, 2000]:

Truthmaker Axiom (TA): For all propositions ϕ, if ϕ is true (holds), then something makes it true (hold), *i.e.*, there is some s such that "$s \models \phi$". (I.4)

Within informatics, truthmakers may be seen as the link between (naïvely conceived) physical reality and models of logics built out of some mathematical representation of that reality. For example, in intuitionistic and modal logics, possible worlds may be seen as truthmakers. More concretely, in Chapter 9 we give an example of a model of **BI** built out of a representation of a computer's memory. The possible worlds may be seen as memory states, with accessibility via computation steps. A computer's memory is a leading example of a consumable entity, *i.e.*, a *resource*.

What, then, is the connection between semantics and resources ?

Resources, Truth and Proof

The notion of *resource* is widely used in common discourse, with the following definition (we quote from the Oxford English Dictionary [OED, 1976]):

> **rèsour′ce** (...) *n.* (usu. in *pl.*) Means of supplying what is needed, stock that can be drawn on, available assets; ...

The *notion* of resource is, however, a primitive in informatics. It refers to any entity which may be referenced by a procedure.[4] Examples of

[4]We use the term procedure, rather broadly, to mean a specified operation, not necessarily a procedure expressed in a programming language.

resources include such basic concepts as the *space* available in a computer's memory or the *time* taken by a program's execution. Other examples are physical components and devices, processes, procedures or functions, data (structures), (computer) memory and information itself. A key characteristic of a resource is that it may be *consumed* and the availability of resources is a significant constraint on the behaviour of systems and must therefore be a significant part of any analyses of their properties. Indeed, the notion of resource has been considered similarly informally but rather carefully by Brinch Hansen, in his text on operating systems [Hansen, 1973]:

> "The word *resource* covers physical components, processes, procedures and data structures; in short, any object referenced by computations."

The use of "referenced" here is particularly significant: a resource may be available to, and hence be referenced by, more than one computation. It follows that a theory of resources must account for the notion of *sharing of resources* (and, of course, *non-sharing*) by the component processes of computations.

In fact, there has recently been much interest in so-called "resource-sensitive" logics, with various interpretations of Girard's linear logic [Girard, 1987] being the main focus. However, linear logic was introduced as a tool for analysing the proof-theoretic properties of classical (and intuitionistic) logic and its analysis of resources, its rôle as meta-language in domain theory, implicit in much of the work of Plotkin and his collaborators (see also [Amadio and Curien, 1998]), notwithstanding, is correspondingly constrained to the so-called "number-of-uses" reading of propositions. Nevertheless, these resource-interpretations provide a useful informal background for our analysis.

We develop an example, following Pym, O'Hearn, and Yang [Pym et al., 2000], in which resources are considered to be coins which may be used to buy chocolates or candy from a vending machine. Here we deliberately develop an example which is similar to examples provided by C.A.R Hoare and J.-Y. Girard in their related settings.

A proposition is a declarative statement about cost and a (hypothetical) judgement of consequence is read declaratively as follows:

$$\phi \models \psi \quad : \quad \begin{array}{l}\text{If I have enough money to make } \phi \text{ true,}\\ \text{then I have enough to make } \psi \text{ true.}\end{array}$$

Note that we have implied here a semantic consequence relation, $\models$, between propositions. As we have seen, such a relation is usually defined, with respect to a model $\mathcal{M}$, by

$$\phi \models \psi \quad \text{iff} \quad \mathcal{M} \models \phi \text{ implies } \mathcal{M} \models \psi$$

with, in our setting, the modelling of resource residing in the model. This is indeed what we intend here but, for this informal example, we project the modelling of resource directly onto the propositions and work with just "$\phi \vdash \psi$". This approach has two immediate advantages. Firstly, it will serve as a way of motivating our subsequent modelling of resource and, secondly, it facilitates a direct comparision with a similar example rendered in linear logic.

Relative to this reading of consequence, we can explain the meanings of two conjunctions and two implications.

We get a resource semantics for the usual intuitionistic, or additive, implication and conjunction as follows:[5]

$\phi \rightarrow \psi$: If I were to obtain enough money to make ϕ true, then I should also have enough to make ψ true;

$\phi \wedge \psi$: The money I have got is both enough to make ϕ true and enough to make ψ true.

$\phi \vee \psi$: The money I have got is enough either to make ϕ true or to make ψ true.

However, from the point of view of resources, we can define corresponding multiplicative connectives which, respectively, combine and divide resources, as follows:

$\phi \mathrel{-\!\!*} \psi$: If you were to give me enough to make ϕ true then, combined with what I have already got in my pocket, I should have enough to make ψ true; and

$\phi * \psi$: I can use part of my money to make ϕ true and have enough left over to make ψ true (and *vice versa*).

These connectives are the core of **BI**, *the logic of bunched implications*.[6] We call $*$ "star" and $\mathrel{-\!\!*}$ "magic wand".

Given these readings, the following judgements say that for one coin I can buy one candy but that to buy one chocolate I must have two coins:

[5]It should be noted that the reading we provide here for the disjunction, $\vee$, is intuitionistic.

[6]In the relevant logic tradition, "multiplicative" is often termed "intensional" and "additive" termed "extensional". The reasons for this terminology are beyond our present scope.

- $coin \vdash candy$, and
- $coin * coin \vdash choc$,

where the basic propositions are

- *coin* : I have (at least) one coin in my pocket,
- *choc* : I have enough to buy a chocolate, and
- *candy* : I have enough to buy a candy.

The most distinctive feature of **BI** is its treatment of the two implications as having equivalent status. We illustrate how this works. We certainly expect

$$coin \vdash coin \mathrel{-\!\!*} choc$$

because if I have a coin in my pocket, and if you give me another, then I will have enough to buy a chocolate but

$$coin \nvdash coin \to choc$$

because a single coin is not enough to buy a chocolate. The difference between the two implications is clear here.

We emphasize here that multiple purchases (vends) are formed using $*$ and not $\wedge$: The ability to purchase two candies, requiring two coins, may be represented as

$$coin * coin \vdash candy * candy$$

but

$$coin \vdash candy \wedge candy$$

asserts that one coin is enough to buy a candy and enough to buy a(nother) candy. This rather strange assertion amounts to nothing more than an instance of the usual rule of $\wedge$I, *i.e.*,

$$\frac{\phi \vdash \psi \quad \phi \vdash \psi}{\phi \vdash \psi \wedge \psi},$$

so that $coin \vdash candy \wedge candy$ amounts to a duplication of the assertion that one coin is enough to buy a candy.

The additive disjunction, $\vee$, is perhaps less strange: it merely asserts that we have enough to buy one of the disjuncts. The reader may be wondering whether there is a form of disjunction which correspsonds to $*$ in a way analogous to that in which $\vee$ corresponds to $\wedge$. There is such a disjunction but it inhabits a slightly different logical setting (which we discuss in Chapter 7) and its interpretation in terms of resources is a more delicate matter.

To understand the two sorts of connectives formally, we must either introduce the idea of a proof theory based on *bunches* or provide a model-theoretic semantics for the formulæ (see Chapter 1). For now, it suffices to remark that we must permit collections of assumptions to be combined in two different ways, one way corresponding to the additives and the other to the multiplicatives.

The reader will detect some similarity here with J.-Y. Girard's well-known "Marlboro's and Camels", or "use once", reading of linear logic. However, the divergences are more interesting than the (somewhat superficial) similarities and illustrate just how great is the difference between **BI** and linear logic .

Firstly, Girard's reading is about "proofs-as-actions" where, for example,

- *choc*: the (type of the) act of buying a chocolate.

In contrast, our reading is not defined in terms of proofs; rather, it arises from a purely declarative reading of formulæ.[7] We emphasize here that we do not regard a proposition as a resource and (so) a proof as a way to manipulate resources. Rather, the reading is completely declarative: a proposition is a statement about the world whose judgement of truth may involve consideration of resources.[8]

Secondly, the difference is not merely one of emphasis but may be seen at the technical level of logical consequence. For instance,

- $coin \mathrel{-\!\!*} choc \not\vdash coin \to choc$

is something we would certainly expect, because $coin \mathrel{-\!\!*} choc$ is true when you have one coin in your pocket, but $coin \to choc$ is certainly not. In linear logic, however, in which $\phi \to \psi$ is rendered, via the exponential , !, as $!\phi \multimap \psi$, one gets

- $coin \multimap choc \vdash coin \to choc \qquad (= !coin \multimap choc)$

[7]This origin notwithstanding, we proceed, in Chapter 9, to provide two interpretations of **BI** in which the access of proof-objects to resources is central.

[8]Note, however, that a reading of propositions-as-resources is in no way precluded although, for reasons which we sketch below, it is essentially different from that available in linear logic. Such a reading may be seen most clearly in the term models required for our various completeness proofs (Chapters 4, 5, 13 and 14) and in the "sharing interpretation" of logic programming provided in Chapters 9 and 16.

no matter what *coin* and *choc* are, because one can compose on the left with dereliction $!\phi \vdash \phi$.

There are other examples in **BI** which violates the "use once" idea from linear logic (here I denotes the unit of the product $*$):

- $I \vdash (coin \wedge (coin \to choc)) \mathrel{-\!\!*} choc$, and
- $I \vdash (coin \mathrel{-\!\!*} ((coin \to coin \to choc) \to choc)$.

Now these judgements seem wrong from the point of view of linear logic because

- $I \not\vdash (coin \& (!coin \multimap choc)) \multimap choc$, and
- $I \not\vdash (coin \multimap !(!(!coin \multimap (!coin \multimap choc)) \multimap choc)$.

The first case would violate the idea that a linear function of type $A \& B \multimap C$ must use one of its input components, but not both, and the second would violate the idea that a linear function cannot use its argument twice. However, if we discard this perspective and think declaratively, using the reading of formulæ advanced in this section, then the truth of **BI**'s judgements is quite straightforward and not at all surprising. In **BI**, the proof of the last judgement, when viewed as a term of **BI**'s λ-calculus, $\alpha\lambda$, introduced in Chapters 2 (see also [O'Hearn and Pym, 1999, O'Hearn, 1999, Pym, 1999]), does indeed use its argument twice. Indeed, in [O'Hearn and Pym, 1999] we gave a resource reading of proofs to justify this judgement; the declarative justification is much more immediate.

Of course, our discussion need not be construed as a criticism of linear logic. The "proofs-as-actions" reading, in particular, is vividly intuitive: It gives a consistent way of understanding the position that linear logic's consequence relation takes on the judgements above.

So, for our present purposes, we have the following axiomatization of resources, via Kripke's idea of possible worlds:

- An underlying *set* of resources, M, here, for simplicity, untyped but not essentially so;
- A representative for *zero* resources, e;
- A way of *combining* resources, $\cdot$, here, for simplicity, commutative but not essentially so;
- A way of *comparing* resources, $\sqsubseteq$.

Mathematically, we recognize that we require a preordered (commutative) monoid

- $\mathcal{M} = (M, \cdot, e, \sqsubseteq)$.[9]

We can then regard a unit of resource $m \in M$ as a truthmaker[10] for propositions ϕ by defining a forcing relation, $\models$, between worlds, or resources, and formulæ

$$m \models \phi,$$

in the style of Kripke semantics, by induction on the structure of formulæ. This forcing relation is read as "m is sufficient to make ϕ true". For example, we have

$$m \models \phi \rightarrow \psi \quad \text{iff} \quad \text{for all } n \sqsubseteq m \ n \models \phi \text{ implies } n \models \psi$$

and

$$m \models \phi \mathrel{-\!\!*} \psi \quad \text{iff} \quad n \models \phi \text{ implies } m \cdot n \models \psi,$$

with the latter clause being read as "m is sufficient to make $\phi \rightarrow \psi$ true just in case if n is sufficient to make ϕ true, then $m \cdot n$ is sufficient to make ψ true".

Our monetary example suggests that the natural numbers may give a model of **BI**, and indeed they do, as follows:

- $\mathcal{N} = (N, +, 0, \leq)$,

where $+$, 0, and $\leq$ have their usual meanings. Then the implication $\mathrel{-\!\!*}$ is modelled by the following, in which m, n, *etc.* range over the set of natural numbers, N:

- $m \models \phi \mathrel{-\!\!*} \psi$ iff for all n, if $n \models \phi$, then $m + n \models \psi$.

Just as a resource semantics allows us to define two forms of implication, so we can use it to define two forms of universal quantifier. The forcing clauses are analagous to those for the corresponding implications, essentially:

$$m \models \forall x.\phi \quad \text{iff} \quad \text{for all } n \sqsubseteq m \text{ and all } t \text{ at } n,\ n \models \phi[t/x]$$

and

$$m \models \forall_{\mathbf{new}} x.\phi \quad \text{iff} \quad \text{for all } n \text{ and all } t \text{ at } n,\ m \cdot n \models \phi[t/x],$$

[9]Such that if $m \sqsubseteq m'$ and $n \sqsubseteq n'$, then $m \cdot n \sqsubseteq m' \cdot n'$.
[10]Recall the Truthmaker Axiom (I.4).

Here again we see the distinction between the conservation of resources captured by the additive $\forall$ and the combination of resources captured by the multiplicative $\forall_{\mathbf{new}}$. Two existential quantifiers, with corresponding semantics, may be defined similarly.

So far, we have been motivated be a very simple notion of resource. In particular, we have paid no attention to the following two key aspects of resource semantics:

- The *location* of a resource; and
- The *ownership* of a resource.

Moreover, we have paid little theoretical attention to the internal structure of resources, *i.e.*, the elements of M. However, the combination of the additive and multiplicative connectives admits a treatment of both *local* and *global* reasoning within a common logical analysis.[11] This local–global distinction is exploited to good effect in Chapter 9, which summarizes ideas presented in [Ishtiaq and Pym, 1998, Pym et al., 2000, Ishtiaq and O'Hearn, 2001].

A more sophisticated model of resources might, for example, use a multi-sorted set of resources, perhaps with elements carrying significant "internal" structure, with correspondingly sorted operations. Such an analysis is beyond our present scope.

Proof-search and Its Operational Semantics

Another view of consequences is provided by the theory of *reductive logic*.[12] Instead of asking whether, in a given model $\mathcal{I}$, a consequence holds semantically, *i.e.*, whether

$$\phi_1; \dots ; \phi_m \models_{\mathcal{I}} \phi \quad \text{iff} \quad \mathcal{I} \models \phi_i, \text{ for each } 1 \leq i \leq m, \text{ implies } \mathcal{I} \models \phi$$

holds, we can ask whether the *syntactic structure* of $\phi_1; \dots ; \phi_m$ and ϕ carries in enough information to try to decide whether ϕ is a consequence of $\phi_1; \dots ; \phi_m$.

Recall that the sequent calculus manipulates directly sequents of the form

$$\phi_1; \dots ; \phi_m \ \vdash \ \psi,$$

which should be understood as saying "if all of the ϕs hold, then the ψs hold". Manipulations of sequents are performed by adding connectives

[11] We remark that both location and ownership may be seen to emerge from our discussion of logic programming for both propositional **BI** (in Chapter 9) and predicate **BI** (in Chapter 16).
[12] So far, we have considered only deductive logic.

either to the left-hand side, the antecendent, or to the right-hand side, the succedent. For example, conjunction has left and right rules

$$\frac{\Gamma;\phi_1;\phi_2 \vdash \phi}{\Gamma;\phi_1 \wedge \phi_2 \vdash \phi}\ \wedge L \quad \text{and} \quad \frac{\Gamma \vdash \phi_1 \qquad \Gamma \vdash \phi_2}{\Gamma \vdash \phi_1 \wedge \phi_2}\ \wedge R,$$

respectively, the right rule, for example, being read as, "if $\Gamma \vdash \phi_1$ and $\Gamma \vdash \phi_2$ are provable, then $\Gamma \vdash \phi_1 \wedge \phi_2$ is provable". Similarly, the implication left- and right-rules are

$$\frac{\Gamma \vdash \phi_1 \qquad \Gamma;\phi_2 \vdash \psi}{\Gamma;\phi_1 \supset \phi_2 \vdash \psi}\ \supset L \quad \text{and} \quad \frac{\Gamma;\phi_1 \vdash \phi_2}{\Gamma \vdash \phi_1 \supset \phi_2}\ \supset R,$$

respectively. The left rule makes a connection between the two premisses via the introduced implication. The right rule moves a proposition from the antecedent to the conclusion by forming the implication. All proofs must start with *axiom sequents* of the form

$$\frac{}{\Gamma;\alpha \vdash \alpha}\ \mathit{Axiom}.$$

Proof-search is the construction of proofs of sequents using inference rules as *reduction operators*, read from conclusion to premisses.[13] We start from a given, putative endsequent and perform a search by successively applying reduction operators. We terminate the search successfully if all branches of the *search tree* so generated have axiom sequents at their leaves.

The significance of the Cut-elimination theorem for proof-search may be seen very clearly:

An inference rule

$$\Downarrow \quad \frac{\text{PREMISS}_1 \ldots \text{PREMISS}_m}{\text{CONCLUSION}}\ R$$

has the *subformula property* just in case each PREMISS_i, $1 \leq i \leq m$, is formed entirely from subformulæ of CONCLUSION.

As reduction operators, read from conclusion to premisses,

$$\Uparrow \quad \frac{\text{SUFFICIENT PREMISS}_1 \ldots \text{SUFFICIENT PREMISS}_m}{\text{PUTATIVE CONCLUSION}}\ R,$$

rules which have this property are much less non-deterministic than those which don't. For example, compare the following two rules, in which we write ?- for the putative consequence:

$$\Uparrow \quad \frac{\Gamma \mathrel{?\text{-}} \phi_1 \qquad \Gamma;\phi_2 \mathrel{?\text{-}} \psi}{\Gamma;\phi_1 \supset \phi_2 \mathrel{?\text{-}} \psi}\ \supset L \qquad \Uparrow \quad \frac{\Gamma,\phi \mathrel{?\text{-}} \psi \qquad \Gamma \mathrel{?\text{-}} \phi}{\Gamma \mathrel{?\text{-}} \psi}\ \mathit{Cut}.$$

[13]Kleene [Kleene, 1968] explains this for the classical predicate calculus.

The Cut requires the generation of the formula ϕ, because it is not detemined by the conclusion. In the $\supset L$ rule, however, each of ϕ_1, ϕ_2 and Γ is determined by the conclusion. Cut is the only rule of the sequent calculus for propositional **CL** or **IL** which fails to have the subformula property.[14]

So, we can consider using the sequent calculus to try to decide whether or not a putative consequence

$$\phi_1; \ldots ; \phi_m \text{ ?- } \phi$$

holds. We do this by using the rules, read as reduction operators from conclusion to premisses, to remove the connectives from each side of the putative consequence, as follows: If we want to find a proof of $\Gamma \vdash \phi$, of what is it sufficient to have proofs ? An example will be instructive.

Consider $\phi; \phi \supset \psi; \phi \supset \chi \text{ ?- } \psi \wedge \chi$. This consequence does indeed hold, and may be established as follows:

$$\dfrac{\dfrac{\vdots}{\phi;\phi\supset\psi;\phi\supset\chi \text{ ?- } \psi}\supset \mathrm{L} \qquad \dfrac{\dfrac{}{\phi;\phi\supset\psi \text{ ?- } \phi}\,\text{Axiom} \qquad \dfrac{}{\phi;\phi\supset\psi;\chi \text{ ?- } \chi}\,\text{Axiom}}{\phi;\phi\supset\psi;\phi\supset\chi \text{ ?- } \chi}\supset \mathrm{L}}{\phi;\phi\supset\psi;\phi\supset\chi \text{ ?- } \psi\wedge\chi}\wedge\mathrm{R}.$$

Here we see an intermediate stage in the construction of a proof. The first step is to break up the conjunction $\psi \wedge \chi$ on the right. Following the right hand branch, we choose to do a $\supset$L, using $\phi \supset \chi$, thereby creating two axioms. The left hand branch, here undeveloped, is completed similarly.

From the computational point of view, we have glossed over several important issues here. For example:

- We chose to develop the right-hand branch first. Alternatively, we could have chosen the left-hand one first, or developed them together, "in parallel". In more complex situations, a parallel execution may be very attractive, yet require *communication* between the *processes* which calculate each branch;

- At each inference, we chose a proposition to reduce: for example, on the right-hand branch, we chose to reduce, using $\supset$L, $\phi \supset \chi$ rather than $\phi \supset \psi$.

[14]On the other hand, proofs using Cut are much shorter than their Cut-free counterparts [Boolos, 1998].

An algorithm to calculate proofs, or decide putative consequences, must make such choices: it must resolve the *non-determinism* that is inherent in the problem. This highlights a view of logic which has emerged from computational concerns, summarized by the slogan

$$\textit{Logic} = \textit{Inference} + \textit{Control}.$$

Thus the nature of the reasoning determined by a system of logic depends not only on the inference rules (or the satisfaction relation) but also on the régime which controls their use. These issues are very clearly seen in the logic programming language Prolog [Gillies, 1996]. Prolog is based on *proof-search*, as we have described, but the behaviour of programs written in Prolog and its variants depends critically on the choice of search strategy.

The degree of non-determinism in logics such as linear logic is greatly increased by the presence of "multiplicative" rules. To see this, consider the $\otimes R$ rule,

$$\Downarrow \quad \frac{\Gamma_1 \vdash \phi_1 \quad \Gamma_2 \vdash \phi_2}{\Gamma_1, \Gamma_2 \vdash \phi_1 \otimes \phi_2}.$$

As a rule of deduction, this presentation is adequate but, as a rule of reduction it must be written as

$$\Uparrow \quad \frac{\Gamma_1 \text{ ?- } \phi_1 \quad \Gamma_2 \text{ ?- } \phi_2}{\Gamma \text{ ?- } \phi_1 \otimes \phi_2} \quad \Gamma = \Gamma_1, \Gamma_2.$$

Now the difficulty is clear: faced with $\Gamma \vdash \phi_1 \otimes \phi_2$ the division of Γ into Γ_1 and Γ_2 must be calculated and for Γ of length n there are 2^n choices.

There is a standard solution to this problem, the so-called *input/output* model [Hodas and Miller, 1994, Winikoff and Harland, 1995, Harland et al., 1996, Harland and Pym, 1997].

- Firstly, pass all of Γ to the left-hand branch of the proof, leaving the right-hand branch to be determined later, as follows:

$$\Uparrow \quad \frac{\Gamma \text{ ?- } \phi_1 \quad ? \text{ ?- } \phi_2}{\Gamma \text{ ?- } \phi_1 \otimes \phi_2}.$$

- Proceed with the left-hand branch until (recursively) it is completed. At this point, calculate which of the formulæ in Γ have been used to complete the left-hand branch and collect into a finite sequence Γ_{left}.

- The remaining, unused, formulæ may now be passed to the right-hand branch of the proof, as follows:

$$\Uparrow \quad \frac{\Gamma_{\text{left}} \text{ ?- } \phi_1 \quad \Gamma \backslash \Gamma_{\text{left}} \text{ ?- } \phi_2}{\Gamma \text{ ?- } \phi_1 \otimes \phi_2} \quad \Gamma = \Gamma_{\text{left}}, \Gamma \backslash \Gamma_{\text{left}}.$$

A similar strategy is required for any multiplicative connective.

In **BI**, we have not only the propositional multiplicatives, $*$ and $-\!*$ (and $\#$ in the classical and related systems) but also multiplicative quantifiers, $\forall_{\mathbf{new}}$ and $\exists_{\mathbf{new}}$. In Chapters 9 and 16, we show how these connectives provide an interpretation of proof-search, and hence of logic programming, based on the sharing and non-sharing of resources.

Logical frameworks

A key concept in linguistics, particularly in the context of computing, is that of a *meta-language*, *i.e.*, a language that is used to describe other languages. The languages that may be described by a given meta-language are called its *object-languages*. A language, be it meta- or object-, may be structured so as to describe a system of *logic*, the essential aspects of such a structure being the isolation of identities such as *propositions* and *inference rules* or *axioms*.

It is important to emphasize that this no idle exercise: the proliferation of useful logical systems is great, not only in mathematics and philosophy, but particularly in areas such as software engineering, linguistics and artificial intelligence. We must have a uniform theory of how to represent them to machines.

A meta-language may be either formal or informal. For example, the informal meta-language usually used for describing systems of logic, *e.g.*, in textbooks such as [Prawitz, 1965, Chellas, 1980], is English. However, in order to describe a system of logic to a computer, perhaps for the purpose of implementing a theorem prover or a logic programming language, English will not do — because it is too imprecise. For this purpose, we need a *formal* meta-language.

So, *logical frameworks* are formal meta-logics which, *inter alia*, provide languages for describing logics in a manner that is suitable for mechanical implementation. The field of logical frameworks began with a seminal paper by Harper, Honsell and Plotkin [Harper et al., 1993] (see also [Avron et al., 1992, Pym, 1990, Pym and Wallen, 1991, Galmiche and Pym, 2000]), although the ideas may be seen in work of Aczel, Prawitz and Martin-Löf (see the references in [Harper et al., 1993]), which themselves may be seen to be in a tradition beginning with Kant's *Logik*. The basic ideas are rather elegant. In order to describe a logical framework, we must [Pym, 1996, Ishtiaq and Pym, 1998, Galmiche and Pym, 2000]:

1 Characterize the class of object-logics to be represented;

2 Give a meta-logic or language, together with its meta-logical status *vis-à-vis* the class of object-logics;

3 Characterize the representation mechanism for object-logics.

The above prescription may be summarized by the slogan

$$Framework = Language + Representation.$$

We restrict our attention to frameworks based on type-theoretic meta-logics.

The leading example of a logical framework is the aptly named LF, due to Harper, Honsell and Plotkin [Harper et al., 1993] (see also [Avron et al., 1992, Pym, 1990, Pym and Wallen, 1991, Galmiche and Pym, 2000]). LF's language is the $\lambda\Pi$-calculus, a minimal system of first-order dependent function types. The language $\lambda\Pi$ may be viewed as presentation of the $(\supset,\forall)$-fragment of minimal predicate logic: this presentation is the meta-logic used by LF. The representation mechanism is *judgements-as-types*, used to describe object-logics presented in Gentzen's natural deduction style.

Recall the $\vee$E and $\vee$I rules found in the natural deduction presentation of **CL** or **IL**:

$$\frac{\phi_1 \vee \phi_2 \quad \begin{array}{c}[\phi_1]\\ \vdots \\ \psi\end{array} \quad \begin{array}{c}[\phi_2]\\ \vdots \\ \psi\end{array}}{\psi}\ \vee\mathrm{E} \qquad \frac{\phi_1}{\phi_1 \vee \phi_2}\ \vee\mathrm{I}_1 \qquad \frac{\phi_2}{\phi_1 \vee \phi_2}\ \vee\mathrm{I}_2\,.$$

This pairing of rules for introducing and elimination connectives is the key characteristic of natural deduction.

Now, suppose that we want to express the rule of $\vee$-elimination in a metalogic, *i.e.*, in a logical framework. We must consider what meta-linguistic and meta-logical constructions are necessary in order to define the rule. Mathematical and computational logic provide us with some hints, however. Specifically, we know that *Horn clauses* are sufficiently expressive to describe *inductive definitions* of the kind which express inference rules. We can see how this works by continuing with our example.

Recall how we read the inference rule $\vee$E.

- Firstly, it is *schematic*: it describes an inference for each choice of ϕ_1, ϕ_2 and ψ.

- Secondly, the rule describes a construction not on propositions but rather on *proofs*: the premisses are that $\phi_1 \vee \phi_2$ has a proof and that there are proofs of ψ from each of ϕ_1 and ϕ_2.
- Thirdly, it forms the *conjunction* of the three premisses.
- Fourthly, the premisses describe the *implications* of ψ by ϕ_1 and ϕ_2, respectively;
- Finally, it describes the *implication* of the conclusion by the premisses.

So, we can annotate the inference rule with the meta-logical connectives **FORALL**, **AND** and **IMPLIES** as follows:

$$\mathbf{FORALL} \quad \phi_1, \phi_2, \psi$$

$$\begin{array}{ccccc} & & [\phi_1] & & [\phi_2] \\ & & \vdots & & \vdots \\ & & \mathbf{IMPLIES} & & \mathbf{IMPLIES} \\ & & \vdots & & \vdots \\ \phi_1 \vee \phi_2 & \mathbf{AND} & \psi & \mathbf{AND} & \psi \\ \hline & & \psi & & \end{array} \quad \mathbf{IMPLIES} \qquad \vee E$$

In LF, $\vee$-elimination is represented by a formal constant symbol, $\vee$E, which has as its type the meta-logical description of the inference rule, corresponding exactly to the annotated figure above:

$$\begin{array}{lll} \vee\mathrm{E} & : & \Pi\phi_1 : o\,.\,\Pi\phi_2 : o\,.\,\Pi\psi : o\,. \\ & & \\ & & (\mathbf{proof}(\phi_1 \vee \phi_2) \\ & & \& \\ & & (\mathbf{proof}(\phi_1) \to \mathbf{proof}(\psi)) \\ & & \& \\ & & (\mathbf{proof}(\phi_2) \to \mathbf{proof}(\psi))) \\ & & \\ & & \to \mathbf{proof}(\psi) \end{array}$$

Here we have introduced some new syntax, that of *type theory*. The $\lambda\Pi$-calculus is an example of such a type theory. The Π is the meta-logical *universal quantifier*, or "for all", but the variable over which it quantifies must be typed. In this case, the type is o, of propositions, so that $\phi : o$ should be read as "ϕ is a proposition". We have also used a meta-logical implication, denoted by $\to$, and conjunction, denoted by &.

Finally, we have used the *judgement* **proof**, which has type $o \to \textbf{Type}$, where **Type** denotes the collection of all types. So, for any proposition ϕ, $\textbf{proof}(\phi)$ is a type. Similarly, the disjunction symbol, $\vee$, has type $(o \,\&\, o) \to o$. With these definitions, we can read the expression above as, "$\vee$E is a constant of the following type: for all propositions ϕ_1, ϕ_2 and ψ, if we have a proof of $\phi_1 \vee \phi_2$, a proof of ψ built out of a proof of ϕ_1 and a proof of ψ built out of a proof of ϕ_2, then we can build a proof of ψ". This illustrates the essence of the judgements-as-types notion of representation. It should be clear that the technique is rather general. It is applicable to a wide, and quantifiable, range of logical systems.

In order to describe a system of logic $\mathcal{L}$ in LF, we must describe all of the syntax and inference rules of the logic in the way we described $\vee$E. Such a collection of definitions is called the LF signature of the logic $\mathcal{L}$ and is usually written $\Sigma_{\mathcal{L}}$. Given such a signature, LF's representation of $\mathcal{L}$'s proofs is constructed by "function application", just as in functional programming using languages such as Lisp, ML, Miranda, Haskell and Scheme. To see how this works, consider again the definition of $\vee$E.

- We can construct an *instance* of the schema by applying the constant $\vee$E to three particular propositions, corresponding to each of the Π-quantified propositional variables ϕ_1, ϕ_2 and ψ, to obtain

$$
\begin{array}{lll}
(\vee\mathrm{E})(p_1)(p_2)(q) & : & (\textbf{proof}(p_1 \vee p_2) \\
 & & \& \\
 & & (\textbf{proof}(p_1) \to \textbf{proof}(q)) \\
 & & \& \\
 & & (\textbf{proof}(p_2) \to \textbf{proof}(q))) \\
 & & \\
 & & \to \textbf{proof}(q)
\end{array}
$$

  by replacing each of ϕ_1, ϕ_2 and ψ by p_1, p_2 and q, respectively. Formally, this construction is effected by three applications of the type-theoretic rule of "Π-application":

$$
\frac{\Gamma \vdash_{\Sigma_{\mathcal{L}}} M : \Pi x : A \,.\, B \quad \Gamma \vdash_{\Sigma_{\mathcal{L}}} N : A}{\Gamma \vdash_{\Sigma_{\mathcal{L}}} MN : B[N/x]}.
$$

- Given such an instance of the $\vee$E-rule, we can apply it. To do so we must have a proof of $p_1 \vee p_2$, a proof of q from p_1 and a proof of q from p_2. Let's call them P_1, P_2 and Q, of type $\textbf{proof}(p_1)$, $\textbf{proof}(p_2)$ and $\textbf{proof}(q)$, respectively. The instance of the rule takes as arguments this triple of proofs and returns a proof of q, *i.e.*,

$$
(\vee\mathrm{E})(p_1)(p_2)(q)(P_1, P_2, Q) : \textbf{proof}(q).
$$

- The term $(\vee\mathrm{E})(p_1)(p_2)(q)(P_1, P_2, Q)$ may be used in LF's representation of $\mathcal{L}$ whenever a term of type $\mathbf{proof}(q)$ is required.

Of course, not all representations of an object-logic are appropriate for our purposes. Typically, we require that a representation be *uniform*. To understand this notion, we must be a bit more precise about the central notion of mathematical logic, *i.e.*, *logical consequence*. Recall that a consequence relation is a relation between finite sequences of propositions (the antecedents or hypotheses) and propositions (the succedents or conclusions). Such a relation is usually written in the form $\phi_1, \ldots, \phi_m \vdash \phi$ (or $\Gamma \vdash \phi$) and in classical and intuitionistic logic is required to satisfy the following three *structural rules*:

- Exchange: if $\Gamma \vdash \phi$, then $\Delta \vdash \phi$, where Δ is a permutation of Γ;
- Weakening: if $\Gamma \vdash \phi$, then $\Gamma; \Gamma' \vdash \phi$;
- Contraction: if $\Gamma; \Gamma \vdash \phi$, then $\Gamma \vdash \phi$.

Given a consequence $\Gamma \vdash \phi$, we can ask for a *realizer* Φ for it. A *realizer* is an operator which transforms Γ into ϕ. In intuitionistic logic, realizers amount to functions and are described by terms of Church's λ-calculus according to the so-called Curry-Howard-de Bruijn correspondence [Howard, 1980, Barendregt, 1992]. In general, realizers are given by the proof trees of the logic in question.

The class of uniform representations is identified by considering *surjective* mappings between the proofs of consequences of the object-logic $\Delta \vdash_L \phi$ and terms M, such as the one constructed above, of the meta-logic such $\Gamma_\Delta \vdash_{\Sigma_L} M : A_\phi$, where Γ_Δ denotes LF's representation of Δ and A_ϕ denotes the LF-type which represents ϕ. So, all judgements in the meta-logic have corresponding judgements in the object-logic. It has been shown that LF may be used to uniformly represent a very wide, and quantifiable, class of logical systems, including natural deduction style presentations (as discussed, by example, above) of:

- Classical first-order predicate logic, in which we have the familar propositional connectives, conjunction, disjunction, implication and negation, used to combine predicates of the form $\phi(x)$, such as $Red(x)$, or, "x is red". We also permit quantification over first-order variables to form. For example, $\forall x \,.\, \phi(x)$, or, "all xs are red";
- Classical higher-order predicate logic, in which we permit quantification not only over variables which stand for terms, such as the names of individuals, but also over predicates themselves. For example, $\exists R \,.\, Transitive(R)$, or, "there is a relation R which is transitive";

- Intuitionistic first-order predicate logic, in which we take weaker axioms for negation than in classical predicate calculus. Specifically, the system is formulated so that "law of the excluded middle",

$$\vdash \phi \vee \neg\phi$$

does not hold. Semantically, this decision takes us into the realm of *possible worlds* and *toposes* [Lambek and Scott, 1986].

- The *modal logic* S4, in which we use the modalities *necessitation* and *possibility*. Necessitation,

$$\Box\phi \qquad \text{or "}\phi\text{ is a necessary truth"},$$

is used to express the fact that ϕ is true in *all* possible worlds, or, roughly, in all models. Possibility,

$$\Diamond\phi \qquad \text{or "}\phi\text{ is a possible truth"},$$

is used to express the fact that ϕ is true is *some* possible world, or, roughly, in some model. See [Chellas, 1980] for more details about modal logics;

- Hoare's program logic [Apt, 1989], in which the propositions are triples of the form

$$\{\phi_{PRE}\}\ \mathcal{P}\ \{\phi_{POST}\}.$$

Here logical assertions ϕ_{PRE} and ϕ_{POST} are used to reason about programs $\mathcal{P}$ by asserting, respectively, pre- and post-conditions which must hold before a program may execute correctly and after it has executed correctly. See [Winskel, 1993] for more details.

Notably absent from the list above are logics from the substructural [Restall, 1999] family, including Lambek's systems [Lambek, 1968, Lambek, 1969, Lambek, 1972], relevant logics [Anderson et al., 1992, Dunn, 1986] and, indeed, linear logic. As we have seen, one of the key characteristics of linear logic is its failure to satisfy the structural properties of Weakening and Contraction,

$$\frac{\Gamma \vdash \psi}{\Gamma, \phi \vdash \psi}\ W \qquad \frac{\Gamma, \phi, \phi \vdash \psi}{\Gamma, \phi \vdash \psi}\ C.$$

In the absence of these structural properties, conjunction (similarly disjunction) decomposes into *additive*, *e.g.*,

$$\frac{\Gamma \vdash \phi_1 \quad \Gamma \vdash \phi_2}{\Gamma \vdash \phi_1 \wedge \phi_2}\ \wedge R$$

and *multiplicative*, *e.g.*,

$$\frac{\Gamma_1 \vdash \phi_1 \quad \Gamma_2 \vdash \phi_2}{\Gamma_1, \Gamma_2 \vdash \phi_1 \otimes \phi_2} \otimes R$$

versions. Note that in the additive version the Γ is shared by the two premisses and propagated to the conclusion, but that in the multiplicative version Γ_is ($i = 1, 2$) from the two premisses must be combined in the conclusion.

In order to uniformly represent substructural logics (and, it seems, some program logics) in a logical framework, it seems we must have a framework based on a substructural type theory. This is the point of departure for the RLF logical framework which is based on a type theory, $\lambda\Lambda$, which corresponds to a structural variant of a fragment of **BI** which is closely related to linear logic. We present RLF and $\lambda\Lambda$ in Chapter 15. RLF's semantics may be seen as a resource semantics and our development is guided by this intuition.

We conclude Chapter 15 with a somewhat speculative discussion of the proof- and model-theoretic issues in the design of a truly bunched dependent type theory, *i.e.*, a type theory which stands in a suitably robust propositions-as-types correspondence with predicate **BI**.

I

PROPOSITIONAL BI

Chapter 1

INTRODUCTION TO PART I

Some of the content of Part I has appeared in [O'Hearn and Pym, 1999, Pym, 1999, Pym et al., 2000, Armelín and Pym, 2001]. References are given in the text as appropriate.

—DJP

1. A Proof-theoretic Introduction

One of the most important outcomes of the study of linear logic, much more than the formal system itself, is its revealing of the computational significance of the structural rules of Weakening and Contraction [Girard, 1987]. Logically, their absence leads to the decomposition of conjunction into *additive* (&) and *multiplicative* ($\otimes$) forms, which may be given a sequential natural deduction presentation as follows:

$$\frac{\Gamma \vdash \phi \quad \Gamma \vdash \psi}{\Gamma \vdash \phi \& \psi}\ \& I \qquad \frac{\Gamma \vdash \phi_1 \& \phi_2}{\Gamma \vdash \phi_i}\ (i = 1, 2) \quad \& E$$

$$\frac{\Gamma \vdash \phi \quad \Delta \vdash \psi}{\Gamma, \Delta \vdash \phi \otimes \psi}\ \otimes I \qquad \frac{\Gamma, \phi, \psi \vdash \chi \quad \Delta \vdash \phi \otimes \psi}{\Gamma, \Delta \vdash \chi}\ \otimes E.$$

If we have the rules of Weakening and Contraction

$$\frac{\Gamma \vdash \psi}{\Gamma, \phi \vdash \psi}\ W \qquad \frac{\Gamma, \phi, \phi \vdash \psi}{\Gamma, \phi \vdash \psi}\ C,$$

then these rules for $\otimes$ and & define the same connective, but without them the connectives are distinct. A similar decomposition obtains for disjunction, the multiplicative version arising in the classical (as opposed to intuitionistic) setting.

The decomposition of the connectives has a long history in the relevant logic tradition but the possibilities revealed by restricting structural rules were given a new perspective by the appealing "resource interpretations" of linear logic. The leading example is perhaps the *number-of-uses* reading in which a proof of a linear implication $\phi \multimap \psi$ determines a function that *uses* its argument exactly once. Like $\otimes$, the linear implication is multiplicative, which is to say that it uses separate contexts in its elimination rule.

$$\frac{\Gamma, \phi \vdash \psi}{\Gamma \vdash \phi \multimap \psi} \multimap I \qquad \frac{\Gamma \vdash \phi \multimap \psi \quad \Delta \vdash \phi}{\Gamma, \Delta \vdash \psi} \multimap E.$$

However, an important message of linear logic is that, in order to obtain an expressive system, one cannot stay in the pretty-but-weak purely multiplicative system: it seems crucial to allow access to structurals in some manner. In linear logic, this is done via the "!" modality, which admits a recovery of intuitionistic (or additive) implication $\phi \to \psi$ as $!\phi \multimap \psi$. The number-of-uses reading of implication is extended to $!\phi$ as "as many ϕs as required".[1]

Access to the structurals may be recovered in another, rather different, way, not involving a modality. Just as we decomposed conjunction directly into multiplicative and additive parts, so we can decompose *implication* directly. The technical cost of this conceptual symmetry is that we must work with a more richly structured notion of sequent, entailing a more delicate analysis of the proof-theoretic relationship between implication and conjunction.

Implication is inextricably bound up with conjunction, or at least with antecedent-forming operations used to formulate sequents. This connection goes so far that it is sometimes said that an introduction rule

$$\frac{\Gamma, \phi \vdash \psi}{\Gamma \vdash \phi \to \psi}$$

may be regarded, proof-theoretically, as defining the *meaning* of $\to$ [Sundholm, 1986]; it is clear that the character of the implication in a logic is, in a sense, determined by that of the comma or conjunction.

If, as is the case in **BI**, we have two forms of implication, then we are faced with the question of which of them to use in the introduction rule.

$$\frac{\Gamma, \phi \vdash \psi}{\Gamma \vdash \phi ? \psi}.$$

[1]The rôle of the modalities, or exponentials, is central to the development of linear logic.

That is, should the conclusion $\phi ? \psi$ in this rule be a multiplicative or additive implication ?

The connection between introduction rules and implications suggests a solution: If an antecedent-forming operation determines the behaviour of an implication, and we have two implications, then we should have two antecedent-forming operations. So, we postulate a new context-forming operation, ";", and stipulate that Contraction and Weakening are permitted for ";" but not for ",". The introduction rules then become

$$\frac{\Gamma , \phi \vdash \psi}{\Gamma \vdash \phi \mathrel{-\!\!*} \psi} \quad \text{and} \quad \frac{\Gamma ; \phi \vdash \psi}{\Gamma \vdash \phi \to \psi} .$$

The antecedents are no longer sequences; rather, they are trees with propositions as leaves and internal nodes labelled with "," or ";", or in short, *bunches* [Dunn, 1975, Belnap, 1982, Read, 1988].

Corresponding to **BI**'s natural deduction system is a simply-typed lambda calculus, $\alpha\lambda$, which gives a representation of **BI**'s natural deduction proof-objects. For example, the typing rules for the two kinds of lambda-abstraction, corresponding to the right rules for $\mathrel{-\!\!*}$ and $\to$ are, respectively

$$\frac{\Gamma , x : A \vdash M : B}{\Gamma \vdash \lambda x : A.M : A \mathrel{-\!\!*} B} \quad \text{and} \quad \frac{\Gamma ; x : A \vdash M : B}{\Gamma \vdash \alpha x : A.M : A \to B} .$$

Here we are working with bunches of typed variables, rather than bunches of formulæ, in which no variable may occur more than once. There are two combinators for application, one for each abstractor.

BI's proof theory may also be presented as a sequent calculus, in which, as explained in the introduction, elimination rules are replaced by rules which introduce connectives to the left-hand sides of sequents.

For example, the rules for eliminating $\to$ and $\mathrel{-\!\!*}$,

$$\frac{\Gamma \vdash \phi \to \psi \quad \Delta \vdash \phi}{\Gamma ; \Delta \vdash \psi} \to E \quad \text{and} \quad \frac{\Gamma \vdash \phi \mathrel{-\!\!*} \psi \quad \Delta \vdash \phi}{\Gamma , \Delta \vdash \psi} \mathrel{-\!\!*} E,$$

are replaced by

$$\frac{\Gamma \vdash \phi \quad \Delta(\psi) \vdash \chi}{\Delta(\Gamma ; \phi \to \psi) \vdash \chi} \to L \quad \text{and} \quad \frac{\Gamma \vdash \phi \quad \Delta(\psi) \vdash \chi}{\Delta(\Gamma , \phi \mathrel{-\!\!*} \psi) \vdash \chi} \to L,$$

respectively.

We establish the Cut-elimination theorem for **BI** and show that the natural deduction system and the sequent calculus are equivalent.

2. A Semantic Introduction

It is all very well to describe proof systems in this way and we can indeed argue that the proof-theoretic "meaning", in the sense of [Sundholm, 1986], is rather clear. However, we must ask what, if any, is the semantic significance of the resulting logic ? We argue herein that **BI** possesses three very natural semantics:

- Algebraic and topological models, in the tradition of Boole, Heyting and Tarski;
- Categorical models, in the tradition of Brouwer, Heyting, Kolmogorov, Dana Scott and Lambek;
- Kripke models, in the tradition of Kripke, Beth, Tarski and Joyal, as represented by [Lambek and Scott, 1986].

Of course, as we explain at the appropriate points in our development, these three semantics are intimately related to one another, being instances of the same abstract construction. However, their motivations and styles are sufficiently different to warrant separate presentations.

2.1 Algebraic and Topological Semantics

The first semantics, described in Chapter 3, follows, on the one hand, in the tradition of Heyting [Girard, 1989] and, on the other, in that of Tarski [Tarski, 1956]. We describe an algebraic structure corresponding to the logical structure of **BI**. Such a structure combines that required for the additive, intuitionistic, part of **BI**, *i.e.*, a Heyting algebra, with that required for the multiplicative part, *i.e.*, the $(\otimes, I, \multimap)$-fragment of intuitionistic linear logic. This latter structure is closely related to the algebraic structures described in [Restall, 1999, Trolestra, 1992].

We also give an associated syntactic calculus, which amounts to a Hilbert-type proof system for **BI**.

2.2 Categorical Semantics

The second semantics is a BHK-style semantics of the proof theory of propositional **BI** which arises from *doubly closed categories* (DCCs), in which a single category admits two closed structures or function spaces. It shows clearly the difference with linear logic, where two closed categories, connected by a monoidal co-monad [Barber, 1996], are usually used.

The purely multiplicative fragment of **BI**, sometimes called BCI logic, sometimes called multiplicative intuitionistic linear logic, has been studied in many different contexts but a categorical semantics is studied in some detail in Lambek's work on "Deductive Systems and Categories" [Lambek, 1968, Lambek, 1969, Lambek, 1972].

Categorical models of the proofs of predicate logics are a more delicate matter. Just as for intuitionistic predicate logic, for predicate **BI** we move to models with indexed, or fibred, structure [Seely, 1983].

2.3 Kripke Semantics

The third semantics is a Kripke-style semantics of formulæ, which combines Kripke's semantics of intuitionistic logic and Urquhart's semantics of multiplicative intuitionistic linear logic based on the idea of "pieces of information" [Urquhart, 1972]. This second semantics gives **BI** more of a genuine status as a logic: It gives us a way to read the formulæ as propositions that are true (or not) relative to a given world. It is motivated by the notion of resource, as discussed in the Introduction.

Recall from the Introduction that we require a resource to be a preordered commutative monoid, informally, a "Kripke resource monoid",

$$\mathcal{M} = (M, e, \cdot, \sqsubseteq),$$

in which we require the following bifunctoriality condition:

$$\text{if} \quad m \sqsubseteq m' \quad \text{and} \quad n \sqsubseteq n', \quad \text{then} \quad m \cdot n \sqsubseteq m' \cdot n'.$$

Whenever we refer to a preordered commutative monoid we assume that this condition holds.

Preordered commutative monoids may be used to provide a truth-functional, Kripke-style semantics for **BI**. The basic idea of **BI** is to allow the connectives of multiplicative intuitionistic linear logic, **MILL**, and intuitionistic logic, **IL**, to exist side-by-side, without recourse to the modalities used to recover intuitionistic strength in linear logic. This may be done with an inductively defined forcing relation $\models$. The clauses for the additives follow the intuitionistic pattern:

$$m \models \phi \wedge \psi \quad \text{iff} \quad m \models \phi \quad \text{and} \quad m \models \psi$$

$$m \models \phi \vee \psi \quad \text{iff} \quad \text{either} \quad m \models \phi \quad \text{or} \quad m \models \psi$$

$$m \models \phi \to \psi \quad \text{iff} \quad \text{for all} \quad n \sqsubseteq m, \quad n \models \phi \quad \text{implies} \quad n \models \psi.$$

The clauses for the multiplicatives are more interesting, following Urquhart's [Urquhart, 1972] semantics for **MILL**:

$$m \models \phi * \psi \quad \text{iff} \quad \text{for some } n, n' \in M,\ m \sqsubseteq n \cdot n'\ ,\ n \models \phi \text{ and } n' \models \psi$$

$$m \models \phi \mathrel{-\!\!*} \psi \quad \text{iff} \quad \text{for all } n \in M,\ n \models \phi \text{ implies } m \cdot n \models \psi.$$

The case for $*$ may be read as requiring that a resource be *partitioned* into components which force the constituents of the proposition. The case for $\mathrel{-\!\!*}$ may be read as combining the cost of a function with the cost of its argument to give the cost of its value. The additive cases may be read as requiring the *conservation* of resources.

We must also take care of the logical units, of $\wedge$, $*$ and $\vee$, with the ones for $\wedge$ and $*$ being unsurprising,

$$m \models \top \quad \text{for all} \quad m \in M$$

$$m \models I \quad \text{iff} \quad m \sqsubseteq e,$$

and have evident resource interpretations. However, the unit of $\vee$, $\bot$, offers a choice.

One possibility is to adopt the following clause for $\bot$:

$$m \not\models \bot \quad \text{for any} \quad m \in M \tag{1.1}$$

This choice allows us to state and prove a soundness theorem and a completeness theorem for **BI** without $\bot$. We discuss another choice below.

All propositions must satisfy the familiar monotonicity property from intuitionistic logic:

Kripke Monotonicity (or Hereditary) : $m \models \phi$ and $n \sqsubseteq m$ implies $n \models \phi$.

The forcing $m \models \phi$ may be read as "ϕ holds with cost m" or "ϕ holds according to the information m". Informally, we call this semantics "Kripke resource semantics".

EXAMPLE 1.1 *A simple example is provided by the monoid*

$$\mathcal{N} = (\mathbb{N}, 0, +, \leq),$$

the lifted natural numbers ordered by less than. If we interpret $\phi \mathrel{-\!\!*} \psi$ as a function from ϕ to ψ, with cost m, and ϕ as the argument of this function, with cost n, then the cost of computing the result of the function application is $m + n$.

Note that the *distributive law*,

$$\phi \wedge (\psi \vee \chi) \;{*\!*}\; (\phi \wedge \psi) \vee (\phi \wedge \chi),$$

holds in Kripke resource semantics. Consequently, there is no hope of completeness for linear logic [Girard, 1987] with respect to Kripke resource semantics: The distributive law fails for linear logic's additives.

For the mutliplicatives, this semantics amounts to the one given by Urquhart [Urquhart, 1972], based on the idea of "pieces of information", for the ($\multimap$, $\otimes$, I)-fragment of intuitionistic linear logic (or **MILL**). It should be clear that such Kripke resource monoids trivially subsume the preorders used in Kripke semantics for intuitionistic logic.

More generally, we remark that an alternative presentation of the semantics of substructural connectives is both possible and commonplace in substructural logic [Restall, 1999, Read, 1988, Anderson and Belnap, 1975, Anderson et al., 1992]. Briefly, our use of a monoidal product $\cdot$ together with an order $\sqsubseteq$ may be replaced by a ternary relation R on a set M of worlds, so that, for example, the forcing relation for $-\!\!*$ is rendered as

$$l \models \phi -\!\!* \psi \quad \text{iff} \quad \begin{array}{l} \text{for all } m, n \in M \text{ such that } R(l, m, n), \\ \text{if } m \models \phi, \text{ then } n \models \psi. \end{array}$$

This point of view is consistent with Day's [Day, 1970] use of *premonoidal categories* in his treatment of tensor products on functor categories. We discuss Day's construction in Chapter 3 and exploit it in the Kripke models we introduce in Chapters 4, 5, 13 and 14. However, the combination of a monoidal product $\cdot$ and an order $\sqsubseteq$ is very naturally motivated by our combination of additive (intuitionistic) and multiplicative (linear) connectives.

As we have already suggested, the handling of inconsistency, $\bot$, is a delicate matter in this setting. Indeed, the elementary "Kripke resource semantics" we have presented so far does not support a completeness theorem (see Chapter 4) in the presence of $\bot$. What are we to make of this ? The answer lies in the degree of internalization of inconsistency in the semantics. Consider that the (complete) algebraic models discussed in Chapter 3 or, more generally, the categorical semantics of proofs, also in Chapter 3, include representatives for inconsistency (the initial object, $\mathbf{0}$, which interprets $\bot$). The elementary forcing semantics, in contrast, may handle inconsistency only by denying the existence of a world at which $\bot$ is forced. Completeness for a monoid-based forcing semantics may be achieved in topological settings in which internal represtentives for inconsistency, together with a non-indecomposable treatment of disjunction [Lambek and Scott, 1986], are available. We develop such a semantics, which is naturally available in a topological setting, in Chapter 5 — for propositional **BI**, using the category of sheaves on a topological space and Grothendieck sheaves on a preordered monoid — and in Chapter 14 — for predicate **BI**, using the category of sheaves on a topological space. Whilst it may well be possible to handle completeness with inconsistency using simpler tools, we suggest that our methods represent the natural level of abstraction for this problem.

It is worth pausing to take stock. We began in the Introduction by presenting an informal cost model and we extrapolated from it to arrive at the elementary monoid semantics. However, we have pointed out that there is a technical problem in the elementary semantics: the treatment of $\perp$ is incomplete.

Of course, there are two mathematical ways of handling this: internalize $\perp$ or make $\cdot$ partial; so that, in the term monoid, $(\phi) \cdot (\phi \mathrel{-\!\!*} \perp)$ would simply be undefined. The former is more appealing from the point of view of our mathematical development. However, our whole approach to **BI**'s Kripke semantics has been inspired by our desire to model resources: From this point of view, why should $\cdot$ be total ? In fact, one of our computational examples, a model based on the computer's store [Pym et al., 2000], described in Chapter 9, is most naturally described in terms of a partial operation. A partial monoid semantics is explored by Galmiche, Méry and Pym in [Galmiche et al., 2002].

3. Towards Classical Propositional BI

So far our discussion has been confined to what we can call "intuitionistic **BI**" or, indeed, "minimal **BI**", to which we can easily add the intuitionistic negation, $\neg\phi = \phi \to \perp$. From the semantic point of view, the key characteristic of a classical system is the strength of negation. Within the single-conclusioned formulation of bunched consequence, we can strengthen the intuitionistic negation, via the addition of RAA, but it is a move to a multiple-conclusioned formulation which suggests a form of multiplicative negation.

In Chapter 7, we give a brief discussion of some classical bunched logics. There are four basic possibilities for combining the intuitionistic and classical additives and multiplicatives:

- Intuitionistic additives and intuitionistic multiplicatives;
- Classical additives and intuitionistic multiplicatives;
- Intuitionistic additives and classical multiplicatives;
- Classical additives and classical multiplicatives.

We consider, informally, algebraic semantics, proof systems and forcing semantics for classical systems. We conclude with a brief discussion of Troelstra's analysis of additive implication [Trolestra, 1992].

From a computational point view, we have several concrete models which are of interest, based on the resource interpretation of **BI**'s classical additives and intuitionistic multiplicatives. These models are described in Chapter 9.

Chapter 7 concludes with a brief discussion of Troelstra's analysis of a classical system which combines additive and multiplicative implications, using neither bunches nor exponentials [Trolestra, 1992].

4. Logical Relations

The basic relationship between proof-theoretic consequences and semantic consequences for a given logic is described by soundness and completeness theorems, the latter usually being effected by a "model existence" lemma in which a model is constructed from the syntax, or proofs, of the logic. For propositional **BI**, we provide such analyses in Chapters 4 and 5.

A useful refinement of this basic framework is, in suitable circumstances, provided by the theory of *logical relations*, [Statman, 1985b, Plotkin, 1980], which is used, among other things, to study λ-definability of functions. We provide an account of *bunched logical relations*, *i.e.*, logical relations for $\alpha\lambda$, based on a set-theoretic notion of "Kripke $\alpha\lambda$-model" , with, for simplicity, just $-\!\!*$- and $\rightarrow$-types.

5. Computational Models

We conclude Part I with a chapter on computational interpretations of **BI**, which arise from the following models of computation:

- Proof-search and (propositional) logic programming;
- Interference and non-interference in imperative programming.

The first of these, discussed in [O'Hearn and Pym, 1999], is an example of what is, perhaps, the most immediate and most basic form of computational interpretation of a logical system: the attempt to calculate proofs by treating the rules of the logic as a reductive system. In the propositional setting, the result of the computation is either failure, or success together with the proof which is calculated; there is no answer substitution to be calculated. In the setting of **BI**, we show that the (bunched) structure of the program determines which program clauses have access to which of the atomic assumptions declared in the program.

The second, also discussed in [O'Hearn and Pym, 1999] but developed at more length in [O'Hearn, 1999], is concerned with the sharing and non-sharing of memory by procedures in imperative programming languages of the kind described by Idealized Algol [Reynolds, 1981] or Reynolds' syntactic control of interference, or SCI. The basic idea is that procedures that are combined using multiplicative combination *cannot* share resources with each other and that procedures combined using additive combination *may*, though need not, share resources.

We go on to give three further examples, which are properly models of **BI** but not of the basic, (Heyting,Lambek) version, which is mainly studied herein, but rather of Boolean **BI**, as described in Chapter 7:

- Petri nets;
- A CCS-like model;
- A pointers model.

Each of these provides support for our interpretation of **BI**'s semantics as an account of resources and their computational properties. Their unifying feature is a reliance on the spatial interpretation of **BI**'s semantics. In particular, they exploit **BI**'s ability to treat local and global reasoning separately — and on equal footings — within a single semantic framework.

A discussion of the value of some aspects of the three models mentioned above may be found in [Pym et al., 2000].

Chapter 2

NATURAL DEDUCTION FOR PROPOSITIONAL BI

1. Introduction

BI has a simple and elegant proof theory, the presentation of which we begin in this section. We start with a definition of **BI** as a system of natural deduction, formulated in the sequential, or linearized, style. Rather than establish the metatheory of this system in the manner of Prawitz [Prawitz, 1965], we first formulate a representation of **BI**'s proofs as a λ-calculus, $\alpha\lambda$, with types given by **BI**'s propositions. We then establish normalization for $\alpha\lambda$. **BI**'s natural deduction system was introduced in [O'Hearn and Pym, 1999, Pym, 1999].

2. A Natural Deduction Calculus

In this section, we give a presentation of **BI** in sequential natural deduction form, *i.e.*, a sequential presentation based on introduction and elimination rules. Let L denote a set of atomic propositional letters and let p, q, *etc.* range over L. The set of **BI** propositions over L, $\mathcal{P}(L)$, is given by the following inductive definition:

PROPOSITIONS

$$
\begin{array}{rclr}
\phi & ::= & \mathrm{p} & \text{atoms} \\
 & | & I & \text{multiplicative unit} \\
 & | & \phi * \phi & \text{multiplicative conjunction} \\
 & | & \phi \mathrel{-\!\!*} \phi & \text{multiplicative implication} \\
 & | & \top & \text{additive unit} \\
 & | & \phi \wedge \phi & \text{additive conjunction} \\
 & | & \phi \rightarrow \phi & \text{additive implication} \\
 & | & \bot & \text{additive disjunctive unit} \\
 & | & \phi \vee \phi & \text{additive disjunction.}
\end{array}
$$

The additive connectives correspond to those of intuitionistic logic (**IL**), whereas the multiplicative connectives correspond to those of multiplicative, intuitionistic linear logic (**MILL**).

As we have seen, the presence of the additive and multiplicative implications as primitives forces to work with contexts which are not merely finite sequences but rather finite trees.

BUNCHES

$$
\begin{array}{lcll}
\Gamma & ::= & \phi & \text{propositional assumption} \\
 & \mid & \emptyset_m & \text{multiplicative unit} \\
 & \mid & \Gamma, \Gamma & \text{multiplicative combination} \\
 & \mid & \emptyset_a & \text{additive unit} \\
 & \mid & \Gamma ; \Gamma & \text{additive combination}
\end{array}
$$

For example, the bunch $\phi_1, ((\phi_3, \phi_4); \phi_2)$ may be drawn as

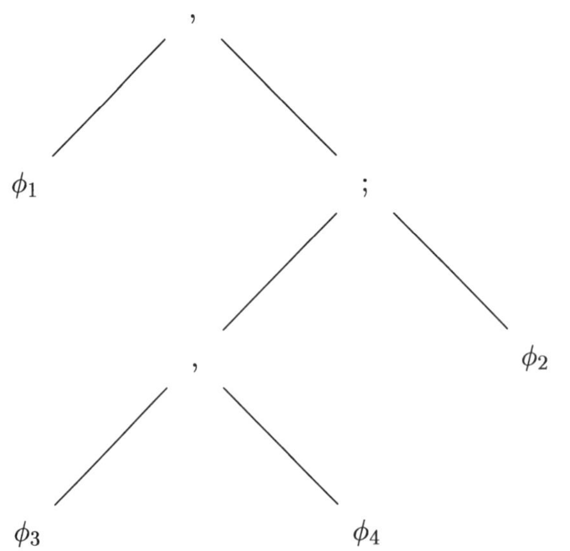

The main point of the definition of bunches is that ";" admits the structural properties of weakening and contraction, whereas "," does not: this distinction allows the correct formulation of the two implications. Bunches are structured as trees, with internal nodes labelled with either "," or ";" and leaves labelled with propositions. Bunches may be represented using lists of lists, *etc.* as described in [Read, 1988]. We write $\Gamma(\Delta)$, and refer to Δ as a *sub-bunch* of Γ, for a bunch Γ in which Δ appears as a sub-tree and write $\Gamma[\Delta'/\Delta]$ for Γ with Δ replaced by Δ'. We write $\Gamma(-)$ to denote a bunch Γ which is incomplete and which may be completed by placing a bunch in its hole, and will use this notation to refer to that part of $\Gamma(\Delta)$ which is not part of Δ. We require that "," and ";" be commutative monoids, giving rise to the coherent equivalence, $\Gamma \equiv \Gamma'$, as follows:

COHERENT EQUIVALENCE: $\Gamma \equiv \Gamma'$

1 Commutative monoid equations for $\emptyset_a$ and ";".

2 Commutative monoid equations for $\emptyset_m$ and ",".

3 Congruence: if $\Delta \equiv \Delta'$ then $\Gamma(\Delta) \equiv \Gamma(\Delta')$.

Note that ";" and "," do not distribute over one another. We use $=$ for syntactic identity of bunches.

Although we have given the basic definition of bunches, a more structured presentation, stratified bunches, is possible. The idea is to stratify bunches into multiplicative and additive sub-bunches. So, if the top-most bunch-former is ",", then we get

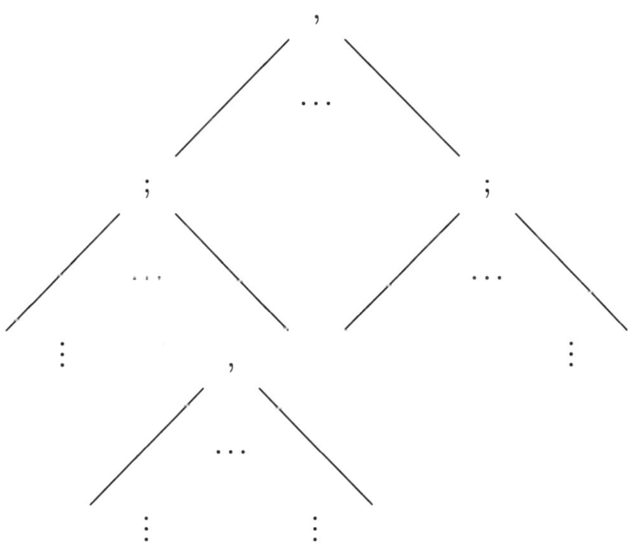

and alternately if the top-most bunch-former is ";".

This presentation is discussed and exploited in Pablo Armelín's work on logic programming with **BI**, where it simplifies the definition of an operational semantics (*q.v.* Chapters 9 and 16).

We call the natural deduction system for propositional **BI**, given in Table 2.1, **NBI**.

IDENTITY AND STRUCTURE

$$\frac{}{\phi \vdash \phi}\ Axiom \qquad \frac{\Gamma \vdash \phi}{\Delta \vdash \phi} \equiv \text{(where } \Delta \equiv \Gamma) \ E$$

$$\frac{\Gamma(\Delta) \vdash \phi}{\Gamma(\Delta; \Delta') \vdash \phi}\ W \qquad \frac{\Gamma(\Delta; \Delta) \vdash \phi}{\Gamma(\Delta) \vdash \phi}\ C$$

MULTIPLICATIVES

$$\frac{}{\emptyset_m \vdash I}\ I\,I \qquad \frac{\Gamma(\emptyset_m) \vdash \chi \quad \Delta \vdash I}{\Gamma(\Delta) \vdash \chi}\ I\,E$$

$$\frac{\Gamma \vdash \phi \quad \Delta \vdash \psi}{\Gamma, \Delta \vdash \phi * \psi}\ *I \qquad \frac{\Gamma(\phi, \psi) \vdash \chi \quad \Delta \vdash \phi * \psi}{\Gamma(\Delta) \vdash \chi}\ *E$$

$$\frac{\Gamma, \phi \vdash \psi}{\Gamma \vdash \phi \mathrel{-\!\!*} \psi}\ {-\!\!*}\,I \qquad \frac{\Gamma \vdash \phi \mathrel{-\!\!*} \psi \quad \Delta \vdash \phi}{\Gamma, \Delta \vdash \psi}\ {-\!\!*}\,E$$

ADDITIVES

$$\frac{}{\emptyset_a \vdash \top}\ \top I \qquad \frac{\Gamma(\emptyset_a) \vdash \chi \quad \Delta \vdash \top}{\Gamma(\Delta) \vdash \chi}\ \top E$$

$$\frac{\Gamma \vdash \phi \quad \Delta \vdash \psi}{\Gamma; \Delta \vdash \phi \wedge \psi}\ \wedge I \qquad \frac{\Gamma(\phi; \psi) \vdash \chi \quad \Delta \vdash \phi \wedge \psi}{\Gamma(\Delta) \vdash \chi}\ \wedge E$$

$$\frac{\Gamma; \phi \vdash \psi}{\Gamma \vdash \phi \to \psi}\ \to I \qquad \frac{\Gamma \vdash \phi \to \psi \quad \Delta \vdash \phi}{\Gamma; \Delta \vdash \psi}\ \to E$$

$$\frac{\Gamma \vdash \bot}{\Gamma \vdash \phi}\ \bot E$$

$$\frac{\Gamma \vdash \phi_i}{\Gamma \vdash \phi_1 \vee \phi_2}\ (i = 1,\, 2) \quad \vee I \qquad \frac{\Gamma \vdash \phi \vee \psi \quad \Delta(\phi) \vdash \chi \quad \Delta(\psi) \vdash \chi}{\Delta(\Gamma) \vdash \chi}\ \vee E$$

Table 2.1. Propositional **NBI**

Notice that the introduction and elimination rules for additive and multiplicative implications, conjunctions and units are identical in form, following Prawitz's prescription [Prawitz, 1971]. The difference between them is the antecedent-combining operations they use. We can replace the ∧E rule with the simpler, and perhaps more familiar, form

$$\frac{\Gamma \vdash \phi_1 \wedge \phi_2}{\Gamma \vdash \phi_i} \quad i = 1, 2.$$

As usual, we have the following easy lemma:

LEMMA 2.1 *1* $\Gamma(\phi_1, \phi_2) \vdash \psi$ *iff* $\Gamma(\phi_1 * \phi_2) \vdash \psi$;

2 $\Gamma(\phi_1; \phi_2) \vdash \psi$ *iff* $\Gamma(\phi_1 \wedge \phi_2) \vdash \psi$.

□

The additive maintenance of bunches may be made explicit by replacing each of the binary and ternary rules by their explicitly additive counterparts, as follows:

$$\frac{}{\phi; \Gamma \vdash \phi} \; \mathit{Axiom}$$

$$\frac{\Gamma \vdash \phi \rightarrow \psi \quad \Gamma \vdash \phi}{\Gamma \vdash \psi} \rightarrow E \qquad \frac{\Gamma \vdash \phi \quad \Gamma \vdash \psi}{\Gamma \vdash \phi \wedge \psi} \wedge I$$

$$\frac{\Gamma \vdash \phi \vee \psi \quad \Gamma; \phi \vdash \chi \quad \Gamma; \psi \vdash \chi}{\Gamma \vdash \chi} \vee E.$$

Call the explicit system so obtained $\mathbf{NBI}^{\heartsuit}$. Then the following lemma follows by a familiar and straightforward induction on the structure proofs, using the admission by ";" of Weakening and Contraction:

LEMMA 2.2 $\Gamma \vdash \phi$ *is provable in* **NBI** *if and only if it is provable in* $\mathbf{NBI}^{\heartsuit}$.

In fact, Weakening and Contraction can be omitted from $\mathbf{NBI}^{\heartsuit}$ by modifying the way which is standard for intuitionistic logic [Troelstra and Schwichtenberg, 1996]. Weakening is pushed to the leaves of proofs via the axiom of the form

$$\frac{}{\Gamma; \phi \vdash \phi} \; \mathit{Axiom},$$

but must also be built into the binary multiplicative rules. For example, $-\!\!* \, E$ must be reformulated as

$$\frac{\Gamma \vdash \phi -\!\!* \psi \quad \Delta \vdash \phi}{(\Gamma, \Delta); \Theta \vdash \psi} -\!\!* \, E.$$

To see why, consider that in a proof

$$\dfrac{\dfrac{\phi \mathrel{-\!\!*} \psi \vdash \phi \mathrel{-\!\!*} \psi \quad \phi \vdash \phi}{\phi \mathrel{-\!\!*} \psi, \phi \vdash \psi}\ {-\!\!*}E}{(\phi \mathrel{-\!\!*} \psi, \phi); \chi \vdash \psi}\ W$$

there is no way, because ";" does not distribute over ",", to push the Weakening above the application. No such amendment to the binary additive rules is necessary because of the associativity of ";"; see also [O'Hearn, 1999].

Contraction is incorporated into the rules for the additives. However, we shall see that Contraction cannot be eliminated from the corresponding calculus of λ-terms for **BI**. We will see this by comparing the realizing terms for

$$(\chi, \chi \mathrel{-\!\!*} (\phi * \psi)); (\chi, \chi \mathrel{-\!\!*} (\phi * \psi)) \vdash \phi * \psi\,, \tag{2.1}$$

inferred from

$$\phi, \psi; \chi, \chi \mathrel{-\!\!*} (\phi * \psi) \vdash \phi * \psi \qquad \text{and} \qquad \chi, \chi \mathrel{-\!\!*} (\phi * \psi) \vdash \phi * \psi$$

using $*E$, and

$$\chi, \chi \mathrel{-\!\!*} (\phi * \psi) \vdash \phi * \psi. \tag{2.2}$$

LEMMA 2.3 (ADMISSIBILITY OF CUT) *The following version of the cut rule is admissible in* **NBI***:*

$$\frac{\Gamma(\phi) \vdash \psi \quad \Delta \vdash \phi}{\Gamma(\Delta) \vdash \psi}\,.$$

PROOF By induction on the structure of proofs in **NBI**. □

Note that the rule given Lemma 2.3 covers both the ";" and "," cases in the construction of bunches.

One important property of **BI** is that it is conservative over both **MILL** and **IL**. In the case of **IL**, this means that $\phi_1; \ldots ; \phi_n \vdash \phi$ is provable in **BI** if and only if $\phi_1, \ldots , \phi_n \vdash \phi$ is provable in **IL**, where each ϕ_i and ϕ is a proposition built up using additive connectives only. For this result, $\emptyset_a$ plays the role of the empty context in **IL**. For **MILL** the statement is similar, using "," and $\emptyset_m$ and propositions built from multiplicatives. That these conservativity properties hold may be seen straightforwardly from a semantics of **BI**'s proof theory.

Finally, note that **BI**'s two implications each give rise to a notion of theorem: a proposition τ is a *theorem* if either

- $\emptyset_a \vdash \tau$ or
- $\emptyset_m \vdash \tau$

is provable. However, we have that if $\emptyset_a \vdash \phi$, then $\emptyset_m \vdash \phi$, so that we need just

- $\emptyset_m \vdash \phi$.

3. The $\alpha\lambda$-calculus

We define a typed λ-calculus whose types are given by **BI**'s propositions. Corresponding to each connective, or type-constructor, there is an operation on terms, which may be considered proof-objects for **NBI**, *i.e.*, **BI** stands in propositions-as-types correspondence [Howard, 1980, Barendregt, 1992] with $\alpha\lambda$.

Contexts are structured as bunches. We call the following system the *simply-typed* $\alpha\lambda$*-calculus*:

BUNCHES

$$\begin{array}{rclr} \Gamma & ::= & x : \phi & \text{variables} \\ & | & \emptyset_m & \text{multiplicative unit} \\ & | & \Gamma, \Gamma & \text{multiplicative combination} \\ & | & \emptyset_a & \text{additive unit} \\ & | & \Gamma ; \Gamma & \text{additive combination} \end{array}$$

We associate distinct variables with each proposition that occurs in a context and adopt the coherent equivalence, $\equiv$, and congruence, $\cong$, as in **BI**. We write $\Gamma \cong \Delta$ to indicate that Γ and Δ are isomorphic as labelled trees. We write $i(\Gamma)$ to denote the set of identifiers (variables) in the bunch Γ. The rules of **NBI** for the $\alpha\lambda$-calculus are given in Table 2.2, in which $\perp_\phi(M)$ is the canonical term of type ϕ constructed from any term t of type $\perp$.

We will usually suppress the explicit $\mathbf{app}_{-*}$ and $\mathbf{app}_{\rightarrow}$ in $\alpha\lambda$-terms, writing just MN for $\mathbf{app}_{\_}(M, N)$, provided there is little likelihood of confusion.

Turning to the equational theory of $\alpha\lambda$-terms, we use the symbol $\succcurlyeq$ to denote one-step reductions, from left to right.

IDENTITY AND STRUCTURE

$$\frac{}{x : \phi \vdash x : \phi}\ Axiom \qquad \frac{\Gamma \vdash M : \phi \quad \Delta(x : \phi) \vdash N : \psi}{\Delta(\Gamma) \vdash N[M/x] : \psi}\ Cut$$

$$\frac{\Gamma(\Delta) \vdash M : \phi}{\Gamma(\Delta; \Delta') \vdash M : \phi}\ W \qquad \frac{\Gamma(\Delta; \Delta') \vdash M : \phi}{\Gamma(\Delta) \vdash M[i(\Delta)/i(\Delta')] : \phi}\ (\Delta' \cong \Delta)\quad C$$

$$\frac{\Gamma \vdash M : \phi}{\Delta \vdash M : \phi}\ (\Delta \equiv \Gamma)\quad E$$

UNITS

$$\frac{}{\emptyset_m \vdash I : I}\ I\,I \qquad \frac{\Gamma(\emptyset_m) \vdash N : \phi \quad \Delta \vdash M : I}{\Gamma(\Delta) \vdash \mathbf{let}\, I\, \mathbf{be}\, M\, \mathbf{in}\, N : \phi}\ I\,E$$

$$\frac{\Gamma \vdash M : \bot}{\Gamma \vdash \bot_\phi(M) : \phi}\ \bot E$$

$$\frac{}{\emptyset_a \vdash \top : \top}\ \top I \qquad \frac{\Gamma(\emptyset_a) \vdash N : \phi \quad \Delta \vdash M : \top}{\Gamma(\Delta) \vdash N[M/\top]\phi}\ \top E$$

MULTIPLICATIVES

$$\frac{\Gamma, x : \phi \vdash M : \psi}{\Gamma \vdash \lambda x : \phi . M : \phi -\!\!* \psi}\ -\!\!* I \qquad \frac{\Gamma \vdash M : \phi -\!\!* \psi \quad \Delta \vdash N : \phi}{\Gamma, \Delta \vdash \mathbf{app}_{-\!*}(M, N) : \psi}\ -\!\!* E$$

$$\frac{\Gamma \vdash M : \phi \quad \Delta \vdash N : \psi}{\Gamma, \Delta \vdash M * N : \phi * \psi}\ *I \qquad \frac{\Gamma(x : \phi, y : \psi) \vdash N : \chi \quad \Delta \vdash M : \phi * \psi}{\Gamma(\Delta) \vdash \mathbf{let}\,(x, y)\, \mathbf{be}\, M\, \mathbf{in}\, N : \chi}\ *E$$

ADDITIVES

$$\frac{\Gamma ; x : \phi \vdash M : \psi}{\Gamma \vdash \alpha x : \phi . M : \phi \to \psi}\ \to I \qquad \frac{\Gamma \vdash M : \phi \to \psi \quad \Delta \vdash N : \phi}{\Gamma; \Delta \vdash \mathbf{app}_{\to}(M, N) : \psi}\ \to E$$

$$\frac{\Gamma \vdash M : \phi \quad \Delta \vdash N : \psi}{\Gamma; \Delta \vdash \langle M, N \rangle : \phi \wedge \psi}\ \wedge I \qquad \frac{\Gamma \vdash M : \phi \wedge \psi \quad \Gamma; x : \phi; y : \psi \vdash N : \chi}{\Gamma \vdash N[\pi_1 M/x, \pi_2 M/y] : \chi}\ \wedge E$$

$$\frac{\Gamma \vdash M : \phi_i}{\Gamma \vdash \mathbf{in}_i(M) : \phi_1 \vee \phi_2}\ (i = 1, 2)\quad \vee I$$

$$\frac{\Gamma \vdash M : \phi \vee \psi \quad \Delta(x : \phi) \vdash N : \chi \quad \Delta(y : \psi) \vdash P : \chi}{\Delta(\Gamma) \vdash \mathbf{case}\, M\, \mathbf{of}\, \mathbf{in}_1(x) \Rightarrow N\, \mathbf{or}\, \mathbf{in}_2(y) \Rightarrow P : \chi}\ \vee E$$

Table 2.2. **NBI** for the $\alpha\lambda$-calculus

β-reductions	η-reductions
$\mathbf{app}_{\rightarrow}(\alpha x : \phi.M, N) \succcurlyeq M[N/x]$	$\alpha x : \phi.Mx \succcurlyeq M \quad (x \notin \mathrm{FV}(M))$
$\mathbf{app}_{-\!*}(\lambda x : \phi.M, N) \succcurlyeq M[N/x]$	$\lambda x : \phi.Mx \succcurlyeq M \quad (x \notin \mathrm{FV}(M))$
$\pi_1\langle M_1, M_2\rangle \succcurlyeq M_1$	$\langle \pi_1 M, \pi_2 M\rangle \succcurlyeq M$
$\pi_2\langle M_1, M_2\rangle \succcurlyeq M_2$	
$(\mathbf{let}\,(x,y)\,\mathbf{be}\,M_1 * M_2\,\mathbf{in}\,N) \succcurlyeq N[M_1/x, M_2/y]$	$(\mathbf{let}\,(x,y)\,\mathbf{be}\,M\,\mathbf{in}\,x * y) \succcurlyeq M$
$(\mathbf{case}\,\mathbf{in}_1(M)\,\mathbf{of}\,\mathbf{in}_1(x) \Rightarrow N_1\,\mathbf{or}\,\mathbf{in}_2(y) \Rightarrow N_2)$ $\succcurlyeq N_1[M/x]$	$M \succcurlyeq \mathbf{case}\,M\,\mathbf{of}$ $\mathbf{in}_1(x) \Rightarrow \mathbf{in}_1(x) \vee \mathbf{in}_2(y) \Rightarrow \mathbf{in}_2(y)$
$(\mathbf{case}\,\mathbf{in}_2(M)\,\mathbf{of}\,\mathbf{in}_1(x) \Rightarrow N_1\,\mathbf{or}\,\mathbf{in}_2(y) \Rightarrow N_2)$ $\succcurlyeq N_2[M/y]$	

Figure 2.1. $\beta\eta$-reductions

REDUCTIONS

We formulate the equational theory of $\alpha\lambda$-terms via a system of reduction rules which is given in Figure 2.1.

We must also take the ζ-reductions for the disjunctive and multiplicative conjunctive terms, $\mathbf{case}\,z\,\mathbf{of}\ \mathbf{in}_1(x) \Rightarrow M\ \mathbf{or}\ \mathbf{in}_2(y) \Rightarrow N$ and $\mathbf{let}\,(x,y)\,\mathbf{be}\,z\,\mathbf{in}\,M$. These reductions correspond to the so-called commuting conversions of natural deduction systems; see [Girard et al., 1989, Benton et al., 1993]. For a concise presentation, we adapt the notion of a *term context* (see, for example, [Barendregt, 1984, Barber, 1996]), defined recursively in Figure 2.2 (for brevity, we omit the additive cases other than those for disjunction).

A given term context *binds* a variable x if that term context has a sub-term of the form $\mathbf{let}\,(x,y)\,\mathbf{be}\,C[-]\,\mathbf{in}\,M$, $\mathbf{let}\,(x,y)\,\mathbf{be}\,M\,\mathbf{in}\,C[-]$, $\lambda x : \phi.C[-]$ or $\alpha x : \phi.C[-]$. If $C[-]$ is a context and M is a term, $C[M]$ is $C[-]$ in which $-$ is replaced with M.

The ζ-reductions may now be stated as in Figure 2.3.[1]

[1]The necessity of the ζ-reductions for, say, disjunction may be understood as a consequence of the failure of $\vee E$ to be suitably "syntax-directed".

$$
\begin{array}{rcl}
C[-] & ::= & - \mid \mathbf{let}\, I\, \mathbf{be}\, C[-]\, \mathbf{in}\, M \mid \mathbf{let}\, I\, \mathbf{be}\, M\, \mathbf{in}\, C[-] \\
& \mid & M * C[-] \mid C[-] * M \\
& \mid & \mathbf{let}\, (x,y)\, \mathbf{be}\, C[-]\, \mathbf{in}\, M \mid \mathbf{let}\, (x,y)\, \mathbf{be}\, M\, \mathbf{in}\, C[-] \\
& \mid & \lambda x : \phi . C[-] \mid C[-]\, M \mid M\, C[-] \\
& \mid & \ldots \quad \text{(some additive cases)} \\
& \mid & \mathbf{in}_1(C[-]) \mid \mathbf{in}_2(C[-]) \\
& \mid & \mathbf{case}\, M\, \mathbf{of}\, \mathbf{in}_1(x) \Rightarrow C[-]\, \mathbf{or}\, \mathbf{in}_2(y) \Rightarrow P \\
& \mid & \mathbf{case}\, M\, \mathbf{of}\, \mathbf{in}_1(x) \Rightarrow N\, \mathbf{or}\, \mathbf{in}_2(y) \Rightarrow C[-]
\end{array}
$$

Figure 2.2. Term Context

ζ-reductions for $*$

$$
\begin{array}{r}
C[\mathbf{let}\, I\, \mathbf{be}\, M\, \mathbf{in}\, N] \succcurlyeq \mathbf{let}\, I\, \mathbf{be}\, M\, \mathbf{in}\, C[N] \\
C[\mathbf{let}\, x,y\, \mathbf{be}\, M\, \mathbf{in}\, N] \succcurlyeq \mathbf{let}\, x,y\, \mathbf{be}\, M\, \mathbf{in}\, C[N] \\
(C \text{ does not bind } x,\ y)
\end{array}
$$

ζ-reductions for $\vee$

$$
\begin{array}{r}
C[\mathbf{case}\, M\, \mathbf{of}\, \mathbf{in}_1(x) \Rightarrow N\, \mathbf{or}\, \mathbf{in}_2(y) \Rightarrow P] \succcurlyeq \\
\mathbf{case}\, M\, \mathbf{of}\, \mathbf{in}_1(x) \Rightarrow C[N]\, \mathbf{or}\, \mathbf{in}_2(y) \Rightarrow C[P]
\end{array}
$$

Figure 2.3. ζ-reductions

Here we follow the orientation of the commuting conversions taken in [Girard et al., 1989]. Note that here, in contrast to [Benton et al., 1993], for example, ζ-reductions are necessary for neither Weakening nor Contraction: This is because we have formulated these rules as direct manipulations of the structure of sequents rather than via a modality such as Girard's [Girard, 1987] "!", which must be treated as a connective.

We take the obvious α-reductions. We also write $M \succcurlyeq M'$ if M reduces to M' in possibly many steps. We write $M = M'$ if either $M \succcurlyeq M'$ or $M' \succcurlyeq M$ and $\Gamma \vdash M = M' : \phi$ for $\Gamma \vdash M : \phi$ and $M = M'$. We shall be concerned throughout with $\beta\eta\zeta$-equality and $\beta\eta\zeta$-normal forms. We say that an $\alpha\lambda$-term is *well-typed* in Γ if there is a ϕ such that $\Gamma \vdash M : \phi$ is provable.

LEMMA 2.4 (ADDITIVE RULES) *The following additive versions of the Axiom, $\wedge I$, $\wedge E$, $\to E$ and $\vee E$ rules are derivable in $\alpha\lambda$:*

$$\frac{}{x:\phi;\Gamma \vdash x:\phi}\ \mathit{Axiom}$$

$$\frac{\Gamma \vdash M:\phi \quad \Gamma \vdash N:\psi}{\Gamma \vdash \langle M,N\rangle : \phi \wedge \psi}\ \wedge I \qquad \frac{\Gamma \vdash M:\phi_1 \wedge \phi_2}{\Gamma \vdash \pi_i M : \phi_i}\ (i=1,2) \quad \wedge E$$

$$\frac{\Gamma \vdash M:\phi \to \psi \quad \Gamma \vdash N:\phi}{\Gamma \vdash MN:\psi}\ \to E$$

$$\frac{\Gamma \vdash M:\phi \vee \psi \quad \Gamma(x:\phi) \vdash N:\chi \quad \Gamma(y:\psi) \vdash P:\chi}{\Gamma \vdash \mathbf{case}\, M\, \mathbf{of}\ \mathbf{in}_1(x) \Rightarrow N\ \mathbf{or}\ \mathbf{in}_2(y) \Rightarrow P:\chi}\ \vee E.$$

PROOF An easy construction using the structurals. □

Note that the system, corresponding to $\mathbf{NBI}^{\heartsuit}$ without W and C, formed by substituting these rules and deleting C and W is *not* equivalent to $\alpha\lambda$. To see this, reconsider (2.1) and (2.2), with $\alpha\lambda$-terms added. We can prove

$$(x:\chi \mathrel{-\!\!*} (\phi * \psi)\,,\, y:\chi)\,;\,(x':\chi \mathrel{-\!\!*} (\phi * \psi)\,,\, y':\chi) \vdash \mathbf{let}\,(x',y')\,\mathbf{be}\,xy\,\mathbf{in}\,x'y' : \phi * \psi$$

but not

$$x:\chi \mathrel{-\!\!*} (\phi * \psi)\,,\, y:\chi \vdash \mathbf{let}\,(x,y)\,\mathbf{be}\,xy\,\mathbf{in}\,xy : \phi * \psi.$$

We see that, for $\alpha\lambda$, Contraction cannot be omitted from the alternative calculus. As for the propositional consequences, Weakening must be incorporated into the binary multiplicative rules. See also [O'Hearn, 1999].

LEMMA 2.5 (ADMISSIBILITY OF CUT) *The following Cut rule is admissible:*

$$\frac{\Gamma(x:\phi) \vdash M:\psi \quad \Delta \vdash N:\phi}{\Gamma(\Delta) \vdash M[N/x]:\psi}.$$

PROOF By induction on the structure of proofs in $\alpha\lambda$. The argument follows the usual pattern, with the induction hypothesis based on *Multicut*,

$$\frac{\Gamma(x_1:\phi_1 \mid \ldots \mid x_m:\phi_m) \vdash M:\psi \quad \Delta_1 \vdash N_1:\phi_1 \ldots \Delta_m \vdash N_m:\phi_m}{\Gamma(\Delta_1 \mid \ldots \mid \Delta_m) \vdash M[N_1/x_1 \ldots N_m/x_m]:\psi}$$

in order to handle Contraction.

Similar arguments have been presented in many of our references; see, for example, [Barber, 1996, Barber and Plotkin, 1997, Benton et al., 1993] □

We can conveniently state the *strengthening* property for $\alpha\lambda$. Note that $x : \phi$ extends the sub-bunch Δ of Γ via ";", *not* ",".

LEMMA 2.6 (STRENGTHENING) *If* $\Gamma(\Delta; x : \phi) \vdash M : \psi$ *is provable in* $\alpha\lambda$ *and if* $x \notin \mathrm{FV}(M)$, *then* $\Gamma(\Delta) \vdash M : \psi$ *is provable in* $\alpha\lambda$.

PROOF By induction on the structure of proofs in $\alpha\lambda$, relying on the restriction of the statement to extensions of Δ via ";" so that the argument proceeds just as for the usual simply-typed λ-calculus; see [Barendregt, 1992] for a general discussion. □

Before proceeding to the proof theory of $\alpha\lambda$, we remark briefly on the resource interpretation of proofs. The linear λ-calculus admits the so-called "use-once" interpretation of linear, not exponentiated, variables. Whilst this interpretation obtains in the purely multiplicative fragment of $\alpha\lambda$, it fails for terms (proofs) which involve both additives and multiplicatives.

For example [O'Hearn and Pym, 1999], the judgement

$$\emptyset_m \vdash \phi \mathrel{-\!\!*} ((\phi \to \phi \to \psi) \to \psi)$$

has a proof that uses the witness for a proof of the premiss ϕ twice:

$$\dfrac{\dfrac{\dfrac{\dfrac{\dfrac{\vdots}{x : \phi; f : \phi \to \phi \to \psi \vdash f\,x : \phi \to \psi} \qquad \dfrac{\vdots}{x : \phi; f : \phi \to \phi \to \psi \vdash x : \phi}}{x : \phi; f : \phi \to \phi \to \psi \vdash (f\,x)\,x : \psi}\; C, \to E}{x : \phi \vdash \alpha f\,.\,(f\,x)\,x : ((\phi \to \phi \to \psi) \to \psi)}\; \to I}{\emptyset_m, x : \phi \vdash \alpha f\,.\,(f\,x)\,x : ((\phi \to \phi \to \psi) \to \psi)}\; \equiv}{\emptyset_m \vdash \lambda x\,.\,\alpha f\,.\,(f\,x)\,x : \phi \mathrel{-\!\!*} ((\phi \to \phi \to \psi) \to \psi)}\; \mathrel{-\!\!*} I.$$

In the key step (top right in the figure), we use the admissible rule for $\to$-elimination (or equivalently we use $\to E$ followed by Contraction). It is at this step that the way that $\to$ admits sharing between $(f\,x)$ and its argument x appears.

The point of this example is the two occurrences in the body $(f\,x)\,x$ of the argument x to a $\mathrel{-\!\!*}$-typed function. This serves to illustrate that the idea that a multiplicative, or "linear", function uses its argument exactly once does not directly carry over to **BI** (apart, of course, from in the purely multiplicative fragment). For this reason, we have chosen not to adopt the symbol $\multimap$, with the associated readings it tends to carry, for **BI**'s multiplicative implication.

If the number-of-uses reading — which is often considered to be characteristic for a logic, like linear logic, which restricts Contraction — does not carry over, then what meaning do the **BI** connectives have ? And what is the justification for this judgement ?

We can answer these questions in two ways. First, we claim that the Kripke resource semantics, given in Chapters 4 and 5, provides a natural truth-based semantics for **BI**. We do not have to appeal to λ-calculus to make sense of it (although we certainly can do so). Secondly, although **BI** can be understood in terms of its truth-functional semantics, we can also offer a resource-based interpretation of proofs which does justify these judgements. We call this the (propositional) *sharing interpretation*, which we develop in Chapter 9.

4. Normalization and Subject Reduction

The strong normalization (SN) theorem for the simply-typed λ-calculus, with the usual functional ($\rightarrow$), conjunctive ($\wedge$ or $\times$) and co-product ($\vee$ or $+$) types is well-documented, with [Prawitz, 1965, Prawitz, 1971] and [Girard et al., 1989] being the most accessible.[2] The basic idea of the proof of SN is Tait's [Tait, 1967] notion of *reducibility*, taken together with Girard's [Girard, 1972] notion of *neutrality*, which facilitates a technical improvement of Tait's proof.

For the remainder of this chapter, we confine our attention to $\alpha\lambda$ without $\vee$ (or $\perp$) and references to $\alpha\lambda$ should be taken to exclude them. However, the methods discussed in [Prawitz, 1965, Ghani, 1995, Ritter et al., 2000, Pym and Ritter, 2001] can be used to extend SN and SR to $\vee$ (and $\perp$).

The basic idea of reducibility is to structure the reductions of terms according to the structure of their types. Without $\vee$, we define the set $\mathbf{Red}(\phi)$, of *reducible terms of type* ϕ, relative to a given context, as follows:

1 If M has atomic type, α, then $M \in \mathbf{Red}(\alpha)$ if it is strongly normalizing (SN);

[2][Girard et al., 1989] provides the details for $\rightarrow$ and $\wedge$; definitions are provided for $\vee$ but for the extension of the proof of SN to $\vee$ one is referred to [Prawitz, 1971], which is formulated in terms of natural deduction proof trees rather than a term calculus.

2 If M has type $\phi_1 \wedge \phi_2$, then $M \in \mathbf{Red}(\phi_1 \wedge \phi_2)$ if both $\pi_1 M \in \mathbf{Red}(\phi_1)$ and $\pi_2 M \in \mathbf{Red}(\phi_2)$;

3 If M has type $\phi \to \psi$, then $M \in \mathbf{Red}(\phi \to \psi)$ if, for every $N \in \mathbf{Red}(\phi)$, $MN \in \mathbf{Red}(\psi)$.

The basic idea of neutrality is to pick out those terms which are *not* immediately constructed by introduction rules. In the absence of $\vee$ (and $\perp$), the neutral terms are those of the form x, $\pi_1 M$, $\pi_2 M$, MN.

The key technical lemma is then that the sets $\mathbf{Red}(\phi)$ satisfy the following conditions:

CR1 If $M \in \mathbf{Red}(\phi)$, then M is SN;

CR2 If $M \in \mathbf{Red}(\phi)$ and $M \succcurlyeq M'$, then $M' \in \mathbf{Red}(\phi)$;

CR3 If M is neutral and every redex in M reduces to a term $M' \in \mathbf{Red}(\phi)$, then $M \in \mathbf{Red}(\phi)$;

CR4 If M is both neutral and normal, then $M \in \mathbf{Red}(\phi)$.

This lemma is proved by induction on the structure of types and on a measure $\nu(M)$, which bounds the length of every reduction sequence beginning with M.

The proof of SN now proceeds, by an induction on the structure of terms, to show that all terms are reducible. The argument uses the following lemma (*q.v.* [Girard et al., 1989]):

> Let M be any, not necessarily reducible, term with free variables among $x_1 : \phi_1, \ldots, x_m : \phi_m$. If $N_1, \ldots, N_m$ are reducible terms of type $\phi_1, \ldots, \phi_m$, then $M[N_1/x_1, \ldots, N_m/x_m]$ is reducible.

The extension of these ideas to $\vee$ is well-understood; see, for example [Prawitz, 1971, Troelstra and Schwichtenberg, 1996].

Given SN for the simply-typed λ-calculus, we can obtain SN for $\alpha\lambda$ by the technique of *translation* as described, for example, in [Troelstra and Schwichtenberg, 1996]. Given systems $\mathbf{S}$ and $\mathbf{S}'$ of terms and reductions, together with a mapping $\mapsto$ from $\mathbf{S}$ to $\mathbf{S}'$ with the property that each reduction step in $\mathbf{S}$ maps to one or more reductions in $\mathbf{S}'$, we can infer SN for $\mathbf{S}$ from SN for $\mathbf{S}'$.

We define a mapping $\mapsto$ from $\alpha\lambda$ to the simply-typed λ-calculus as follows:

$$\begin{array}{ccccccc} \to & \mapsto & \to & \quad & -\!\!* & \mapsto & \to \\ \wedge & \mapsto & \wedge & & * & \mapsto & \wedge \\ \top & \mapsto & \top & & I & \mapsto & \top. \end{array}$$

Considering the reductions, we must first recall that the elimination rule for $\wedge$ may be written in the form[3]

$$\frac{\phi \wedge \psi \qquad \begin{array}{c}[\phi]\,;\,[\psi]\\ \vdots \\ \chi\end{array}}{\chi}\ \wedge\mathrm{E}.$$

We then have that a reduction of the form

$$\frac{\dfrac{\begin{array}{c}\vdots\\ \phi\end{array} \quad \begin{array}{c}\vdots\\ \psi\end{array}}{\phi * \psi}\ {*}\mathrm{I} \qquad \begin{array}{c}[\phi]\,,\,[\psi]\\ \vdots \\ \chi\end{array}}{\chi}\ {*}\mathrm{E} \quad \succcurlyeq \quad \begin{array}{c}\begin{array}{c}\vdots\\ \phi\end{array} \qquad \begin{array}{c}\vdots\\ \psi\end{array}\\ \vdots \\ \chi\end{array} \quad ,$$

in which we omit the $\alpha\lambda$-terms, maps to the reduction of the form

$$\frac{\dfrac{\begin{array}{c}\vdots\\ \phi\end{array} \quad \begin{array}{c}\vdots\\ \psi\end{array}}{\phi \wedge \psi}\ \wedge\mathrm{I} \qquad \begin{array}{c}[\phi]\,;\,[\psi]\\ \vdots \\ \chi\end{array}}{\chi}\ \wedge\mathrm{E} \quad \succcurlyeq \quad \begin{array}{c}\begin{array}{c}\vdots\\ \phi\end{array} \qquad \begin{array}{c}\vdots\\ \psi\end{array}\\ \vdots \\ \chi\end{array} \quad .$$

Other reductions behave similarly under $\mapsto$. Note that the choice of the generalized form of $\wedge$E ensures that the commuting conversion for $*$ has an image under $\mapsto$.

THEOREM 2.7 (STRONG NORMALIZATION) *All well-typed $\alpha\lambda$-terms are strongly normalizing,* i.e., *all reduction sequences terminate.*

PROOF By induction on the structure of well-typed terms, using the translation to simply-typed λ-calculus via $\mapsto$. □

LEMMA 2.8 (INVERSION) *The $\to I$ and $-\!\!* I$ rules are invertible,* i.e., *the following are admissible in* **NBI***:*

- *Inverse of $\to I$:*

$$\frac{\Gamma \vdash \alpha x : \phi . M : \psi}{\Gamma \,;\, x : \phi \vdash M : \psi};$$

[3]We adopt the (informal) graphical representation of natural deduction for **BI** for clarity. It is essentially equivalent to our sequential presentation of natural deduction for **BI**. Recent work by Jules Bean has begun to develop a version of Fitch-style box proofs [Prawitz, 1965] for **BI**, based on the geometic/combinatorial idea of "ribbons".

- *Inverse of* $-\!\!*\, I$:
$$\frac{\Gamma \vdash \lambda x : \phi . M : \psi}{\Gamma, x : \phi \vdash M : \psi}.$$

PROOF SKETCH By induction on the structure of proofs. Each of the premisses may be inferred by either by an application of the corresponding introduction rule, $\to I$ or $-\!\!*\, I$, respectively, or by a structural rule. □

THEOREM 2.9 (SUBJECT REDUCTION) *If* $\Gamma \vdash M : \phi$ *is provable in* $\alpha\lambda$ *and* $M \succcurlyeq M'$, *then* $\Gamma \vdash M' : \phi$ *is provable in* $\alpha\lambda$.

PROOF SKETCH By induction on the structure of the derivaton of the reduction $M \succcurlyeq M'$, using Lemmas 2.8 and 2.5.

For example, the base case for β-reduction (for example, for $-\!\!*$) has $-\!\!*\, E$,
$$\frac{\Gamma \vdash \lambda x : \phi . M : \phi -\!\!* \psi \quad \Delta \vdash N : \phi}{\Gamma, \Delta \vdash (\lambda x : \phi . M) N : \psi}$$
as the last inference in the derivation (possibly followed by instances of structural rules). By Lemma 2.8, we have $\Gamma, x : A \vdash M : \psi$ and so, by Lemma 2.5, $\Gamma, \Delta \vdash M[N/x] : \psi$.

The other cases are similar. □

We briefly remark on the question of confluence. In [Barber, 1996], the failure of confluence for **DILL**'s λ-calculus is illustrated by the remark that the term $\mathbf{let}\, I\, \mathbf{be}\, x\, \mathbf{in}\, I * I$ reduces to both $x * I$ and $I * x$. The concerns of [Barber, 1996] notwithstanding, we conjecture that it is possible, perhaps using techniques similar to those adopted in [Pym and Ritter, 2001], which exploit work of Neil Ghani [Ghani, 1995], to set up $\alpha\lambda$ with $*$ and $\vee$ (and their units) as a confluent system.

5. Structural Variations on BI and $\alpha\lambda$

5.1 Affinity and Relevance

So far we have consider just two bunch-forming operations:

- The *linear* ",", which admits neither Weakening nor Contraction; and
- The *intuitionistic* ";", which admits both.

However, two further operators also arise naturally:

- The *affine* ",", which admits Weakening but not Contraction; and
- The *relevant* ",", which admits Contraction but not Weakening.

The affine variation of $\alpha\lambda$, and hence of propositional **BI**, is obtained as follows:

AFFINE COHERENT EQUIVALENCE

We add

4 $\emptyset_m = \emptyset_a$.

to Coherent Equivalence.

It then follows that Weakening for ",",

$$\frac{\Gamma(\Delta) \vdash M : \phi}{\Gamma(\Delta, \Delta') \vdash M : \phi} \text{ W},$$

is admissible. To see this, consider the following derivation:

$$\dfrac{\dfrac{\dfrac{\dfrac{\phi \vdash \chi}{\phi\,,\,\emptyset_m \vdash \chi}\ \emptyset_m \text{ is unit of ","}}{\phi\,,\,\emptyset_a \vdash \chi}\ \emptyset_m = \emptyset_a}{\phi\,,\,(\emptyset_a\,;\,\psi) \vdash \chi}\ \text{W}}{\phi\,,\,\psi \vdash \chi}\ \emptyset_a \text{ is unit of ";"}.$$

Alternatively, we can choose to define the affine system by the addition of Weakening for "," and then derive the logical equivalence of $\emptyset_a$ and $\emptyset_m$.

REDUCTIONS

In $\alpha\lambda$, we must add reductions which provide the projections for $*$:

$$(\mathbf{let}(x, y) = M * N \,\mathbf{in}\, x) \succcurlyeq x \qquad (\mathbf{let}(x, y) = M * N \,\mathbf{in}\, y) \succcurlyeq y\,.$$

LEMMA 2.10 (AFFINE CONVERTIBILITY) *In the affine calculus, the following is admissible:*

$$\frac{\Gamma(\Delta; \Delta') \vdash M : \phi}{\Gamma(\Delta, \Delta') \vdash M : \phi}.$$

The relevant variation is obtained by adding Contraction for ", ":

$$\frac{\Gamma(\Delta, \Delta') \vdash M : \phi}{\Gamma(\Delta) \vdash M[i(\Delta)/i(\Delta')] : \phi} \quad (\Delta' \cong \Delta).$$

This choice gives rise to a version of $*$ which admits Contraction and which, within the relevant logic tradition, is called *fusion* and which is usually denoted by $\circ$.

Contraction for "," implies the bunched logic form of *Dereliction*.

5.2 Dereliction

The presence of bunched structure admits the possibility of a rule of *dereliction* of the following form:

$$\frac{\Gamma(\Delta, \Delta') \vdash \phi}{\Gamma(\Delta; \Delta') \vdash \phi} \; D.$$

Note that this form of dereliction does *not* rely on the presence of a modality or exponential, such as linear logic's !. A consequence of Dereliction is that we get Contraction for ",": suppose $\Delta' \cong \Delta$, then

$$\frac{\dfrac{\Gamma(\Delta, \Delta') \vdash M : \phi}{\Gamma(\Delta; \Delta') \vdash M : \phi} D}{\Gamma(\Delta) \vdash M[i(\Delta)/i(\Delta')] : \phi} (\Delta' \cong \Delta) \quad C.$$

5.3 Non-commutativity

Whilst intuitionistic conjunction is necessarily commutative, monoidal conjunctions need not be. We can take a non-commutative bunch-former, "|", together with its unit $\emptyset_n$ by taking the clauses $\emptyset_n$ and $\Gamma \mid \Gamma$ in the definition of bunches.[4]

Along with "|" come the non-commutative conjunction, $\diamond$ with unit H, and the "left" and "right" implications, respectively $\multimap\!\!\lhd$ and $\multimap\!\!\rhd$, which stand in the usual adjunctive relationship:

$$\frac{x : \phi \diamond \Gamma \vdash M : \psi}{\Gamma \vdash \lambda_{-\lhd} x : \phi . M : \phi \mathbin{-\lhd} \psi} \mathbin{-\lhd} I \qquad \frac{\Gamma \vdash M : \phi \mathbin{-\lhd} \psi \quad \Delta \vdash N : \phi}{\Delta \diamond \Gamma \vdash MN : \psi} \mathbin{-\lhd} E$$

and

$$\frac{\Gamma \diamond x : \phi \vdash M : \psi}{\Gamma \vdash \lambda_{-\rhd} x : \phi . M : \phi \mathbin{-\rhd} \psi} \mathbin{-\rhd} I \qquad \frac{\Gamma \vdash M : \phi \mathbin{-\rhd} \psi \quad \Delta \vdash N : \phi}{\Gamma \diamond \Delta \vdash MN : \psi} \mathbin{-\rhd} E.$$

We refrain from syntactically distinguishing the two applications.

We must also extend Coherent Equivalence with monoid equations for $\emptyset_n$ and | and **NBI** with suitable rules.

[4]Note that non-commutative bunching may be taken either in addition to or instead of the commutative ",".

Related non-commutative systems have been considered by Lambek [Lambek, 1968, Lambek, 1969, Lambek, 1972], Yetter [Yetter, 1990], Ruet and Fages [Ruet and Fages, 1998] and Retore [Retoré, 1998]. Just the last two of these employ forms of bunching.

5.4 More Combinators

We remark that there is no technical obstruction to taking as many bunch-forming operators as one desires. Proof systems may be given in the evident way and Day's construction provides a systematic construction of monoidal closed structures in functor categories.

Chapter 3

ALGEBRAIC, TOPOLOGICAL, CATEGORICAL

1. An Algebraic Presentation

In this section, we are primarily concerned with truth and provability, rather than the structure of proofs and so present a simple-minded algebraic semantics and associated calculus for **BI**. This presentation of **BI** makes little or no explicit use of bunches, *i.e.*, **BI**'s tree-structured contexts.

In order to motivate the models, however, it is useful to sketch briefly the categorical interpretation which lies at the core of **BI** and which we describe in more detail in the sequel.

Suppose, recalling our proof-theoretic introduction in Chapter 1, that we are to have a logic with two implications. Then, categorically, the natural notion of consequence arises from doubly closed categories, which are categories that possess two closed structures or function spaces. That is, we have a single category with two adjunctions

$$[A * B, C] \cong [A, B \mathrel{-\!\!*} C] \qquad \text{and} \qquad [A \wedge B, C] \cong [A, B \to C]$$

which determine the properties of $\mathrel{-\!\!*}$ and $\to$.[1] The algebraic models are collapsed versions of these structures, where the additive implication $\to$ corresponds to that of intuitionistic logic and the multiplicative $\mathrel{-\!\!*}$ to that of a basic substructural logic.

To describe the algebraic models, we recall firstly that Heyting algebras are the algebraic models of intuitionistic propositional logic. A

[1] These ideas are developed in § 3.

Heyting algebra is a lattice with greatest and least elements in which the meet $a \wedge b$ is *residuated*, *i.e.*, there is an implication operator, $\rightarrow$, satisfying

$$a \wedge b \leq c \quad \text{iff} \quad a \leq b \rightarrow c.$$

Secondly, an algebraic model of a basic substructural logic containing conjunction $*$, unit I and implication $-\!\!*$ is similar, except that $*$ is not required to be idempotent, *i.e.*, have the properties of meet, and I is not required to be top. That is, we would require a preorder with a (monotone) commutative monoid structure that is residuated, so that

$$a * b \leq c \quad \text{iff} \quad a \leq b \mathrel{-\!\!*} c.$$

Because we have all of the connectives of intuitionistic logic and the basic substructural logic at the same time, we ask for an algebra that has both kinds of structure simultaneously.

> *A* **BI**-*algebra is a Heyting algebra equipped with an additional residuated commutative monoid structure.*

It is important to note that the same underlying order is used to describe the residuated structure in both cases. Categorically, this corresponds to the two closed structures being defined with respect to a single class of arrows.

From this notion of **BI**-algebra, it is straightforward to derive a collection of axioms and rules — a Hilbert-type system, equivalent to **NBI** — for proving judgements $\phi \vdash \psi$, where the formulæ ϕ and ψ are built from propositional variables, the additive connectives $\rightarrow$, $\wedge$, $\top$, $\vee$ and $\bot$, and the multiplicative connectives I, $*$ and $-\!\!*$. The axioms and rules of this system are those for (some presentation of) intuitionistic propositional logic, together with the ones for the multiplicatives given in Table 3.1.

We say that "$\psi \vdash \phi$ is derivable or provable" to indicate that $\psi \vdash \phi$ may be proven using this system or, equivalently, when $[\![\psi]\!] \leq [\![\phi]\!]$ for all interpretations $[\![-]\!]$ in **BI**-algebras: **BI**-algebras obviously give sound and complete models of the proof system just given.

2. A Topological Presentation

A (commutative) topological monoid is a (commutative) monoid in the category **Top** of topological spaces and continuous maps between them, *i.e.*, a topological space $\mathcal{X}$, with open sets $\mathcal{O}(\mathcal{X})$, together with two arrows, a tensor product $* : \mathcal{X} \times \mathcal{X} \longrightarrow \mathcal{X}$ and its unit $e : 1 \longrightarrow \mathcal{X}$ such that the usual monoidal diagrams commute [Mac Lane, 1971].

We need to interpret a formula $\phi * \psi$ as the tensor product, $U * V$ of the interpretations, respectively U and V, of ϕ and ψ. The tensor product of

$$\phi * (\psi * \chi) \dashv\vdash (\phi * \psi) * \chi \qquad \phi * I \dashv\vdash \phi \dashv\vdash I * \phi$$

$$\frac{\chi \vdash \phi \quad \eta \vdash \psi}{\chi * \eta \vdash \phi * \psi} \qquad \phi * \psi \vdash \psi * \phi$$

$$\frac{\chi * \phi \vdash \psi}{\chi \vdash \phi \mathrel{-\!\!*} \psi} \qquad \frac{\chi \vdash \phi \mathrel{-\!\!*} \psi \quad \eta \vdash \phi}{\chi * \eta \vdash \psi}$$

Table 3.1. Hilbert-type **BI**

two open sets is not necessarily open, however. Consequently, we must require that the monoidal structure be defined by open maps, *i.e.*, which map open sets to open sets.

An *open* topological monoid is one in which the maps $*$ and e, which define the monoidal structure, are open.

LEMMA 3.1 (DISTRIBUTIVITY) *Let $(\mathcal{X}, *, e)$ be a topological monoid. For all open sets U, V_i, $i \in \mathcal{I}$, where $\mathcal{I}$ is some indexing set,*

$$U * (\bigcup_i V_i) = \bigcup_i (U * V_i).$$

PROOF We have that $z \in U * (\bigcup_i V_i)$ iff there exist $x \in U$ and $y_j \in V_j$, for some j, such that $z = x * y_j$ iff $z \in \bigcup_i (U * V_i)$. □

The interpretation of **BI** in an open commutative topological monoid now follows exactly as for the interpretation of intuitionistic logic in a topological space, *i.e.*, with $[\![\bot]\!] = \emptyset$, with the addition of the following:

$$\begin{array}{rcl} [\![\phi * \psi]\!] & = & [\![\phi]\!] * [\![\psi]\!] \\ [\![I]\!] & = & e(1) \end{array}$$

and

$$[\![\phi \mathrel{-\!\!*} \psi]\!] \quad = \quad \bigcup_{i \in \mathcal{I}} \{U_i \mid U_i \text{ is open and } U_i * [\![\phi]\!] \subseteq [\![\psi]\!]\},$$

where $\mathcal{I}$ is an indexing set. This interpretation is well-defined:

LEMMA 3.2 (MULTIPLICATIVE FUNCTION SPACE) $[\![\phi \mathrel{-\!\!*} \psi]\!] * [\![\phi]\!] \subseteq [\![\psi]\!]$.

PROOF We have that $\bigcup_{i \in \mathcal{I}} (U_i * [\![\phi]\!]) \subseteq [\![\psi]\!]$, so that $(\bigcup_{i \in \mathcal{I}} U_i) * [\![\phi]\!] \subseteq [\![\psi]\!]$, by distributivity. □

We can obtain soundness and completeness for these models just as for **BI**-algebras.

3. A Categorical Presentation

The BHK semantics of proofs of **BI** rests directly on a class of models based on *doubly closed categories*, or *DCCs*. We refer to [Mac Lane, 1971] for basic categorical notions, such as cartesian, monoidal and closed structure. See also Lambek's related work [Lambek, 1968, Lambek, 1969, Lambek, 1972, Lambek, 1993], arising from his early work in mathematical linguistics [Lambek, 1958].

DEFINITION 3.3 (DCC) *A* doubly closed category *is a category equipped with two monoidal closed structures. A DCC is* cartesian *if one of the closed structures is cartesian and the other is symmetric monoidal and* bi-cartesian *(sometimes written bi-CDDC) if it also has finite co-products.* □

To see how (bi-C)DCC structure corresponds to **BI**, consider the two adjunctions[2]

$$[F \otimes D, E] \;\cong\; [F, D \multimap E] \qquad [F \times D, E] \;\cong\; [F, D \rightarrow E],$$

where $\otimes$ is a symmetric monoidal product and $\times$ a cartesian product. In the proof theory, these adjunctions correspond to having one context combination corresponding to $*$ and another to $\wedge$. This leads directly to the rules

$$\frac{\Gamma\, ,\, \phi \vdash \psi}{\Gamma \vdash \phi \mathrel{-\!\!*} \psi} \qquad \frac{\Gamma\, ;\, \phi \vdash \psi}{\Gamma \vdash \phi \rightarrow \psi}$$

and to the tree-like structure of antecedents.

A DCC alone does not constitute a definition of a *model* of **BI**, for which we must also have an *interpretation* of **BI**'s syntax. Such an interpretation is a function from **BI**'s language of propositions to the objects of a DCC, defined by induction on the structure of propositions.

The interpretation of **BI** in a bi-cartesian DCC, with the two closed structures $(\times, 1, \rightarrow)$ and $(\otimes, I, \multimap)$ and co-product $(+, 0)$, is given by a

[2]In the sequel, we drop "cartesian" wherever no confusion is likely.

function $\llbracket - \rrbracket$ such that:

$$\begin{aligned} \llbracket \phi \vee \psi \rrbracket &= \llbracket \phi \rrbracket + \llbracket \psi \rrbracket \\ \llbracket \bot \rrbracket &= 0 \\ \llbracket \phi \wedge \psi \rrbracket &= \llbracket \phi \rrbracket \times \llbracket \psi \rrbracket \\ \llbracket \top \rrbracket &= 1 \\ \llbracket \phi \to \psi \rrbracket &= \llbracket \psi \rrbracket \to \llbracket \phi \rrbracket \end{aligned} \qquad \begin{aligned} \llbracket \phi * \psi \rrbracket &= \llbracket \phi \rrbracket \otimes \llbracket \psi \rrbracket \\ \llbracket I \rrbracket &= I \\ \llbracket \phi \mathrel{-\!\!*} \psi \rrbracket &= \llbracket \psi \rrbracket \multimap \llbracket \phi \rrbracket \end{aligned} \tag{3.1}$$

We interpret a bunch Γ by replacing each "," with $*$ and each ";" with $\wedge$. We write $\llbracket - \rrbracket_{\mathcal{D}}$ when we want to indicate that the interpretation is in the (bi-C)DCC $\mathcal{D}$.

One point which deserves comment here concerns disjunction. To interpret the elimination rule for $\vee$ we need to use distributivity of $+$ over both $\otimes$ and $\times$. To see why we get the needed distributivities in bi-cartesian DCCs note first that, since $E \otimes (-)$ and $E \times (-)$ are both left adjoints, they both preserve all colimits. Second, $+$ is a co-product. It follows that we have the isomorphisms

$$\llbracket \phi \rrbracket \times (\llbracket \psi \rrbracket + \llbracket \chi \rrbracket) \cong (\llbracket \phi \rrbracket \times \llbracket \psi \rrbracket) + (\llbracket \phi \rrbracket \times \llbracket \psi \rrbracket)$$

$$\llbracket \phi \rrbracket \otimes (\llbracket \psi \rrbracket + \llbracket \chi \rrbracket) \cong (\llbracket \phi \rrbracket \otimes \llbracket \psi \rrbracket) + (\llbracket \phi \rrbracket \otimes \llbracket \chi \rrbracket).$$

The first of these two laws shows that DCCs are not models of linear logic: distributivity fails for linear logic's additives.

For the remainder of this chapter, we confine our attention to $\alpha\lambda$ without $\vee$ (or $\bot$) and references to $\alpha\lambda$ should be taken to exclude these connectives.

DEFINITION 3.4 (MODELS OF **BI** IN DCCS) *A categorical model of* **BI** *is a pair* $\langle \mathcal{D}, \llbracket - \rrbracket_{\mathcal{D}} \rangle$, *where* $\mathcal{D}$ *is a DCC and* $\llbracket - \rrbracket_{\mathcal{D}}$ *is an interpretation satisfying (3.1).* □

We will often omit the subscript, writing just $\llbracket - \rrbracket$, when no confusion is likely.

PROPOSITION 3.5 (WEAK SOUNDNESS FOR DCCS) *If* $\Gamma \vdash \phi$ *is provable in* **NBI** *and* $\llbracket \Gamma \rrbracket_{\mathcal{D}}$ *and* $\llbracket \phi \rrbracket_{\mathcal{D}}$ *are defined, then* $[\llbracket \Gamma \rrbracket_{\mathcal{D}}, \llbracket \phi \rrbracket_{\mathcal{D}}] \neq \emptyset$.

PROOF SKETCH By a routine induction on the structure of proofs in **NBI**. □

The assignment of morphisms to derivations is straightforward, with $\alpha\lambda$-terms being interpreted in the usual way, the additives following the pattern for simply-typed λ-calculus with products and sums [Girard et al., 1989, Lambek and Scott, 1986], and the multiplicatives following the pattern for the linear λ-calculus with a tensor product [Benton et al., 1993].

Rather than give a detailed semantics of proofs here, we defer such an account to Chapter 8, in which we give an elementary definition of Kripke $\alpha\lambda$-models together with a categorical treatment via DCCs.

DEFINITION 3.6 (MODELS OF $\alpha\lambda$ IN DCCS) *A categorical model of $\alpha\lambda$ is a categorical model of* **BI** *extended to interpret the terms of $\alpha\lambda$.* □

In models of systems which admit the Cut rule,

$$\frac{\Gamma(x:\phi) \vdash M:\psi \quad \Delta \vdash N:\phi}{\Gamma(\Delta) \vdash M[N/x]:\psi},$$

we must show that a corresponding *substitutivity* property holds. In the case of **BI**, the cut rule is interpreted in DCCs by composition. Let $\bullet$ stand for either $\otimes$ or $\times$ and note that the Cut rule may be expressed, simplifying, for clarity, to the case in which $x:\phi$ is at the left-hand end of the bunch, by the two rules

$$\frac{x:\phi\,,\Gamma \vdash M:\psi \quad \Delta \vdash N:\phi}{\Delta\,,\Gamma \vdash M[N/x]:\psi} \quad \text{and} \quad \frac{x:\phi\,;\Gamma \vdash M:\psi \quad \Delta \vdash N:\phi}{\Delta\,;\Gamma \vdash M[N/x]:\psi}.$$

Suppose

$$[\![M]\!] = [\![\phi]\!] \bullet [\![\Gamma]\!] \xrightarrow{f} [\![\psi]\!]$$

and

$$[\![N]\!] = [\![\Delta]\!] \xrightarrow{g} [\![\phi]\!].$$

Then the Cut rule is interpreted by the composition

$$[\![\Delta]\!] \bullet [\![\Gamma]\!] \xrightarrow{g \bullet 1_{[\![\Gamma]\!]}} [\![\phi]\!] \bullet [\![\Gamma]\!] \xrightarrow{f} [\![\psi]\!].$$

Thus we have the following:

LEMMA 3.7 (SUBSTITUTION) *The cut rule is sound in models of $\alpha\lambda$ in DCCs.* □

We can now give a reasonable careful sketch of the soundness of **BI**, viewed as $\alpha\lambda$, for models in DCCs.

PROPOSITION 3.8 (SOUNDNESS FOR DCCS) *If* $\Gamma \vdash M = M' : \phi$ *is provable in* $\alpha\lambda$, *then* $[\![\Gamma \vdash M : \phi]\!]_{\mathcal{D}} = [\![\Gamma \vdash M' : \phi]\!]_{\mathcal{D}}$.

PROOF SKETCH By induction on the structure of proofs in $\alpha\lambda$, using Lemma 3.7. The two closed structures in a DCC are defined independently and so there is no interference between $\alpha\lambda$'s reduction rules for $(\otimes, I, -\!\!\circ)$ and $(\times, 1, \rightarrow)$. Consequently, the coherence properties of the CCC and SMCC structure, via the naturality of the interpretations of the proofs, are not affected by their combination. (This property may be seen easily in the case Kripke $\alpha\lambda$-models, which we develop in detail in Chapter 8: see Definition 8.4.) Checking each of the equalities for $(\otimes, I, -\!\!\circ)$ and $(\wedge, 1, \rightarrow)$ yields the result. See [Troelstra and Schwichtenberg, 1996, Lambek and Scott, 1986, Szabo, 1978, Benton et al., 1993] for similar arguments for the interpretations of the additives in CCCs and the multiplicatives in SMCCs. □

The soundness result can be extended to include the interpretation of $\vee$ (and $\bot$) in bi-DCCs.

The DCC semantics illustrates some, at first sight, unusual properties. In particular, a morphism from E to F may variously be viewed as a map $I \longrightarrow E -\!\!\circ F$ or $1 \longrightarrow E \rightarrow F$ using the adjunctions and, indeed, we have the following isomorphisms of hom sets:

$$[1, E \rightarrow F] \;\cong\; [E, F] \;\cong\; [I, E -\!\!\circ F]$$

In **BI**, these isomorphisms are realized by the following deductions:

$$\dfrac{\dfrac{\phi \vdash \psi}{\emptyset_a ; \phi \vdash \psi}}{\emptyset_a \vdash \phi \rightarrow \psi} \qquad \dfrac{\dfrac{\phi \vdash \psi}{\emptyset_m , \phi \vdash \psi}}{\emptyset_m \vdash \phi -\!\!* \psi}$$

$$\dfrac{\dfrac{\emptyset_a \vdash \phi \rightarrow \psi \quad \overline{\phi \vdash \phi}}{\emptyset_a ; \phi \vdash \psi}}{\phi \vdash \psi} \qquad \dfrac{\dfrac{\emptyset_m \vdash \phi -\!\!* \psi \quad \overline{\phi \vdash \phi}}{\emptyset_m , \phi \vdash \psi}}{\phi \vdash \psi} \; .$$

Although these observations may make the difference between $-\!\!*$ and $\rightarrow$ appear rather slight, the two implications are not interconvertible:

$$\phi -\!\!* \psi \not\vdash \phi \rightarrow \psi \qquad \text{and} \qquad \phi \rightarrow \psi \not\vdash \phi -\!\!* \psi \, .$$

Some examples of DCCs, which appeared in [O'Hearn and Pym, 1999, O'Hearn, 1999], will illustrate these remarks.

EXAMPLE 3.9 (SEE [O'HEARN AND PYM, 1999]) **Set**×**Set** *is bi-cartesian closed, with co-products and cartesian closed structure defined pointwise from their corresponding versions in* **Set**. *A symmetric monoidal closed structure is given by*

$$\begin{aligned} I &= (1,0) \\ (E_0,E_1)\otimes(F_0,F_1) &= ((E_0\times F_0)+(E_1\times F_1),\,(E_0\times F_1)+(E_1\times F_0)) \\ (E_0,E_1)\multimap(F_0,F_1) &= ((E_0\to F_0)\times(E_1\to F_1),\,(E_0\to F_1)\times(E_1\to F_0)) \end{aligned}$$

More generally, if $\mathcal{M}$ is a commutative monoid, considered as a discrete commutative monoidal category, then $\mathbf{Set}^{\mathcal{M}}$ *is a bi-cartesian DCC; in this example $\mathcal{M}$ is the two-element commutative monoid $\{0,1\}$ with addition modulo 2.*

This example does not appear to convey any particularly useful computational ideas, but we can use it to make a few remarks.

1 It is a non-degenerate model, in that I is not a terminal object and $*$ is not cartesian product. So the definition of DCCs does not induce a collapse of the specified structure.

2 There are no maps in the model from 1 to I.

3 $((0,1)\to(1,0)) = (1,0)$ and $((0,1)\multimap(1,0)) = (0,1)$. This, combined with the fact that there are no maps between $(0,1)$ and $(1,0)$ in either direction, implies that there are no maps from $((0,1)\to(1,0))$ to $((0,1)\multimap(1,0))$ or back, confirming the remark above that $\to$ and $\multimap$ are not convertible to one another in the linear version of the bunched language.

4 There is no functor $!$: **Set** × **Set** → **Set** × **Set** admitting an isomorphism $!E\multimap F \cong E\to F$, corresponding to Girard's [Girard, 1987] translation of intuitionistic logic into linear logic, using the exponential, !, thus indicating that a DCC is not simply a model of linear logic in disguise. To understand this remark, consider that $(1,0)\to(2,2) = (2,1)$ but that, for any C, $C\multimap(2,2) = (X,Y)$, for sets X and Y of the same cardinality. Therefore, for any "!" we try to pick, $!E\multimap(2,2)$ cannot be $(2,1)$.

Example 3.9 and our subsequent remarks about the exponential, !, suggest the it is worthing pausing for a brief comparison between the categorical semantics of **BI** and the categorical semantics of intuitionisitic linear logic. A useful summary of the latter has been provided by

Bierman [Bierman, 1995]. The essential structure of a semantics of (the proofs of) propositional intuitionistic linear logic with the connectives[3]

$$\otimes,\ I,\ \multimap,\ \wedge,\ \top,\ \vee \text{ and } !.$$

is given by a *linear category*, consisting of a category $\mathcal{C}$ such that (see [Benton et al., 1992] for the details):

1 $\mathcal{C}$ is symmetric monoidal closed;

2 There is a symmetric monoidal co-monad $(!, \epsilon, \delta, m_{A,B}, m_I)$ on $\mathcal{C}$ such that:

  (a) For every free !-co-algebra, $(!A, \delta_A)$, there are two distinguished monoidal natural transformations, with components $\epsilon_A :!A \to I$ and $d_A :!A \to !A \otimes !A$ (we must check that $! - \otimes ! -$ and I are monoidal functors), which form a commutative co-monoid are which are co-algebra morphisms;

  (b) Whenever $f : (!A, \delta_A) \to (!B, \delta_B)$ is a co-algebra morphism between free-co-algebras, then it is also a co-monoid morphism;

3 $\mathcal{C}$ has finite (cartesian) products and co-products.

It is informative, however, to pay close attention to the interpretation of formulæ and consequences (sequents) within this structure. Tensor product and linear implication are interpreted by the SMCC structure and the additive conjunction and disjunction are interpreted by product and co-product, respectively. The sequential antecedent-constructor, ",", is interpreted by the monoidal structure, with the consequence that Weakening and Contraction are not available (sound) for arbitrary formulæ. However, the co-monad structure, used to interpret ! in the evident way, ensures that these rules are available for formulæ with ! as outermost connective. Now, the co-Kleisli category associated with a linear category is a CCC, and so a model intuitionistic logic, with the consequence that Girard's translation of intuitionisitic implication, $\phi \to \psi$, as $!\phi \multimap \psi$, is sound. Note, however, that we have obtained the adjunction

$$\frac{\Gamma,\ !\phi \vdash \psi}{\Gamma \vdash !\phi \multimap \psi}$$

[3] It is common in the world of linear logic to use & to denote the additive conjunction, here denoted by $\wedge$ with unit $\top$, and $\oplus$ to denote additive conjunction, here denoted here by $\vee$ with unit $\bot$.

and that to get an additive implication *without* the exponential ! we must take an additive antecedent constructor, *viz.* ";",

$$\frac{\Gamma ; \phi \vdash \psi}{\Gamma \vdash \phi \to \psi},$$

interpreted by cartesian *closed* structure, as in a DCC.

An alternative perspective is provided by dual intuitionistic linear logic, **DILL**, [Barber, 1996]. There the organization of the recovery of the intuitionistic structure is rather different. Syntactically, the antecedents of **DILL**'s sequents are divided into two zones using a "stoup", as follows:[4]

$$\Gamma \mid \Delta \vdash \phi .$$

The left-hand zone contains the intuitionistic assumptions and the right-hand the linear ones. The exponential, !, is introduced and eliminated on the right of the sequent, in the natural deduction style and a key feature of the system is the invertible rule

$$\frac{\Gamma, \phi \mid \Delta \vdash \psi}{\Gamma \mid !\phi, \Delta \vdash \psi},$$

so that, once again, we see the soundness of the translation of $\phi \to \psi$ as $!\phi \multimap \psi$.

Categorically, **DILL**'s proofs are modelled by a pair of categories $(\mathcal{S}, \mathcal{C})$, in which $\mathcal{S}$ is an SMCC and $\mathcal{C}$ is a CCC, together with an adjoint pair of symmetric monoidal functors $G \vdash F : \mathcal{C} \to \mathcal{S}$. Tensor product and linear implication are interpreted in the obvious way, in $\mathcal{S}$, and the interpretation of ! is given by

$$[\![!\phi]\!] = FG([\![\phi]\!]),$$

i.e., the exponentiation is interpreted in the CCC and then mapped back to $\mathcal{S}$.

We remark that these structures provide frameworks within which it is possible to analyse the addition of an exponential, !, to **BI**. One takes a bi-cartesian DCC together with a monoidal co-monad *with respect to tensor (not the cartesian) product in the DCC.* An example of such a model, with a computational motivation and based on presheaves of posets ordered by subset-inclusion, $\mathbf{Set}^{\mathcal{FS}}$, has been given by by O'Hearn

[4]Barber uses ";" to denote the stoup (*cf.* [Girard, 1993]) but we prefer | to avoid confusion with bunches.

in [O'Hearn, 2000]. The logical properties of such structures remain to be explored.

EXAMPLE 3.10 *The cartesian closed structure of* **Cat** *is the familiar one: finite products of categories and the one-object category give the additive product and the functor category* B^A *gives the additive exponent. Another closed structure on* **Cat** *is given as follows: the symmetric monoidal structure is given by Gray's tensor product [Gray, 1974] with the one-object category as unit;* $A \multimap B$ *is the category whose objects are functors and arrows are "transformations",* i.e., *families of maps similar to natural transformations but lacking the naturality constraints. It follows that* **Cat** *is an* affine $\alpha\lambda$*-model,* i.e., *in which we have* $I \cong 1$. *Foltz, Lair and Kelly [Foltz et al., 1980] have shown that these are the only symmetric monoidal closed structures on* **Cat**.

It is fairly straightforward to show that if a cartesian DCC is *well-pointed*, *i.e.*, for any pair of parallel maps $f, g : C \longrightarrow D$ there is a map $e : 1 \longrightarrow C$ such that $e; f \neq e; g$, and is not simply a preorder, then the two units I and 1, are isomorphic.

EXAMPLE 3.11 *Generalizing Example 3.9, if* $\mathcal{M}$ *is a commutative monoid, considered as a discrete monoidal category, then* $\mathbf{Set}^{\mathcal{M}}$ *is a* $\alpha\lambda$*-model. In Example 3.9,* $\mathcal{M}$ *is the two-element commutative monoid* $\{0, 1\}$ *with addition modulo* 2.

EXAMPLE 3.12 *Let* $\mathcal{M}$ *be the commutative monoid* $\{0, 1\}$ *with multiplication. Regard* $\mathcal{M}$ *as a one-object category with object* $*$. *There is a symmetric monoidal structure on* $\mathbf{Set}^{\mathcal{M}}$ *given by*

$$\begin{aligned}(A \otimes B)* &= \{(a, b) \in A* \times B* \mid A0a = a \vee B0b = b\} \\ (A \multimap B)* &= \mathbf{Set}^{\mathcal{M}}[A, B]\end{aligned}$$

A non-example is also informative. Again, we follow [O'Hearn and Pym, 1999, O'Hearn, 1999].

EXAMPLE 3.13 *It follows from well-known results in domain theory,* e.g., *see [Plotkin, 1978, Amadio and Curien, 1998, O'Hearn et al., 1995], that CPOs (pointed,* ω*-complete partial orders) admit constructions which closely resemble the connectives of* **BI**. *Suppose* E *and* F *are CPOs. The "multiplicative" constructions, the strict function space* $E \multimap F$ *and the smash product* $E \otimes F$*, and the "additive" constructions, the continuous function space* $E \rightarrow F$ *and the cartesian product* $E \times F$*,* *are all CPOs. However, CPOs are* not *an* $\alpha\lambda$*-model. Whilst* $(\times, 1, \to)$ *is the cartesian closed structure in the category of CPOs and continuous functions,* $(\otimes, I, -\!\!\circ)$ *is the monoidal closed structure in the category of strict continuous functions. For an* $\alpha\lambda$*-model, we require a single category that admits both structures. CPOs interpret in this way not* **BI** *but intuitionistic linear logic, in which the monoidal structure* $(\otimes, I, -\!\!\circ)$ *interprets the multiplicatives, as in* **BI***, and the cartesian structure* $(\times, 1, \to)$ *interprets intuitionistic logic. The two categories are related by a comonad which interprets the exponential* !. *CPOs, and indeed any cartesian closed category, provide a degenerate* $\alpha\lambda$*-model in which the two closed structures are taken to be the same.*

That, for example, the coherence space model does not work for **BI** is rather more than just a technical curiosity. The reasons why it does not work run rather deep and relate to the resource interpretations of the connectives. We have already seen this point from a syntactic perspective in Chapter 2.

We have a straightforward completeness theorem (*cf.* [Barber, 1996, Barber and Plotkin, 1997]).

PROPOSITION 3.14 (COMPLETENESS FOR DCCS) *There is a DCC,* $\mathcal{T}$*, such that if* $m : E \to F$ *is a morphism in* $\mathcal{T}$*, then there exist* Γ*,* ϕ *and* M *such that* $[\![\Gamma]\!]_{\mathcal{T}} = E$*,* $[\![\phi]\!]_{\mathcal{T}} = F$*,* $[\![M]\!]_{\mathcal{T}} = m$ *and* $\Gamma \vdash M : \phi$ *is provable in* $\alpha\lambda$*. Moreover, if* $m' : E \to F$*,* $m = m'$ *and* $[\![M']\!]_{\mathcal{T}} = m'$*, then* $\Gamma \vdash M = M' : \phi$ *in* $\alpha\lambda$*.*

PROOF SKETCH We exhibit such a DCC, $\mathcal{T}$, constructed from the types and terms of $\alpha\lambda$, as follows:

- Objects are $\alpha\lambda$-types (*i.e.*, **BI** propositions) ϕ;
- Arrows between ϕ and ψ are given the equivalence class of pairs (x, M), in which x is a variable and M is an $\alpha\lambda$-term such that $x : \phi \vdash M : \psi$ is provable in $\alpha\lambda$, over the equivalence $\equiv$ defined by

$$(x, M) \equiv (x, N) \quad \text{if } x : \phi \vdash M = N : \psi \text{ in } \alpha\lambda \quad (\beta\eta\zeta\text{-equality})$$

$$(x, M) \equiv (y, M[y/x]) \qquad (\alpha\text{-equality})$$

The interpretation $[\![-]\!]_{\mathcal{T}}$ sends each ϕ to itself and each M to is $\beta\eta\zeta$-normal form.

It is straightforward, though lengthy, to verify that $\mathcal{T}$ is a DCC. The identity $\phi \xrightarrow{1_\phi} \phi$ is given by (x, x). Given $(x, M) : \phi \longrightarrow \psi$ and $(y, N) : \psi \longrightarrow \chi$, their composition $(x, M); (y, N)$ is given by $(x, N[M/y])$.

We must establish that $\mathcal{T}$ has both cartesian closed structure and symmetric monoidal closed structure. These two structures are defined independently and there is no interference between $\alpha\lambda$'s reduction rules for them.

The cartesian closed structure is established in the usual way; *q.v.* [Lambek and Scott, 1986, Seely, 1983].

The symmetric monoidal closed structure is established using the techniques found in [Barber, 1996, Bierman, 1995, Benton et al., 1993]. Note that the ζ-rules for **let** must be employed in order to establish the coherence of the monoidal structure. □

We conclude by remarking that the construction given in our model existence lemma for Kripke models (Lemma 4.6) determines a cartesian DCC and that it is possible to extend our technique to $\vee$ (and $\perp$), interpreted in bi-DCCs, using the techniques described in [Ghani, 1995, Pym and Ritter, 2001].

3.1 Day's Construction

We can generate a rich class of DCCs using a general construction due to Brian Day, [Day, 1970], which we first mentioned in Chapter 1. Day shows that any monoidal (not necessarily closed) category $(\mathcal{C}, \otimes, I)$ induces a monoidal closed structure on the functor category $\mathbf{Set}^{\mathcal{C}^{op}}$, and that if $(\mathcal{C}, \otimes, I)$ is symmetric monoidal, then so is $\mathbf{Set}^{\mathcal{C}^{op}}$. (Day describes his results much more generally, in the context of enriched categories [Day, 1970, Kelly, 1982].) This, combined with the standard fact that $\mathbf{Set}^{\mathcal{C}^{op}}$ is bi-cartesian closed, yields a host of bi-cartesian DCCs.

The construction is as follows. The unit I of the monoidal structure is $\mathcal{C}[-, I]$. Given functors E and F, the formula for the tensor product is written using co-ends:

$$(E \otimes F)X \;=\; \int^{Y,Y'} EY \times FY' \times \mathcal{C}[X, Y \otimes Y'].$$

The formula for $-\!\!\circ$ uses an end:

$$(E -\!\!\circ F)X \;=\; \int_Y \mathbf{Set}[EY, F(X \otimes Y)] \;\cong\; \mathbf{Set}^{\mathcal{C}^{op}}[E(-), F(X \otimes -)].$$

The formulæ for $(E \otimes F)X$ and $(E -\!\!\circ F)X$ are both contravariant in Z, giving the morphism parts of the functors.

Day's construction gives us a way of embedding any symmetric monoidal closed category into a bi-cartesian DCC: we apply the Yoneda embedding $\mathcal{C} \longrightarrow \mathbf{Set}^{\mathcal{C}^{op}}$, which sends X to $\mathcal{C}[-, X]$. It is standard that Yoneda is full and faithful, and it is not difficult to show that it preserves monoidal closed structure [O'Hearn, 1999]. From this we can conclude that **BI** is conservative over **MILL**, again not only on the level of provability but also on that of the semantics of proofs.

Two observations are useful for working with the tensor product. The first is that we have a form of pairing operation: given $a \in EY$ and $b \in FY'$ we can form an element $[a, b] \in (E \otimes F)(Y \otimes Y')$. To see how this element is defined, consider that the co-end $(E \otimes F)X$ may be described as a quotient of quintuples $(Y, Y', f : X \to_{\mathcal{C}} Y \otimes Y', a \in EY, b \in FY')$. The pair $[a, b]$ is then the equivalence class of $(Y, Y', id_{Y \otimes Y'}, a, b)$.

The second is a representation result which characterizes maps out of a tensor products: natural transformations $E \otimes F \longrightarrow G$ are in bijective correspondence with families of functions

$$EX \times FY \longrightarrow G(X \otimes Y)$$

natural in X and Y. To see why this is true, consider the definition of $-\!\!\circ$, and the isomorphism $[E \otimes F, G] \cong [E, F -\!\!\circ G]$: the multi-map characterization is essentially forced by $-\!\!\circ$.

These observations may be put together to describe the functorial action of $\otimes$. Given natural transformations $\eta : E \to E'$ and $\mu : F \to F'$, we obtain a natural family of maps $EX \times FY \to (E' \otimes F')(X \otimes Y)$ by composing $\eta_X \times \mu_Y$ with pairing. Then we obtain a natural transformation $E \otimes F \to E \otimes F$ by the representation of maps out of a tensor in terms of multi-maps.

Finally, we note that it is often possible to give an explicit description of the tensor product without using co-ends at all. In fact, we have already done this with the **Set** $\times$ **Set** example above.

3.2 Conservativity

We have observed **BI**'s conservativity over **IL** and **MILL** from a proof-theoretic point of view. From a semantic point of view, the conservativity of **BI** over **IL** may be seen immediately in terms of DCCs. For suppose $\mathcal{C}$ is a categorical model of **IL**, *i.e.*, a bi-cartesian closed category. Then we can regard it as a bi-cartesian DCC in which the two closed structures are the same. When we restrict to the additive fragment of **BI**, this shows that the denotations of **BI** proofs are exactly the same as those in the model of **IL**.

The Yoneda functor fully and faithfully embeds a CCC in its corresponding category of presheaves. Indeed, for a small $\mathcal{C}$, $\mathbf{Set}^{\mathcal{C}^{op}}$ is bi-cartesian closed. Similarly to O'Hearn's use [O'Hearn, 1999] of a result due to Day [Day, 1973], we can get a similar result for monoidal closed categories using Day's construction. To formulate this result, we recall a few definitions.

- A *symmetric monoidal closed functor* between symmetric monoidal closed categories $(\mathcal{C}, \multimap, \otimes, I)$ and $(\mathcal{C}', \multimap', \otimes', I')$ is a functor $F : \mathcal{C} \longrightarrow \mathcal{C}'$, together with a map $t_0 : I' \to F(I)$ and natural transformations $t_1 : F(A) \otimes' F(B) \longrightarrow F(A \otimes B)$ and $t_2 : F(A \multimap B) \longrightarrow F(A) \multimap' F(B)$, subject to some coherence conditions [Eilenberg and Kelly, 1965].

- A functor is *strong monoidal* if t_0 and t_1 are isomorphisms, subject to some coherence conditions.

- A functor is said to *preserve SMCC structure* if it is an SMCC functor in which each t_i is an isomorphism.

PROPOSITION 3.15 (EMBEDDING SMCCS) *Suppose $\mathcal{C}$ is a small symmetric monoidal closed category. Then there is a fully faithful embedding $\mathcal{C} \longrightarrow \overline{\mathcal{C}}$, where $\overline{\mathcal{C}}$ is a bi-cartesian DCC, which preserves the symmetric monoidal closed structure. (If the unit in $\mathcal{C}$ is terminal, then $\overline{\mathcal{C}}$ is an affine bi-cartesian DCC.)*

PROOF We know ([Im and Kelly, 1986]) that Yoneda, $Y : \mathcal{C} \longrightarrow \mathbf{Set}^{\mathcal{C}^{op}}$, is strong monoidal. For the closed part, we exhibit t_2 and its inverse. Let $s \in \mathcal{C}[W, X \multimap Y]$. Then the element $t_2 s \in (\mathcal{C}[-, X] \multimap \mathcal{C}[-, Y])W$ is defined as follows: We require a map $(t_2 s[W'](f : W' \to X)) : W \otimes W' \to Y$; this is $(W \otimes f); s'$, where $s' : W \otimes X \to Y$ is the map corresponding to s via the defining adjunction. Its inverse is defined as follows: Given $r \in (\mathcal{C}[-, X] \multimap \mathcal{C}[-, Y])W$, we get $r' : W \to X \multimap Y$ as the map obtained by currying from $r[X]id_X : W \otimes X \to Y$. □

It follows that **BI** is a conservative extension of both **IL** and **MILL**.

3.3 Structural Variations

Categorical models of the affine and relevant versions of **BI** and $\alpha\lambda$ may be obtained by imposing some conditions on DCCs.

For affine models we require a DCC such that $1 \cong I$. The soundness of the Weakening rule for ",",

$$\frac{\phi \vdash \chi}{\phi, \psi \vdash \chi},$$

is a consequence of this isomorphism. To see this, consider that we must demonstrate the existence of an arrow w the diagram

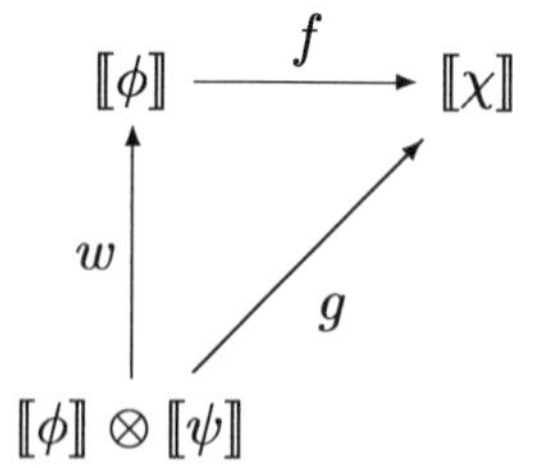

and so, given the arrow f, construct the arrow g. The existence of w follows from $1 \cong I$. We have

$$[\![\phi]\!] \otimes [\![\psi]\!] \overset{1_{[\![\phi]\!]} \otimes \bigcirc_{[\![\psi]\!]}}{\longrightarrow} [\![\phi]\!] \otimes 1.$$

Therefore, since $1 \cong I$, we have

$$[\![\phi]\!] \otimes [\![\psi]\!] \overset{1_{[\![\phi]\!]} \otimes \bigcirc_{[\![\psi]\!]}}{\longrightarrow} [\![\phi]\!] \otimes I = [\![\phi]\!].$$

A computationally-motivated example of an affine model is developed in some detail by O'Hearn in [O'Hearn, 1999].

For relevant models, we must ask for canonical arrows $E \overset{c}{\longrightarrow} E \otimes E$, so that the existence of an arrow $E \otimes E \overset{f}{\longrightarrow} F$ implies the existence of an arrow $E \overset{g}{\longrightarrow} F$.

Models which admit Dereliction,

$$\frac{\phi, \psi \vdash \chi}{\phi ; \psi \vdash \chi},$$

require canonical arrows $[\![\phi]\!] \times [\![\psi]\!] \longrightarrow [\![\phi]\!] \otimes [\![\psi]\!]$.

The non-commutative variations, though unproblematic, require change to the underlying categorical structure. A non-commutative monoidal conjunction, together with its two associated implications, may be interpreted in a category which carries *monoidal biclosed* structure, *i.e.*, a non-commutative monoidal product, $\diamond$ with unit H, together with two adjoints, $-\!\!\triangleleft$ and $-\!\!\triangleright$. Similar structures were considered by Lambek in [Lambek, 1968, Lambek, 1969, Lambek, 1972]. In such a structure, the two function spaces satisfy the isomorphisms

$$[B, A -\!\!\triangleleft C] \cong [A \diamond B, C] \cong [A, B -\!\!\triangleright C].$$

By freely combining (bi)cartesian closed, symmetric monoidal closed, and monoidal biclosed structure, we can provide models for logics with combinations of (i) intuitionistic, (ii) commutative linear, and (iii) non-commutative linear connectives. Note that we can take all three together but that the system we obtain remains quite different from that presented in [Polakow and Pfenning, 1999], which uses three distinct zones, separated by stoups, for the three fragments, just as $\alpha\lambda$ (**BI**) differs from **DILL** [Barber, 1996, Barber and Plotkin, 1997], which uses two zones for the linear and intuitionistic fragments.

A computationally-motivated example of a non-commutative model, based on presheaves over a category of worlds built out of memory locations is developed in [Pym et al., 2000].

Chapter 4

KRIPKE SEMANTICS

1. Kripke Models of Propositional BI

The elementary account of a truth-functional semantics for **BI** that we gave in the Introduction, using "Kripke resource monoids", is, in its conceptual simplicity, rather appealing. However, its simplicity somewhat obscures an important aspect of the semantics of **BI**, namely the structure of the multiplicatives, $*$ and $-\!\!*$, which we must understand in the same (set-theoretic or, rather, categorical) terms as we understand the structure of the additives. In this chapter, we develop the elementary Kripke semantics in a categorical setting and establish a soundness theorem and a completeness theorem (restricted to exclude inconsistency, $\bot$).

Whilst the structure of $\wedge$ and $\rightarrow$ may be seen very clearly from the definition of the forcing relation, $\models$ (they are simply the usual set-theoretic product and implication, respectively) the structure of $*$ and $-\!\!*$ is, perhaps, less clear. To see their structure, we shall need to move away from the category of sets to the category (topos) of presheaves over a symmetric monoidal category and exploit Day's tensor product construction [Day, 1970], which will form the basis of the multiplicative part of our definition of Kripke models.

Let $(\mathcal{C}, I, \cdot)$ be a small monoidal category. Following [Day, 1970], we consider $\mathbf{Set}^{\mathcal{C}^{op}}$. The monoidal structure on $\mathcal{C}$ induces the following monoidal closed structure on $\mathbf{Set}^{\mathcal{C}^{op}}$: for $A, B : \mathcal{C}^{op} \longrightarrow \mathbf{Set}$, the co-end

$$(A \otimes B)m = \int^{n,n'} A(n) \times B(n') \times \mathcal{C}[m, n \cdot n'] \qquad (4.1)$$

defines the multiplicative product, $\otimes$, with the unit given by $\mathcal{C}[-, I]$ and the end

$$\begin{aligned}(A \multimap B)m &= \int_n \mathbf{Set}[A(n), B(m \cdot n)] \\ &\cong \mathbf{Set}^{\mathcal{C}^{op}}[A(-), B(m \cdot -)]\end{aligned} \tag{4.2}$$

defines the multiplicative function space, $\multimap$. The formulæ for

$$(A \otimes B)m \text{ and } (A \multimap B)m$$

are both contravariant in m, giving the morphism parts of the functors; see [Day, 1970, O'Hearn et al., 1995]. Moreover, if the monoidal structure on $\mathcal{C}$ is symmetric, then so is the induced one on $\mathbf{Set}^{\mathcal{C}^{op}}$. These definitions are elegantly explained in Ambler's thesis [Ambler, 1992], which considers the semantics of first-order linear logic in symmetric monoidal closed categories.

Recall from Chapter 3 that two general observations are useful for working with the tensor product given by Day's construction. The first is that we have a form of pairing operation: given $a \in EY$ and $b \in FY'$ we can form an element $[a, b] \in (E * F)(Y * Y')$. To see how this element is defined, consider that the co-end $E * F(X)$ may be described as a quotient of quintuples $(Y, Y', f : X \to_{\mathcal{C}} Y * Y', a \in EY, b \in FY')$. The pair ("Day's pairing operation") $[a, b]$ is then the equivalence class of $(Y, Y', 1_{Y * Y'}, a, b)$.

The second is a representation result which characterizes maps out of a tensor products: natural transformations $E * F \longrightarrow G$ are in bijection with families of functions

$$EY \times FY' \longrightarrow G(Y * Y')$$

natural in Y and Y'. To see why this is true, consider the definition of $-\!\!*$, and the isomorphism $[E * F, G] \cong [E, F -\!\!* G]$: the multi-map characterization is, essentially, forced by $-\!\!*$.

The usual categorical product and exponential lift, respectively pointwise and via Yoneda's lemma, to $\mathbf{Set}^{\mathcal{C}^{op}}$, which also has co-products. So, $\mathbf{Set}^{\mathcal{C}^{op}}$ has, using Day's construction, enough structure to interpret all of the connectives of propositional **BI**.[1]

[1]In Chapter 13, we shall see that we can also interpret predicate **BI** in presheaf categories, $\mathbf{Set}^{\mathcal{C}^{op}}$, in which $\mathcal{C}$ is small monoidal.

We can now define *Kripke models* of propositional **BI**. Our definition lives in the category $\mathbf{Set}^{\mathcal{M}^{op}}$, where $\mathcal{M} = (M, e, \cdot, \sqsubseteq)$ is a preordered commutative monoid, considered as a preordered monoidal category with an arrow from m to n if and only if $m \sqsubseteq n$.

Our definition of satisfaction is consistent with the definitions of the structure of $\otimes$ and $\multimap$ provided by (4.1) and (4.2) but enforces the structure they define via a forcing relation, $\models$. This setting is consistent with the BHK semantics for **BI**, developed in Chapter 3, in which we require that $*$ and $-\!\!*$ *be interpreted* using the (4.1) and (4.2). In the sequel, we will see that the basic truth-functional semantics described in the Introduction may be recovered as a simple restriction of this semantics. For now, it is sufficient to require that propositions be interpreted as objects of $\mathbf{Set}^{\mathcal{M}^{op}}$, a framework which ensures that we have enough structure to define an interpretation function and a satisfaction relation for propositions containing $*$ and $-\!\!*$.

DEFINITION 4.1 (KRIPKE MODELS) *Let $\mathcal{M} = (M, e, \cdot, \sqsubseteq)$ be a preordered commutative monoid, considered as a preordered commutative monoidal category with an arrow from m to n if and only if $m \sqsubseteq n$, and let $\mathcal{P}(L)$ denote the collection of* **BI** *propositions over a language L of propositional letters. A* Kripke model *is a triple*

$$\langle [\mathcal{M}^{op}, \mathbf{Set}], \models, [\![-]\!] \rangle,$$

where $[\mathcal{M}^{op}, \mathbf{Set}]$ is the category of presheaves over the preorder category $\mathcal{M}$ to **Set**, *$\models \subseteq M \times \mathcal{P}(L)$ is a satisfaction relation satisfying the constraints in Table 4.1 and $[\![-]\!] : \mathcal{P}(L) \rightharpoonup obj([\mathcal{M}^{op}, \mathbf{Set}])$ is a partial function from the* **BI** *propositions over L to the objects of $[\mathcal{M}^{op}, \mathbf{Set}]$ such that:*[2]

Kripke monotonicity (or Hereditary): If $n \sqsubseteq m$, then, for each $\phi \in \mathcal{P}(L)$, $m \models \phi$ implies $n \models \phi$.

Wherever no confusion will arise, we shall refer to a model

$$\langle [\mathcal{M}^{op}, \mathbf{Set}], \models, [\![-]\!] \rangle$$

simply as $\mathcal{M}$. □

To see that this definition is consistent with the sub-object classifier semantics of intuitionistic logic [Lambek and Scott, 1986] consider the pullback diagram in $\mathbf{Set}^{\mathcal{M}^{op}}$,

[2]We use a partial function because the interpretation of any given proposition need not be defined in all models.

$m \models \mathrm{p}$	iff	$[\![\mathrm{p}]\!](m) \neq \emptyset$, for $\mathrm{p} \in L$
$m \models \phi * \psi$	iff	for some $n, n' \in M$, $m \sqsubseteq n \cdot n'$, $n \models \phi$ and $n' \models \psi$
$m \models \phi \mathrel{-\!\!*} \psi$	iff	for all $n \in M$ $n \models \phi$ implies $m \cdot n \models \psi$
$m \models \phi \wedge \psi$	iff	$m \models \phi$ and $m \models \psi$
$m \models \phi \vee \psi$	iff	$m \models \phi$ or $m \models \psi$
$m \models \phi \to \psi$	iff	for all $n \sqsubseteq m$, $n \models \phi$ implies $n \models \psi$
$m \models \top$	for all	$m \in M$
$m \models I$	iff	$m \sqsubseteq e$
$m \not\models \bot$	for any	m

Table 4.1. Kripke Semantics

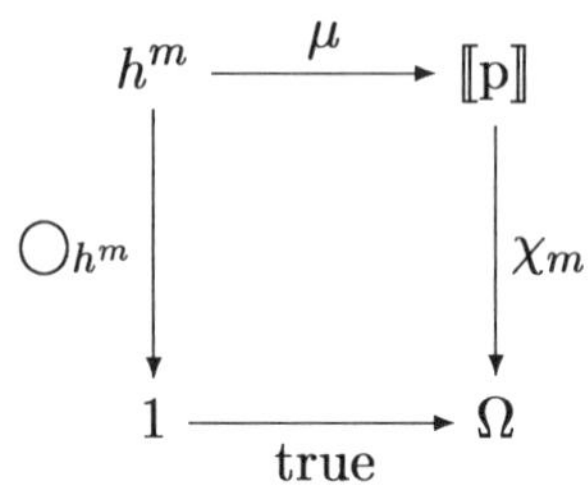

and note that an arrow $h^m \xrightarrow{\mu} [\![\mathrm{p}]\!]$ is, by the Yoneda lemma, determined uniquely by an element $\check{\mu} \in [\![\mathrm{p}]\!](m)$ and so $[\![\mathrm{p}]\!](m)$ is non-empty just in case the square commutes. Note that it is conceptually important, in our setting, that propositions ϕ are interpreted as $[\![\phi]\!] \in obj([\mathcal{M}^{op}, \mathbf{Set}])$, thereby giving access to Day's constructions.

Let $\langle [\mathcal{M}^{op}, \mathbf{Set}], \models, [\![-]\!] \rangle$ be a Kripke model. If $m \models \phi$, then *ϕ is true at m in $\mathcal{M}$*. Where necessary, the forcing relation for a model $\mathcal{M}$ may be distinguished as $\models_{\mathcal{M}}$. We write

$$m \models_{\mathcal{M}} \Gamma \qquad \text{iff} \qquad m \models_{\mathcal{M}} \phi_{\Gamma},$$

where ϕ_Γ is the formula obtained from Γ by replacing each semicolon with $\wedge$ and each comma with $*$, with association respecting the tree structure of Γ. Then $m, \Gamma \models_{\mathcal{M}} \phi$ if $m \models_{\mathcal{M}} \Gamma$ implies $m \models_{\mathcal{M}} \phi$. This defines the truth of ϕ relative to Γ at m . Similarly, if, for all m in all $\mathcal{M}$, $m \models \phi$, then ϕ *is valid*. We write $\Gamma \models \phi$, if and only if, for all m in all $\mathcal{M}$, $m \models \Gamma$ implies $m \models \phi$. This defines the validity of ϕ relative to Γ.

It is a straightforward matter to generalize Definition 4.1 by replacing the category $[\mathcal{M}^{op}, \mathbf{Set}]$ with the category $[\mathcal{C}^{op}, \mathbf{Set}]$, where $\mathcal{C}$ is any symmetric monoidal category. The form of the necessary changes to the definition of $\models$ may be seen in [Lambek and Scott, 1986].

2. Soundness and Completeness for BI without $\perp$

We establish the soundness of **BI** without $\perp$ for elementary Kripke models. It can be argued that soundness holds even in the presence of $\perp$.

THEOREM 4.2 (SOUNDNESS OF **BI** FOR KRIPKE MODELS) *If* $\Gamma \vdash \phi$ *is provable in* **NBI** *without* $\perp$, *then* $\Gamma \models \phi$.

PROOF We pick an arbitrary KRM, $\mathcal{M}$, and proceed by induction on the structure of the proof Φ of $\Gamma \vdash \psi$ in **NBI**. We do just a few illustrative cases.

Axiom: If the last inference in Φ is an axiom inference, then it is immediate that, for any world m in any model $\mathcal{M}$, $m \models \phi$ implies $m \models \phi$.

$-\!\!*$ I: By the induction hypothesis, we have that, for all m, $m \models \Gamma, \phi$ implies $m \models \psi$. We must show that if $m \models \Gamma$, then $m \models \phi -\!\!* \psi$. We have that $m \models \phi -\!\!* \psi$ iff, for all n, $n \models \phi$ implies $m \cdot n \models \psi$. If $n \models \phi$, then $m \cdot n \models \Gamma, \phi$. By the induction hypothesis, $m \cdot n \vdash \psi$. Therefore $m \models \phi -\!\!* \psi$.

$-\!\!*$ E: Follows easily from the definition of $\models$: we must show that if $m \models \Gamma, \Delta$, then $m \models \phi$. We have that $m \models \Gamma, \Delta$ iff there are n and n' such that $m \sqsubseteq n \cdot n'$ and $n \models \Gamma$ and $n' \models \Delta$. By the induction hypothesis, $n \models \phi -\!\!* \psi$ and $n' \models \phi$. Therefore $n \cdot n' \models \psi$ and so, by monotonicity, $m \models \psi$.

$\to$I: The usual intuitionistic argument, *q.v.* [van Dalen, 1983].

$\to$E: Immediate from the definition of $\models$.

The arguments for the other cases are variations on these. □

In the presence of $\bot$, we must argue that $\bot$E is sound. Suppose $m \models \Gamma$ implies $m \models \bot$. Now, $m \models \bot$ for no m. Therefore there is no m such that $m \models \Gamma$. Therefore, we have, vacuously, that if $m \models \Gamma$, then $m \models \phi$.

Turning to completeness, we remark that the absence of $\bot$ will turn out to be critical, *q.v.* the discussion following Theorem 4.7. This deficiency is corrected using topological methods in Chapter 5.

In order to establish elementary completeness, we must first establish the existence of a model which corresponds exactly to the consequences provable in **NBI**. We must begin by developing the idea of a *prime theory* for **BI**. The basic idea is the same as the one used in the proof of the completeness theorem for intuitionistic logic [van Dalen, 1983], adapted to handle the multiplicatives. The rôle of prime theories is to have exactly the structure required by the semantic clauses for the connectives. In particular, the structure required by the semantic clauses for $\vee$ and $*$. With respect to a prime theory, the proof-theoretic meaning of a connective is determined by its introduction rule.

DEFINITION 4.3 (PRIME BUNCH) *A bunch* Γ *is* prime *if*

1 Γ *is closed under consequences* $\vdash$ *generated by* $-\!\!*\,E$ *and* $\to$*E,* i.e.,

(a) Γ *is of the form* $\Gamma(\phi\,,\ \phi -\!\!* \psi)$ *implies* Γ *is also of the form* $\Gamma(\psi)$*;*

(b) Γ *is of the form* $\Gamma(\phi\,;\ \phi \to \psi)$ *implies* Γ *is also of the form* $\Gamma(\psi)$*;*

(c) Γ *is of the form* $\Gamma(\tau\text{-?}\phi)$*, where* τ *is a theorem and -? denotes either* $\to$ *or* $-\!\!*$ *implies* Γ *is also of the form* $\Gamma(\phi)$*;*[3]

2 Γ *is of the form* $\Gamma(\phi * \psi)$ *implies* Γ *is also of the form* $\Gamma(\phi, \psi)$*;*

3 Γ *of the form* $\Gamma(\phi \wedge \psi)$ *implies* Γ *is also of the form* $\Gamma(\phi; \psi)$*;*[4]

4 Γ *of the form* $\Gamma(\phi \vee \psi)$ *implies* Γ *is also either of the form* $\Gamma(\phi)$ *or of the form* $\Gamma(\psi)$.

Here we consider the form of bunches up to coherent equivalence, $\equiv$.
□

DEFINITION 4.4 (EVALUATION) *Let* Γ *be a bunch. Each of the following, of the form* redex $\rightsquigarrow$ reduct, *is called an* evaluation *in* Γ*:*

[3]Alternatively, these cases may be seen as being generated by $-\!\!*$L and $\to$L reductions.
[4]This case is not necessary for the subsequent argument to go through but seems conceptually appropriate at this point.

1(a) $\Gamma(\phi, \phi \mathrel{-\!*} \psi) \leadsto \Gamma(\psi)$;

1(b) $\Gamma(\phi; \phi \rightarrow \psi) \leadsto \Gamma(\phi; \phi \rightarrow \psi; \psi)$;

1(c) $\Gamma(\tau \mathrel{-\!*} \phi) \leadsto \Gamma(\phi)$ *and* $\Gamma(\tau \rightarrow \phi) \leadsto \Gamma(\tau \rightarrow \phi; \phi)$;

2 $\Gamma(\phi * \psi) \leadsto \Gamma(\phi, \psi)$;

3 $\Gamma(\phi \wedge \psi) \leadsto \Gamma(\phi; \psi)$;

4 $\Gamma(\phi \vee \psi) \leadsto \Gamma(\phi)$ *or* $\Gamma(\phi \vee \psi) \leadsto \Gamma(\psi)$.

We say that Γ' *is an* evaluation *of* Γ *and write* $\Gamma \leadsto \Gamma'$, *if* Γ' *is obtained by evaluating any of the redexes in* Γ. *If* Γ' *is also prime, we say that* Γ' *is a* prime *evaluation of* Γ.

We say that $\Gamma'(\Delta') \preceq \Gamma(\Delta)$ *if* $\Gamma' = \Gamma[\Delta'/\Delta]$ *for some* Δ, Δ' *such that* Δ' *is a reduct of* Δ. □

We will refer to the *principal formula* of a redex as determined by the inference used to evaluate it.

A prime evaluation $\lceil \Gamma \rceil$ of a given bunch Γ may be constructed inductively, respecting the structure of Γ. For example, suppose

$$\Gamma = (\phi\, ,\, \phi \mathrel{-\!*} (\psi * \chi))\, ,\, (\phi\, ,\, \phi \mathrel{-\!*} (\psi \vee \chi)).$$

Firstly, we perform all of the possible evaluations, to get to $\overset{\leadsto}{\Gamma}$. Evaluating for $\mathrel{-\!*}$, we get the bunch

$$(\psi * \chi)\, ,\, (\psi \vee \chi).$$

Now, evaluating $*$, we reach

$$(\psi, \chi)\, ,\, (\psi \vee \chi).$$

Evaluating $\vee$, we reach either

$$(\psi, \chi)\, ,\, (\psi) \quad \text{or} \quad (\psi, \chi)\, ,\, (\chi)$$

Each of the steps we have performed has the property that it introduces no new consequences. Finally, we set $\lceil \Gamma \rceil = \Gamma; \overset{\leadsto}{\Gamma}$ (so that $\lceil \Gamma \rceil \vdash \phi_\Gamma$).

Note that evaluations are not unique. Consider, for example, the bunch

$$\phi, \phi \mathrel{-\!*} \psi, \phi \mathrel{-\!*} \chi$$

which may be evaluated to either

$$\psi, \phi \mathrel{-\!*} \chi \quad \text{or} \quad \phi \mathrel{-\!*} \psi, \chi$$

but note that both possible evaluations will be available with our construction of a term model (Lemma 4.6). Note also that, in the presence of $\vee$, evaluation generates a set of bunches.

LEMMA 4.5 (PRIME EVALUATION) *If* $\Gamma \not\vdash \phi$ *in* **NBI** *without* $\bot$, *then there is a prime evaluation,* $\lceil \Gamma \rceil$, *of* Γ *such that* $\lceil \Gamma \rceil \not\vdash \phi$ *in* **NBI** *without* $\bot$.

PROOF The method presented in [van Dalen, 1983], similar to the construction in [Dummett, 1977], in which $\lceil \Gamma \rceil$ is constructed as the co-limit of a sequence,

$$\Gamma = \Gamma_0 \preceq \Gamma_1 \preceq \Gamma_2 \preceq \ldots \preceq \lceil \Gamma \rceil,$$

of extensions of Γ in which we successively treat the connectives that are to be reduced, may be adapted to our setting by constructing a co-limit of evaluations. In our setting, $\Gamma' \preceq \Gamma$ is the "evaluation ordering", as defined in Definition 4.4.

Closing under $\vdash$ and the clause for $\vee$ are essentially identical to the intuitionistic case, *q.v.* [van Dalen, 1983, Dummett, 1977], although note that we have restricted the closure to that generated by evaluating $-\!\!* E$ and $\to E$ and have added the necessary reductions for the remaining connectives.

We successively reduce the remaining unreduced redexes: $\wedge$, close, reduce $\vee$, close, reduce $*$, close, and so on, respecting the original structure of the bunch.[5] We discuss just some of the cases, numbered as in Definition 4.4, in detail.

- Evaluating a redex of the form (1(a)) at stage k, we look for the first redex $\Gamma_k(\phi_1, \phi_1 -\!\!* \phi_2)$ such that $\Gamma \vdash \phi_2$ which has yet to be reduced. It cannot be that $\Gamma_k(\phi_2) \vdash \phi$, for then we should have $\Gamma_k(\phi_1, \phi_1 -\!\!* \phi_2) \vdash \phi$, so we can define

$$\Gamma_{k+1} = \Gamma_k(\phi_2).$$

- Case (1(b)) is similar, with the variation to ϕ_1 and $\phi_1 \to \phi_2$ requiring that we mark the redex in Γ_k as *reduced* in Γ_k, so that looping does not occur. Note that a 1(a) redex cannot create an ill-formed 1(b) redex. For example, if

$$\Gamma = (\phi, \phi -\!\!* \psi), (\psi \to \chi)$$

[5]Note that we intend no fixed order of reductions.

we first, by case 1(a), get to $(\psi), (\psi \to \chi)$ but note that this does *not* form a redex of the form 1(b) with $\psi \to \chi$ (and, indeed, $\Gamma \vdash \chi$ is not provable).

- Evaluating a redex of the form (1(c)) at stage k, we look, for example, for the first redex $\Gamma_k(\tau \twoheadrightarrow \psi)$, where τ is a theorem, such that $\Gamma \vdash \phi$, and put $\Gamma_{k+1} = \Gamma_k(\psi)$. Similarly, the case for $\to$ sets $\Gamma_{k+1} = \Gamma_k(\tau \twoheadrightarrow \psi; \psi)$, with the variation that we mark the redex in Γ_k as *reduced* in Γ_k, so that looping does not occur.

- Evaluating a redex of the form (2) at stage k, we look for the first redex $\Gamma_k(\phi_1 * \phi_2)$ such that $\Gamma_k(\phi_1 * \phi_2) \vdash \phi_1 * \phi_2$ which has yet to be evaluated. It cannot be that $\Gamma_k(\phi_1, \phi_2) \vdash \phi$, for then we should have $\Gamma_k(\phi_1 * \phi_2) \vdash \phi$, so we can define

$$\Gamma_{k+1} = \Gamma_k(\phi_1\,,\, \phi_2).$$

  Note that we have *not* evaluated Γ_k non-trivially in this case: we have simply added to a dependency on $\phi * \psi$ a dependency on the bunch ϕ, ψ, *i.e.*, we have, essentially, performed a $*$L-reduction (*q.v.* Chapter 6) and a contraction, thereby preserving consequences.

The other cases are similar.

We take $\overset{\leadsto}{\Gamma}$ to be the limit of the Γ_ks over $k \geq 0$. Formally, we must consider an inductive definition over trees ordered by $\preceq$ and show that the co-limit exists; the argument is evident, although note that looping is prevented by marking the additive implicational redexes as reduced. Then we set $\lceil \Gamma \rceil = \Gamma\,;\, \overset{\leadsto}{\Gamma}$.

Given this construction, we must then check that (i) $\lceil \Gamma \rceil \nvdash \psi$, and (ii) $\lceil \Gamma \rceil$ is a prime evaluation of Γ.

For (i), we show, by induction on i, that $\Gamma_i \nvdash \phi$, starting with our assumption that $\Gamma \nvdash \psi$. For example, for the $*$ case, suppose that $\Gamma_{i+1} \vdash \phi$, *i.e.*, that $\Gamma_i(\phi_1 * \phi_2) \vdash \phi$. Then, by Lemma 2.1, it must be that $\Gamma_i(\phi_1, \phi_2) \vdash \phi$, a contradiction. The other cases are similar.

For (ii), that $\lceil \Gamma \rceil$ is prime, we proceed just as in [van Dalen, 1983] but taking account of our additional cases: each redex is reduced at some finite stage, thereby satisfying Definition 4.3. For termination, note that we visit the principal formula of each redex only once and that this is sufficient. To see this, consider that in this propositional setting, looping may occur only by the repeated generation of a given proposition. However, the definition of primality is satisfied by generating the given proposition just once. For example, given the bunch $\phi; \phi \to \psi; \psi \to \phi$, we satisfy the definition of primality after a single evaluation of the redex $(\phi, \phi \to \psi)$ (although our procedure subsequently evaluates the redex $(\psi; \psi \to \phi)$ as well). Finally, note that each bunch gives rise to a finite collection of redexes. Similar analyses have been treated formally in, for example, [Dummett, 1977, Wainer and Wallen, 1992, Pinto and Dyckhoff, 1985]. □

Note that, in the presence of disjunction, prime evaluations are not unique: evaluating a redex of the form $\Gamma(\phi \vee \psi)$ yields the choice between $\Gamma(\phi \vee \psi; \phi)$ and $\Gamma(\phi \vee \psi; \psi)$. Formally, this choice may be handled by working with finite sets of evaluations.

The last construction we require is that of a model which corresponds exactly to the proof theory of **BI**.

A bunch Γ is *consistent* just in case $\Gamma \nvdash \bot$; otherwise it is inconsistent. Thus $\bot$, the unit of $\vee$, internalizes inconsistency in **BI**. Our semantics does not account for inconsistency: $\bot$ is *nowhere* forced. Accordingly, for completeness, we must exclude $\bot$ from **BI**. This point is discussed further following the proof of completeness, Theorem 4.7.

LEMMA 4.6 (MODEL EXISTENCE) *There is a Kripke model*

$$\langle [\mathcal{T}^{op}, \mathbf{Set}], \models, [\![-]\!] \rangle,$$

where $\mathcal{T} = (T, e, \cdot, \sqsubseteq)$, *and a world* $t \in T$ *such that if* $\Gamma \vdash \phi$ *is not derivable in* **NBI** *without* $\bot$, *then* $t \models \Gamma$ *and* $t \not\models \phi$.

PROOF We construct, using the normal proofs defined by **NBI** without $\bot$, a term model with the desired property. We begin by noting that, by Lemma 4.5, any bunch Γ may be evaluated to a prime bunch $\lceil \Gamma \rceil$ such that if $\Gamma \nvdash \phi$, then $\lceil \Gamma \rceil \nvdash \phi$.

Let $\mathcal{T} = (T, e, \cdot, \sqsubseteq)$ be the monoid of finite sets of prime evaluations of bunches over $\mathcal{P}(L)$ defined as follows:

- T is $B/\dashv\vdash$, where B is the set of finite sets of prime evaluations of bunches and $\dashv\vdash$ is the evident equality generated by derivability;
- $\cdot$ is given by the prime evaluation of the combination of bunches using the comma, ",": $\Gamma \cdot \Delta \cong \lceil \Gamma , \Delta \rceil$, so that

$$\{\Gamma_1, \ldots, \Gamma_m\} \cdot \{\Delta_1, \ldots, \Delta_n\} = \{ \begin{array}{ccccc} \Gamma_1 \cdot \Delta_1 & , & \Gamma_1 \cdot \Delta_2 & , & \ldots \\ \Gamma_2 \cdot \Delta_1 & , & \Gamma_2 \cdot \Delta_2 & , & \ldots \\ \vdots & , & \vdots & , & \ldots \end{array} \};$$

- e is $\{\emptyset_m\}$;

- $\sqsubseteq$ is given extension of bunches by semicolon, ";": $\Gamma' \sqsubseteq \Gamma$ iff Γ' is of the form $\Gamma; \Delta$, so that

$$\{ \Gamma_1, \ldots, \Gamma_m \} \sqsubseteq \{ \Delta_1, \ldots, \Delta_n \}$$

iff, for all $1 \leq i \leq m$ and all $1 \leq j \leq n$, $\Gamma_i \sqsubseteq \Delta_j$.

We write $\lceil \Gamma \rceil \vdash \phi$ to denote that $\Delta \vdash \phi$, for some $\Delta \in \lceil \Gamma \rceil$.

Now define $[\![-]\!] : \mathcal{P}(L) \rightharpoonup obj([\mathcal{T}^{op}, \mathbf{Set}])$ as follows, in which proofs in **NBI** without $\perp$ are represented as $\alpha\lambda$-terms:

- $[\![I]\!](\Gamma) = \mathcal{T}(\Gamma, \emptyset_m)$;
- $[\![\top]\!](\Gamma) = \{ * \}$;
- For non-unit propositions,

$$[\![\phi]\!](\{\Gamma_1, \ldots, \Gamma_m\}) = \{ \Phi \mid \Phi \text{ is a normal proof in } \mathbf{NBI} \text{ without } \perp \text{ of } \Gamma_i \vdash \phi, \text{ for some } 1 \leq i \leq m \}.$$

(Note that here we presume the extension of SN to $\vee$ using techniques using the methods mentioned in Chapter 2.) Henceforth, we will abuse this notation and write just $[\![\phi]\!](\Gamma)$, reflecting our use of sets of bunches only to formalize the choices generated disjunctive evaluation redexes.

We define $\models$ by

$$\Gamma \models \phi \quad \text{iff} \quad [\![\phi]\!](\Gamma) \text{ is non-empty.}$$

We must check that the various conditions for a Kripke model are satisfied. Firstly, we observe that $\mathcal{T}$ is a commutative monoid with T, $\cdot$, e and $\sqsubseteq$ being well-defined. Secondly, we must check that the definition of $\models$ satisfies the inductive clauses of Definition 4.1. We give a few cases, the remaining ones being similar — note that Γ is read here as an element of the monoid $\mathcal{T}$:

Axiom: $\Gamma \models \mathrm{p}$ if and only if $[\![\mathrm{p}]\!](\Gamma) \neq \emptyset$. Note that since Γ is prime, $[\![\mathrm{p}]\!](\Gamma) \neq \emptyset$ just in case Γ is of the form $\Gamma_1; \mathrm{p}; \Gamma_2$;

I: $\Gamma \models I$ if and only if $[\![I]\!](\Gamma) \neq \emptyset$ if and only if $\mathcal{T}(\Gamma, \emptyset_m) \neq \emptyset$ if and only if $\Gamma \simeq \emptyset_m$;

$\top$: $\Gamma \models \top$ if and only if $[\![\top]\!](\Gamma) \neq \emptyset$ if and only if $\{*\} \neq \emptyset$;

$\vee$: $\Gamma \models \phi \vee \psi$ if and only if $[\![\phi \vee \psi]\!](\Gamma) \neq \emptyset$ if and only if, since Γ is prime, either $[\![\phi]\!](\Gamma) \neq \emptyset$ or $[\![\psi]\!](\Gamma) \neq \emptyset$;

$*$: $\Gamma \models \phi * \psi$ if and only if $[\![\phi * \psi]\!](\Gamma) \neq \emptyset$ if and only if, since Γ is prime, $[\![\phi]\!](\Delta) \neq \emptyset$ and $[\![\psi]\!](\Theta) \neq \emptyset$ and $\Gamma \sqsubseteq \Theta \cdot \Delta$;

$-\!\!*$: $\Gamma \models \phi -\!\!* \psi$ if and only if $[\![\phi -\!\!* \psi]\!](\Gamma) \neq \emptyset$ if and only if, for all Δ, $[\![\phi]\!](\Delta) \neq \emptyset$ implies $[\![\psi]\!](\Gamma \cdot \Delta) \neq \emptyset$.

We note that $\Gamma \sqsubseteq \Gamma'$ if $\Gamma' \equiv \Gamma; \phi_1; \ldots ; \phi_m$, for some $\phi_1, \ldots , \phi_m$. Therefore, adopting the derived form of the axiom rule given in Lemma 6.1, we see that for all $\phi \in \mathcal{P}(L)$, if $\Gamma' \sqsubseteq \Gamma$, then $[\![\phi]\!](\Gamma) \subseteq [\![\phi]\!](\Gamma')$. Note that this containment is well-defined: because we have represented proofs in **NBI** without $\perp$ as $\alpha\lambda$-terms, a Φ that is well-typed in Γ is also well-typed in Γ'. The monoidal isomorphisms are given by equalities.

Finally, we must check that satisfaction corresponds to derivability as required. We claim that $\Gamma \models \phi$ holds — note, as above, that Γ is read here as an element of the monoid $\mathcal{T}$ — if and only if $\Gamma \vdash \phi$ is derivable in **NBI** without $\perp$. Now, $\Gamma \models \phi$ if and only if

$$\{ \Phi \mid \Phi \quad \text{is a normal proof in } \mathbf{NBI} \text{ without } \perp \text{ of } \quad \Gamma \vdash \phi \}$$

is non-empty, which holds if and only if there is a proof of $\Gamma \vdash \phi$. So, given the data $\Gamma \not\vdash \phi$, we set $t = \lceil \Gamma \rceil$, and observe that clearly $\lceil \Gamma \rceil \models \Gamma$, since the construction of $\lceil \Gamma \rceil$ performs — essentially — left reductions for $\wedge$, $\vee$ and $*$ and closes using extension via ";", whilst $\lceil \Gamma \rceil \not\models \phi$. □

The technique used in the proof of Lemma 4.6 is widely applicable. In our present setting, the construction may be seen as justifying the widely adopted view in which proofs in, for example, **MILL** are interpreted as manipulating "propositions-as-resources": the "resources" in such a reading of a proof are the propositions occurring in an antececent, or bunch. In this view, the additive rules *conserve* the resources.

Note that we have used equivalence classes of bunches over the evident relation given by $\vdash$ (*cf.* Propositions 3.14 and 5.15). Such a choice is necessary if it be used to interpret not merely **BI**'s propositional consequences but also $\alpha\lambda$-terms (*i.e.*, the realizers for the propositional consequences).

THEOREM 4.7 (COMPLETENESS OF **BI** WITHOUT $\perp$) **BI** *without* $\perp$ *is complete for Kripke models: if* $\Gamma \models \phi$ *in* **BI** *without* $\perp$, *then* $\Gamma \vdash \phi$ *in* **NBI** *without* $\perp$.

PROOF Suppose $\Gamma \not\vdash \phi$ in **NBI** without $\perp$. Then Lemma 4.6 yields a contradiction. □

We remark that the statement of completeness could, because of our use of proof-objects in the construction of $\mathcal{T}$, be expressed positively and that we adopt the negative formulation merely for consistency with traditional presentations [van Dalen, 1983].

We have remarked that our Kripke semantics does not account for inconsistency. In the light of the constructions we have made, the significance of our choice may be seen clearly. Our semantics requires that

$$m \not\models \bot \quad \text{for any} \quad m \in M$$

and, in order to keep completeness, we have had to exclude $\bot$ from the language. Intuitively, the problem is that its presence permits the formation of inconsistent bunches, even by combining consistent bunches. For example, if we were to permit $\bot$ in the construction for model existence, then we should have that

$$\phi \models \phi \qquad \text{and} \qquad \phi \mathrel{-\!\!*} \bot \models \phi \mathrel{-\!\!*} \bot$$

with the consequence that

$$(\phi) \cdot (\phi \mathrel{-\!\!*} \bot) \models \bot;$$

but our definition of $\models$ requires that there be no m such that $m \models \bot$. Indeed, we have the following, *q.v.* [Pym et al., 2000]:

PROPOSITION 4.8 (INCOMPLETENESS FOR INCONSISTENCY) *For all well-formed ϕ and ψ,*

$$(\phi \mathrel{-\!\!*} \bot) \to \bot ; (\psi \mathrel{-\!\!*} \bot) \to \bot \models ((\phi * \psi) \mathrel{-\!\!*} \bot) \to \bot$$

is valid in the elementary Kripke semantics but

$$(\phi \mathrel{-\!\!*} \bot) \to \bot ; (\psi \mathrel{-\!\!*} \bot) \to \bot \vdash ((\phi * \psi) \mathrel{-\!\!*} \bot) \to \bot$$

is not provable in **BI** *(see also [Pym et al., 2000]).*

PROOF It is a routine calculation to verify that

$$m \models (\phi \mathrel{-\!\!*} \bot) \to \bot \quad \text{iff} \quad \text{there is an } n \text{ such that} \quad n \models \phi.$$

Note that there is no dependency of n upon m, so that the world m required by the definition of validity is unaffected by the worlds required to witness the left- and right-hand sides of the judgement. It follows that the given judgement is correctly one of validity. However, it is straightforward to check (easiest using the sequent calculus given in Chapter 6) that the corresponding sequent is not provable in **BI**. □

Note again that the incompleteness here arises from the interaction between multiplicative implication, $\mathrel{-\!\!*}$, and the unit of (additive) disjunction, $\bot$. However, had we *also* excluded disjunction itself, $\vee$, then we should have been able to use a very direct completeness argument, based on the Yoneda lemma.

Briefly, the term monoid has the set of formulæ as its underlying set and the order and monoid structure are given by

- $\phi \sqsubseteq \psi$ iff $\phi \vdash \psi$,
- $\phi \cdot \psi \ = \ \phi * \psi$ and
- $e \ = \ I$.

As usual, the main subsidiary lemma is model existence:

$$\phi \vdash \psi \textit{ is derivable iff } \phi \models \psi \textit{ in the term model.}$$

This lemma is established by a routine induction on ψ but it is easy to see that the proof breaks down when we encounter $\vee$ or $\perp$. For example, from $\phi \vdash \psi_0 \vee \psi_1$ it does not follow that $\phi \vdash \psi_0$ or $\phi \vdash \psi_1$, as would be needed for the result: take $\phi = \psi_0 \vee \psi_1$. Similarly, the monoid semantics would require that $\phi \vdash \perp$ never holds: but this is not the case when $\phi = \perp$.

This failure for $\perp$ amounts to that which we have already discussed. The failure for $\vee$ amounts to the failure of this Yoneda-based argument to account for the disjunctive choice that is recorded in the construction of a prime extension. However, the simplicity of such argument is very attractive and, pleasingly, it is recovered for the full propositional language, albeit relying on a rather more complex underlying construction, in Chapter 5.

Finally, note that although intuitionistic linear logic is sound for Kripke models,[6] it is *not* complete because of the absence of weakening for the additives. To see this, consider that it is an immediate consequence of the definition of $\models$ that the distributive law,

$$\phi \wedge (\psi \vee \chi) \twoheadleftarrow\!\!\!* \, (\phi \wedge \psi) \vee (\phi \wedge \chi),$$

is valid in all Kripke models. However,

$$\phi \wedge (\psi \vee \chi) \vdash (\phi \wedge \psi) \vee (\phi \wedge \chi)$$

is not a derivable sequent in intuitionistic linear sequent calculus.

[6] Interpret the multiplicatives of linear logic, $\otimes$ and $\multimap$, as our multiplicatives, $*$ and $-\!*$, and the additives of linear logic, & and $\oplus$, as our additives, $\wedge$ and $\vee$, neglecting "!".

3. Kripke Models Revisited

Recall that the Tarski-style version of Kripke semantics for intuitionistic logic may be recovered from the BHK-style, topos-theoretic, version by requiring that a proposition be interpreted as a sub-object of the terminal object (in the chosen category of values), $[\![\phi]\!] \longrightarrow 1$. Intuitively, the restriction to 1, as opposed, say, to an arbitrary set, picks out the situation in which we know just that a proposition holds, and not that it holds with a particular value.

A similar trick works here; our situation is similar to topos-theoretic semantics though we make essential use of the functor-category structure. We interpret **BI**-propositions in the presheaf topos $\mathbf{Set}^{\mathcal{C}^{op}}$, we must exploit the functor-category structure, using Day's tensor product, even to define the interpretation of the connectives.

Recall that for a **BI**-proposition ϕ, $[\![\phi]\!] : [\mathcal{M}^{op}, \mathbf{Set}]$. If $[\![\phi]\!]$ and $[\![\psi]\!]$ are subobjects of 1, then so are $[\![\phi \wedge \psi]\!]$ and $[\![\phi \rightarrow \psi]\!]$. (The situation for $[\![\phi \vee \psi]\!]$ is more delicate and is not discussed here.) However, it does not necessarily follow that $[\![\phi * \psi]\!]$ and $[\![\phi \mathrel{-\!\!*} \psi]\!]$ are sub-objects of 1. To see this, suppose that $\mathcal{M}$ is not connected and consider the definition of $[\![\phi * \psi]\!]$:

$$[\![\phi * \psi]\!](m) = \int^{n,n'} [\![\phi]\!](n) \times [\![\psi]\!](n') \times \mathcal{C}[m, n \cdot n'].$$

Even if $[\![\phi]\!]$ and $[\![\psi]\!]$ are sub-objects of 1, the co-end "sums up" all of the points — roughly, triples of the form $\langle p, q, f\rangle \in [\![\phi]\!](n) \times [\![\psi]\!](n') \times \mathcal{C}[m, n \cdot n']$ — at which the satisfaction relation holds, possibly leading, given that $\mathcal{M}$ is not connected, to a set bigger than the one-point set, $\{*\}$. The situation with $[\![\phi \mathrel{-\!\!*} \psi]\!]$ is similar. The solution to this is quite straightforward.

Let $F : \mathcal{C}^{op} \longrightarrow \mathbf{Set}$ be a functor. Define the *support* of F, *supp*(F), as follows:

$$supp(F)(c) = \begin{cases} \{*\} & \text{if } F(c) \neq \emptyset \\ \emptyset & \text{otherwise.} \end{cases}$$

We can now modify our definition of BHK-interpretation to recover our Tarski-style semantics, with interpretation now denoted by $[\![-]\!]^{\mathrm{T}}$, as follows:

$$[\![\phi \# \psi]\!]^{\mathrm{T}} = \begin{cases} supp([\![\phi \# \psi]\!]) & \text{if } \# \text{ is } * \text{ or } \mathrel{-\!\!*} \\ [\![\phi \# \psi]\!] & \text{otherwise.} \end{cases}$$

Thus we recover our truth-functional semantics within our general categorical framework.

Chapter 5

TOPOLOGICAL KRIPKE SEMANTICS

1. Topological Kripke Models of Propositional BI with $\perp$

BI's Kripke semantics may be adapted to account for $\perp$ by moving from presheaves (or **Set**-valued functor categories) to sheaves on a topological space. Such a move permits a semantics in which we take an inconsistent world, at which $\perp$ is forced, together with a treatment of disjunction that exploits the structure of a topological space which admits a non-indecomposable[1] treatment of disjunction [Lambek and Scott, 1986]. Although this topological semantics is, perhaps, relatively obscure, our subsequent generalization, in § 3, to a semantics based on Grothendieck sheaves recovers the basis of the semantics in preordered monoids.

To understand the apparent need for non-indecomposability, consider a semantics based on a preordered monoid of worlds with an initial object, 0, used to interpret $\perp$, *i.e.*,

$$m \models \perp \quad \text{iff} \quad m = 0,$$

but with the indecomposable treatment of $\vee$,

$$m \models \phi \vee \psi \quad \text{iff} \quad m \models \phi \text{ or } m \models \psi,$$

i.e., rather than require (*cf.* [Lambek and Scott, 1986]) a cover of the world m by worlds n and n' and that both $n \models \phi$ and $n' \models \psi$, we simply

[1] An object C in a topos is ***indecomposable*** if, for all arrows $k : D \longrightarrow C$ and $l : E \longrightarrow C$ such that $[k + l] : D + E \longrightarrow C$ is an epimorphism, either k or l is an epimorphism.

require that m force ϕ or ψ. Then we find that

$$e \models ((\phi \multimap \bot) \multimap \bot) \vee (\phi \multimap \bot) \tag{5.1}$$

holds in this semantics even though

$$((\phi \multimap \bot) \multimap \bot) \vee (\phi \multimap \bot)$$

is not a theorem of **BI**. The failure of theoremhood may be seen from **BI**'s cut-free sequent calculus. To see that (5.1) holds, consider that $e \models \phi \multimap \bot$ either holds or not. If it does, then we are done immediately by the definition of $\models$. If not, then, by the definition of models, we can show that

$$\text{there exists a world } n \text{ such that } n \models \phi \text{ and } n \neq 0. \tag{5.2}$$

Now, suppose $e \models (\phi \multimap \bot) \multimap \bot$. By the definition of $\models$, this is the case if and only if, for all m, if $m \models \phi \multimap \bot$, then $m = 0$. A straightforward calculation reduces this to:

$$\text{for all } m \text{ there exists } n \text{ such that } n \models \phi \text{ and } n \neq 0, \text{ or } m = 0.$$

The result now follows from (5.2).

We remark that our topological interpretations of **BI** suggest a "spatial" interpretation of **BI**'s multiplicatives. We return to this remark in Chapters 9 and 16.

Recall, from Chapter 3, that a *topological monoid* $(\mathcal{X}, *, e)$ is a monoid in the category **Top** of topological spaces, *i.e.*, a topological space

$$(\mid \mathcal{X} \mid, \mathcal{O}(\mathcal{X}))$$

on which is defined a tensor product $* : \mathcal{X} \times \mathcal{X} \longrightarrow \mathcal{X}$, with unit $e : 1 \longrightarrow \mathcal{X}$, such that usual monoidal diagrams commute [Mac Lane, 1971]. Recall also, from Lemma 3.1, that, in any topological monoid, the following distributive law holds: for all open sets U, V_i, for $i \in \mathcal{I}$, where $\mathcal{I}$ is some index set,

$$U * (\bigcup_i V_i) = \bigcup_i (U * V_i).$$

If $(\mid \mathcal{X} \mid, \Omega(\mathcal{X}))$ is a topological space and if $(*, e)$ are defined as above, then we refer to the *topological monoid* $(\mathcal{X}, *, e)$ *on* $(\mid \mathcal{X} \mid, \Omega(\mathcal{X}))$ *and* $(*, e)$. If $*$ and e are open, then we speak of the *open topological monoid* $(\mathcal{X}, *, e)$ *on* $(\mid \mathcal{X} \mid, \Omega(\mathcal{X}))$ *and* $(*, e)$.

LEMMA 5.1 *Let* $(\mathcal{X}, *, e)$ *be a topological monoid and let* $U = \bigcup_i U_i$ *and* $V = \bigcup_j V_j$ *be open covers. Then*

$$U * V \quad = \quad \bigcup_i \bigcup_j (U_i * V_j)\,.$$

PROOF An immediate consequence of the distributivity of $*$ over arbitrary unions. □

We must develop a few properties of sheaves. We follow the terminology and notation for sheaves used in [Lambek and Scott, 1986]. Similar structure is discussed in [Warner, 1983]. In particular, given a sheaf $F : \Omega(\mathcal{X})^{op} \longrightarrow \mathbf{Set}$, we have a mapping $F_{UV} : F(U) \longrightarrow F(V)$ just in case $V \subseteq U$. An $s \in F(U)$ is called *section over* U and we write $F_{UV}(s) = s\,|_V$ to denote the *restriction of* s *to* V.

LEMMA 5.2 (DAY'S PRODUCT FOR SHEAVES) *Let* F *and* G *be sheaves on a topological space* $(|\,\mathcal{X}\,|, \Omega(\mathcal{X}))$. *Let* $(\mathcal{X}, *, e)$ *be the open topological monoid on* $(|\,\mathcal{X}\,|, \Omega(\mathcal{X}))$ *and* $(*, e)$. *Then the functor* $F \otimes G : \Omega(\mathcal{X})^{op} \longrightarrow$ **Set** *defined by the co-end*

$$(F \otimes G)W = \int^{U,V} F(U) \times G(V) \times \Omega(\mathcal{X})^{op}(U * V, W),$$

i.e., *Day's tensor product, is a sheaf, as is the unit of* $\otimes$, $\Omega(\mathcal{X})^{op}(-, I)$.

PROOF Firstly, the product. Let $\bigcup_i U_i$ be an open cover of U and $\bigcup_j V_j$ be an open cover of V. By hypothesis, we have that there is a unique s in $F(U)$ such that $s\,|_{U_i} = s_i$, for all i, and that there is a unique t in $G(V)$ such that $t\,|_{V_j} = t_j$, for all j. Let $\bigcup_k W_k$ be an open cover of $W \subseteq U * V$ (so that $\bigcup_k W_k \subseteq (\bigcup_i U_i) * (\bigcup_j V_j)$). Since $\Omega(\mathcal{X})$ is a preorder, we require a unique r in $(F \otimes G)W$ such that

$$r\,|_{W_k} = r_k$$

for all i, j. Each $(F \otimes G)W_k$ is some set of pairs $[a, b]$ in which a is an element of some U_i and b is an element of some V_j. So we set $r = [s, t]$, Day's pairing of s and t.

Secondly, the unit. We must show that $\Omega(\mathcal{X})^{op}(-, e)$ is a sheaf. Let $\bigcup_i U_i$ be an open cover of U. We must show that there is a unique s in $\Omega(\mathcal{X})^{op}(U, e)$ such that $s\,|_{U_i} = s_i$, for all i. But $\Omega(\mathcal{X})^{op}$ is a preorder, so

$$\Omega(\mathcal{X})^{op}(U, I) \quad = \quad \begin{cases} \{*\} & \text{if } e \subseteq U \\ \emptyset & \text{otherwise,} \end{cases}$$

where $\{*\}$ denotes the one-element set. The result follows. □

LEMMA 5.3 (DAY'S FUNCTION SPACE FOR SHEAVES) *Let F and G be presheaves on a topological space $(|\mathcal{X}|, \Omega(\mathcal{X}))$. Let $(\mathcal{X}, *, e)$ be the open topological monoid on $(|\mathcal{X}|, \Omega(\mathcal{X}))$ and $(*, e)$. Then the functor $F \multimap G : \Omega(\mathcal{X})^{op} \longrightarrow \mathbf{Set}$ defined by the end*

$$\begin{array}{rcl} (F \multimap G)V & = & \int_U \mathbf{Set}[F(U), G(V * U)] \\ & \cong & \mathbf{Set}^{\Omega(\mathcal{X})^{op}}[F, G(V * -)], \end{array}$$

i.e., *the right adjoint of $\otimes$, determines a sheaf.*

PROOF From the isomorphism given above, we can see that the proof must be similar to that for the intuitionistic function space.

We define

$$(F \multimap G)V \quad = \quad \mathbf{Set}^{\Omega(\mathcal{X})^{op}}[F(-), G(V * -)]$$

so that $(F \multimap G)V$ is the set of families $v(U) : F(U) \longrightarrow G(V * U)$ of functions, indexed by open sets U, such that, for all $U' \subseteq U$, the following naturality diagram commutes:

$$\begin{array}{ccc} F(U) & \xrightarrow{v(U)} & G(V * U) \\ {\scriptstyle F(U' \subseteq U)}\big\downarrow & & \big\downarrow{\scriptstyle G(1_V * (U' \subseteq U))} \\ F(U') & \xrightarrow[v(U')]{} & G(V * U') \end{array}$$

Let $\bigcup_j V_j$ be an open cover of V. We must show that there is a unique v in $(F \multimap G)(V)$ such that $v\,|_{V_j} = v_j$, for all j. This follows immediately from the following commuting naturality diagram, for each j,

$$\begin{array}{ccccccc} U & F(U) & \xrightarrow{v(U)} & G(V * U) & & s & V \\ {\scriptstyle U' \subseteq U}\big\uparrow & \big\downarrow & {\scriptstyle G((V_j \subseteq V) * (U' \subseteq U))} & \big\downarrow & & \big\downarrow & \big\uparrow{\scriptstyle V_j \subseteq V} \\ U' & F(U') & \xrightarrow[v_j(U')]{} & G(V_j * U') & & s_j & V_j \end{array}$$

which exists because $V_j \cdot U' \subseteq V * U$ and, since G is a sheaf, there is a unique $s \in G(V * U)$ such that $s\,|_{V_j * U} = s_j$ $(\in G(V_j * U))$.

The morphism part of the construction, as well as application, are defined in the usual way (see, for example, [O'Hearn et al., 1995]). □

Let $\mathbf{Sh}(\mathcal{X})$ denote the category of sheaves on the (open) topological monoid $(\mathcal{X}, *, e)$.

DEFINITION 5.4 (TOPOLOGICAL KRIPKE MODELS) *Let $(\mathcal{X}, *, e)$ be an open topological monoid and let $\mathcal{P}(L)$ denote the collection of* **BI** *propositions over a language L of propositional letters. A* topological Kripke model *is a triple*

$$\langle \mathbf{Sh}(\mathcal{X}), \models, [\![-]\!] \rangle,$$

where $\models \subseteq \mathcal{O}(X) \times \mathcal{P}(L)$, satisfying the conditions in Table 5.1 and $[\![-]\!] : \mathcal{P}(L) \rightharpoonup \mathbf{Sh}(\mathcal{X})$ is a partial function from the **BI** *propositions over L to the objects of $\mathbf{Sh}(\mathcal{X})$ such that:*

Kripke monotonicity: If $V \subseteq U$, then, for each $\phi \in \mathcal{P}(L)$, $U \models \phi$ implies $V \models \phi$.

As before, wherever no confusion will arise, we shall refer to a model

$$\langle \mathbf{Sh}(\mathcal{X}), \models, [\![-]\!] \rangle$$

simply as $\mathcal{X}$. □

We define truth and validity for topological Kripke models just as in Chapter 4.

To see that this definition is consistent with the sub-object classifier semantics of intuitionistic logic [Lambek and Scott, 1986] consider the pullback diagram in $\mathbf{Sh}(\mathcal{X})$,

$$\begin{array}{ccc} h^U & \xrightarrow{\;\mu\;} & [\![\mathrm{p}]\!] \\ {\scriptstyle \bigcirc_{h^U}}\downarrow & & \downarrow{\scriptstyle \chi_U} \\ 1 & \xrightarrow[\;\top\;]{} & \Omega \end{array}$$

and note that an arrow $h^U \xrightarrow{\mu} [\![\mathrm{p}]\!]$ is, by the Yoneda lemma, determined uniquely by an element $\check{\mu} \in [\![\mathrm{p}]\!](U)$.

2. Soundness and Completeness for BI with ⊥

THEOREM 5.5 (SOUNDNESS) *If $\Gamma \vdash \phi$ is provable in* **NBI**, *then $\Gamma \models \phi$.*

$U \models \mathrm{p}$	iff	$[\![\mathrm{p}]\!](U) \neq \emptyset$, for $\mathrm{p} \in L$
$U \models \phi * \psi$	iff	for some $V, V' \in \mathcal{O}(\mathcal{X})$, $U \subseteq V \cdot V'$ and $V \models \phi$ and $V' \models \psi$
$U \models \phi \mathrel{-\!\!*} \psi$	iff	for all $V \in \mathcal{O}(\mathcal{X})$, $V \models \phi$ implies $U \cdot V \models \psi$
$U \models \phi \wedge \psi$	iff	$U \models \phi$ and $U \models \psi$
$U \models \phi \vee \psi$	iff	for some $V, V' \in \mathcal{O}(\mathcal{X})$ such that $U = V \cup V'$, $V \models \phi$ and $V' \models \psi$
$U \models \phi \to \psi$	iff	for all $V \subseteq U$, $V \models \phi$ implies $V \models \psi$
$U \models \top$	for all	$U \in \mathcal{O}(\mathcal{X})$
$U \models I$	iff	$U \subseteq e$
$U \models \bot$	iff	$U = \emptyset$

Table 5.1. Semantics in Sheaves

PROOF The proof proceeds by induction on the structure of proofs in **NBI**. We illustrate the argument with a few cases.

$\bot$E: The induction hypothesis is that, for all U, if $U \models \Gamma$, then $U \models \bot$. But $U \models \bot$ iff $U = \emptyset$, so, by the hereditary condition, we must have that for all $U \subseteq \emptyset$, if $U \models \Gamma$, then $U \models \phi$.

$\vee$I: Without loss of generality, the induction hypothesis is that $\Gamma \models \phi_1$ and we must establish that $\Gamma \models \phi_1 \vee \phi_2$. Suppose that $U \models \phi_1$. If $V \models \phi_2$, where V is possibly empty, then $U \cup V \models \phi_1 \vee \phi_2$. But $U \subseteq U \cup V$, so $U \models \phi_1 \vee \phi_2$. The symmetric case is similar.

$\mathrel{-\!\!*}$I: By the induction hypothesis, we have that, for all U, $U \models \Gamma, \phi$ implies $U \models \psi$. We must show that if $U \models \Gamma$, then $U \models \phi \mathrel{-\!\!*} \psi$. We have that $U \models \phi \mathrel{-\!\!*} \psi$ iff, for all V, $n \models \phi$ implies $U \cdot V \models \psi$. If $V \models \phi$, then $U \cdot V \models \Gamma, \phi$. By the induction hypothesis, $U \cdot V \models \psi$. Therefore $U \models \phi \mathrel{-\!\!*} \psi$.

□

Turning to completeness, we follow the same line of argument as in Chapter 4 but construct from the syntax and proofs of propositional **BI** not a Kripke model but rather a topological Kripke model. Just as in Chapter 4, we need to construct the prime evaluation of a bunch but the same construction, Definition 4.3, works in this setting. Moreover, Lemma 4.5 is unchanged. So it remains only to construct a commutative topological monoid from (prime) bunches.

DEFINITION 5.6 (TERM TOPOLOGICAL KRIPKE MODEL) *We can define a term topological Kripke* **BI**-*model, in which we suppress the routine definition of* $[\![-]\!]$*, as follows:*

- $\mid \mathcal{X} \mid$ *is* $B/\dashv\vdash$*, where* B *is the set of sets of consistent bunches and where* $\dashv\vdash$ *is the evident equality generated by derivability,* i.e., *if* S *and* S' *are sets of consistent bunches, then* $S \vdash S'$ *iff, for any* $\Gamma \in S$*, there exists* $\Gamma' \in S'$ *such that* $\phi_\Gamma \vdash \phi_{\Gamma'}$*;*
- *Open sets are subsets of* $\mathcal{X}$ *closed under prime evaluation of bunches. As usual, prime evaluation generates a set of bunches;*
- *The monoid operation,* $\cdot$*, is given by the consistent prime evaluation of the combination of bunches using the comma, "*,*":* $\Gamma * \Delta \cong \lceil \Gamma, \Delta \rceil$*, where* $\cong$ *denotes isomorphism of labelled trees, so that*

$$\{\Gamma_1, \ldots, \Gamma_m\} \cdot \{\Delta_1, \ldots, \Delta_n\} = \left\{ \begin{array}{ccc} \Gamma_1 * \Delta_1, & \Gamma_1 * \Delta_2, & \ldots \\ \Gamma_2 * \Delta_1, & \Gamma_2 * \Delta_2, & \ldots \\ \vdots, & \vdots, & \ldots \end{array} \right\} \setminus \bot(\Gamma, \Delta)$$

where $\bot(\Gamma, \Delta) = \{ \Gamma_i * \Delta_j \mid \Gamma_i * \Delta_j \vdash \bot \}$*;*

- *The unit,* e*, is given by* $\{ \emptyset_m \}$*, where* $\emptyset_m$ *is the unit of "*,*";*
- $[\![I]\!](\Gamma) = \mathcal{T}(\Gamma, \emptyset_m)$*;*
- $[\![\top]\!](\Gamma) = \{ * \}$*;*
- *For non-unit propositions, but including* $\bot$*, the unit of* $\vee$*,*

$$[\![\phi]\!](\{\Gamma_1, \ldots, \Gamma_m\}) = \{ \Phi \mid \Phi \text{ is a normal proof in } \mathbf{NBI} \text{ of } \Gamma_i \vdash \phi, \text{ for some } 1 \le i \le m \}.$$

Here we intend a restriction of Φ to normal proofs in **NBI** *and presume the extension of SN to $\vee$ (and $\bot$) using the methods mentioned in Chapter 2;*

- *Define the forcing relation $\models$, for non-empty worlds Γ, by $\Gamma \models \phi$ iff $[\![\phi]\!](\Gamma) \neq \emptyset$.*

□

Notice that we remove the inconsistent bunches and that we will always be left with at least the empty set. This property of the term model, together with the appropriate modificaton of the forcing clause for $\bot$, yields completeness. To see this, consider the following example: Let $\mathrm{p} \in L$. Then both p and $\mathrm{p} \mathrel{-\!\!*} \bot$ are consistent bunches but their monoidal combination is not. However, it is easy to see that $\lceil \mathrm{p} \cdot \mathrm{p} \mathrel{-\!\!*} \bot \rceil = \emptyset$ and $\emptyset \models \bot$. How is this to be seen as being consistent with the Kripke models described in Chapter 4, in which $\bot$ is never forced ? One answer is simply that, in order for completeness to go through in the presence of inconsistency, we must use a setting in which "never" is part of the model: the empty set fills exactly this rôle: just as in the Kripke semantics, models inhabit functor categories $\mathbf{Set}^{\mathcal{M}^{op}}$ but for completeness with $\bot$, we must refine this setting to that of sheaves, with the corresponding modification of the forcing relation, on an $\mathcal{M}$ which is a topological space.

An alternative way of handling inconsistency would be to abandon our commitment to a totally defined algebra of worlds and work instead with partial monoids, $\mathcal{M} = (M, \cdot, e, \sqsubseteq)$, in which the composition

$$\cdot : M \times M \rightharpoonup M$$

is a partial function. The basic theory of such models is discussed by Galmiche, Méry and Pym in [Galmiche et al., 2002].

LEMMA 5.7 (MODEL EXISTENCE) *The construction given in Definition 5.6 determines a topological Kripke model, $\mathcal{T}$, such that if $\Gamma \vdash \phi$ is not provable in* **NBI**, *then $\Gamma \not\models \phi$ in $\mathcal{T}$.*

PROOF A routine calculation verifies that we have defined a commutative topological monoid of bunches. The completeness property, if $\Gamma \vdash \phi$ is not provable in **NBI**, then $\Gamma \not\models \phi$ in $\mathcal{T}$, follows by induction on the structure of formulæ.

Paying particular attention to inconsistency, suppose $U \models \bot$ for some non-empty U. $U \models \bot$ iff $[\![U]\!](\bot)$ is non-empty iff some Γ in U is inconsistent. Contradiction. □

Lemma 5.7 may be seen as showing that the syntactic monoid operation is compatible with the evident Scott topology on bunches.

THEOREM 5.8 (COMPLETENESS) **BI** *is complete for topological Kripke models: if* $\Gamma \models \phi$ *in* **BI**, *then* $\Gamma \vdash \phi$ *in* **NBI**.

PROOF Suppose $\Gamma \not\vdash \phi$ in **NBI**. Then Lemma 5.7 yields a contradiction. □

An alternative construction of a term model would follow the pattern of Proposition 5.15 and use open sets of formulæ, closed under $\vdash$. In the more general setting of Grothendieck sheaves, for which we give detailed soundness and completeness proofs, disjunction is handled via the Grothendieck topology. The construction of prime bunches (*cf. prime theories* [van Dalen, 1983]) is of independent interest, however [van Dalen, 1983].

We conclude this section with a remark about our definition of the forcing relation, $\models$.[2] There are two different ways in which to give an interpretation of atomic propositions and how to define $\models$ accordingly.

1 In the first view, $[\![\mathrm{p}]\!]$ is a sheaf $\mathcal{O}(\mathcal{X})^{op} \longrightarrow \mathbf{Set}$ and the interpretation of $- \models \mathrm{p}$ is

$$U \models \mathrm{p} \quad \text{iff} \quad [\![\mathrm{p}]\!](U) \neq \emptyset$$

2 In the second view, we have that

(i) $[\![\mathrm{p}]\!]$ is a sheaf $\mathcal{O}(\mathcal{X})^{op} \longrightarrow \mathbf{Set}$, and

(ii) $[\![\mathrm{p}]\!]$ is a subobject of 1 (note that 1 is also a sheaf).

The interpretation of $- \models \mathrm{p}$ is given by

$$\begin{array}{lll} U \models \mathrm{p} & \text{iff} & [\![\mathrm{p}]\!](U) \neq \emptyset \\ & \text{iff} & U \in [\![\mathrm{p}]\!](U) \end{array}$$

Notice here that since $[\![\mathrm{p}]\!]$ is a subobject of 1 and so $[\![\mathrm{p}]\!](V)$ is either $\{*\}$ or $\emptyset$, we can consider $[\![\mathrm{p}]\!]$ as a collection of open sets and the second "iff" follows.

Now, consider the relationship between (1) and (2). The interpretation of $- \models \mathrm{p}$ in (1) may be described in two stages:

(i) Find a sheaf F such that F is a subobject of 1 and there is an epimorphism from $[\![\mathrm{p}]\!]$ to F:

$$[\![\mathrm{p}]\!] \twoheadrightarrow F \rightarrowtail 1;$$

[2]We are grateful to Hongseok Yang in connection with this remark.

(ii) Give the interpretation of $- \models \mathrm{p}$ according to (2) for F. Then

$$\begin{array}{lll} U \models \mathrm{p} & \text{iff} & F(U) = \{*\} \\ & \text{iff} & U \in F \quad \text{(since } F \text{ is a subobject of 1).} \end{array}$$

Notice here that F is nothing but

$$F(U) = \begin{cases} \{*\} & \text{if } [\![\mathrm{p}]\!] \neq \emptyset \\ \emptyset & \text{if } [\![\mathrm{p}]\!] = \emptyset \end{cases}$$

and that the fact that F is a sheaf follows from the so-called epi-monic factorization theorem in a topos of sheaves.

In both cases, we are using a complete lattice of subobjects of 1 in the category of sheaves.

3. Grothendieck Sheaf-theoretic Models

In this section, which quite closely follows Pym, O'Hearn and Yang [Pym et al., 2000],[3] we give a class of models which generalizes the ones we have so far described and in which we give a detailed proof of completeness. We work with Grothendieck topologies [Mac Lane and Moerdijk, 1992], the algebraic generalization of topological spaces, on preordered commutative monoids. This setting allows us to recover the appealing simplicity of the elementary preordered commutative monoid semantics whilst retaining the topological treatment of inconsistency, via the empty set, necessary for completeness in the presence of $\bot$. The connection between the two topological formulations is the usual one [Mac Lane and Moerdijk, 1992].

DEFINITION 5.9 (GTM) *A* Grothendieck topological preordered commutative monoid, *in short, a* Grothendieck Topological Monoid, *or* GTM, *is a preordered commutative monoid $\mathcal{M} = (M, \cdot, e, \sqsubseteq)$ together with a map $J : M \to \wp(\wp(M))$ satisfying the axioms (1) – (5) discussed below. Such a J is usually called a* Grothendieck topology.[4] □

There are two classes of axioms for GTMs: four axioms for Grothendieck topology and one axiom for the "continuity" of "·". The four axioms for a Grothendieck topology are as follows:

1 *Sieve*: for any $m \in M$ and $S \in J(m)$, "$S \sqsubseteq m$", *i.e.*, for any $m' \in S$, $m' \sqsubseteq m$;

[3] We most particularly acknowledge Yang is this respect and discussions with Didier Galmiche and Daniel Méry have been most helpful.

[4] $\wp(S)$ denotes the power set of the set S.

2 *Maximality*: for any n' such that $n' = n$, $\{n'\}$ is in $J(n)$ (If we were to work with partially ordered monoids, then we should need the following maximality condition instead: for any $m \in M$, $\{m\} \in J(m)$.);

3 *Stability*: for any $m, n \in M$ and $S \in J(m)$ such that $n \sqsubseteq m$, there exists $S' \in J(n)$ such that "$S' \sqsubseteq S$": for any $n' \in S'$, there exists $m' \in S$ such that $n' \sqsubseteq m'$; and

4 *Transitivity*: for any $m \in M$, $S \in J(m)$ and $\{S_{m'} \in J(m')\}_{m' \in S}$, $\bigcup_{m' \in S} S_{m'} \in J(m)$.

We require also an axiom for the "continuity" of "·", as follows:

5 *Continuity*: for any $m, n \in M$ and $S \in J(m)$, "$S \cdot n \in J(m \cdot n)$", *i.e.*, $\{m' \cdot n \mid m' \in S\} \in J(m \cdot n)$.

In the elementary Kripke semantics, based on a preordered commutative monoid, the interpretation of propositions satisfies the Kripke monotonicity condition:

(K): for any $m, n \in M$ such that $n \sqsubseteq m$, if $m \models \phi$, then $n \models \phi$.

Let $\mathcal{M}$ be a GTM and let $\mathcal{P}(L)$ denote the collection of **BI** propositions over a language L of propositional letters. Then a *Grothendieck Resource Model*, or GRM, is a triple $\langle \mathcal{M}, \models, [\![-]\!] \rangle$, satisfying the conditions given in Table 5.2 and not only (K) but also a sheaf-like condition, which is sometimes called *locality* [Mac Lane and Moerdijk, 1992]:

(Sh): for any $m \in M$ and $S \in J(m)$, if "$S \models \phi$", then $m \models \phi$. More precisely, if, for all $m' \in S$, if $m' \models \phi$, then $m' \models \phi$.

The semantics of propositions is parametrized by the interpretation $[\![-]\!]$ of atomic propositions which makes (K) and (Sh) hold for atomic propositions.

Following our earlier pattern of development, the semantics of **BI**'s propositions is given, in Table 5.2, by the forcing relation $m \models \phi$. Formally, we define a model to consist of a GTM together with a satisfaction (or forcing) relation, $\models$, subject to the appropriate formulations of conditions (6) and (7).

DEFINITION 5.10 (GTI) *Let $\mathcal{M}$ be a GTM and let $\mathcal{P}(L)$ denote the collection of* **BI** *propositions over a language L of propositional letters. Then a* Grothendieck Topological Interpretation, *or GTI, is a partial function $[\![-]\!] : L \rightharpoonup \wp(M)$ satisfying (K) and (S):*

$m \models \mathrm{p}$	iff	$m \in [\![p]\!]$
$m \models \top$	iff	always
$m \models \phi \wedge \psi$	iff	$m \models \phi$ and $m \models \psi$
$m \models \phi \to \psi$	iff	for any $n \sqsubseteq m$, if $n \models \phi$, then $n \models \psi$
$m \models \phi \vee \psi$	iff	there exists $S \in J(m)$ such that for any $m' \in S$, $m' \models \phi$ or $m' \models \psi$
$m \models \bot$	iff	$\emptyset \in J(m)$
$m \models I$	iff	there exists $S \in J(m)$ such that for any $m' \in S$, $m' \sqsubseteq e$
$m \models \phi * \psi$	iff	there exists $S \in J(m)$ such that for any $m' \in S$, there exist $n_\phi, n_\psi \in M$ such that $m' \sqsubseteq n_\phi \cdot n_\psi$, $n_\phi \models \phi$ and $n_\psi \models \psi$
$m \models \phi \mathrel{-\!\!*} \psi$	iff	for any $n \models \phi$, $n \cdot m \models \psi$

Table 5.2. Semantics in Grothendieck Sheaves

6 *(K): For any $m, n \in M$ such that $n \sqsubseteq m$, if $m \in [\![\mathrm{p}]\!]$, then $n \in [\![\mathrm{p}]\!]$;*

7 *(Sh): For any $m \in M$ and $S \in J(m)$, if, for all $m' \in S$, $m' \in [\![\mathrm{p}]\!]$, then $m \in [\![\mathrm{p}]\!]$.* □

DEFINITION 5.11 (GRM) *A* Grothendieck Resource Model, *or* GRM *for short, is a triple*

$$\langle \mathcal{M}, \models, [\![-]\!] \rangle$$

in which $\mathcal{M}$ is a GTM, $[\![-]\!]$ is a GTI and $\models \subseteq M \times \mathcal{P}(L)$ is a satisfaction relation satisfying the conditions given in Table 5.2. □

Let $\mathcal{G} = \langle \mathcal{M}, \models, [\![-]\!] \rangle$ be a GRM and, as usual, let ϕ_Γ be the proposition obtained from the bunch Γ by replacing each "," by $*$ and each ";" by $\wedge$. Then the sequent $\Gamma \vdash \phi$ is *valid* in $\mathcal{G}$, written $\Gamma \models_{\mathcal{G}} \phi$, if and only if, for all $m \in M$, $m \models \phi_\Gamma$ implies $m \models \phi$. The sequent $\Gamma \vdash \phi$ is *valid*, written $\Gamma \models \phi$, if and only if, for all GRMs $\mathcal{G}$, $\Gamma \models_{\mathcal{G}} \phi$.

The first two results give the well-definedness of the Grothendieck semantics.

LEMMA 5.12 *Let* $[\![-]\!]$ *be a GTI,* i.e., *an interpretation which makes* (K) *and* (Sh) *hold for atomic propositions, then* (K) *holds for the interpretation* $[\![\chi]\!]$ *of any* **BI** *proposition* χ *containing just propositional letters interpreted by* $[\![-]\!]$.

PROOF For any $m, n \in M$ such that $n \sqsubseteq m$ and $m \models \chi$, we must show $n \models \chi$. The proof proceeds by the induction on the structure of the proposition χ. In most of the cases, the inductive step is immediate. We give just those cases which differ from the corresponding ones in the preordered commutative monoid semantics.

- $\chi = \phi \vee \psi$: since $m \models \phi \vee \psi$, there exists $S_m \in J(m)$ such that for all $m' \in S_m$, $m' \models \phi$ or $m' \models \psi$. By the stability axiom, there exists $S_n \in J(n)$ such that for all $n' \in S_n$, $n' \sqsubseteq m'$ for some $m' \in S_m$. Then, by the induction hypothesis, $n' \models \phi$ or $n' \models \psi$ for any $n' \in S_n$.
- $\chi = \bot$: since $m \models \bot$, $\emptyset \in J(m)$. By the stability axiom, $\emptyset \in J(n)$.
- $\chi = I$: since $m \models I$, there exists $S_m \in J(m)$ such that $m' \sqsubseteq e$ for all $m' \in S_m$. By the stability axiom, there is $S_n \in J(n)$ such that for any $n' \in S_n$, $n' \sqsubseteq m'$ for some $m' \in S_m$. Then, for any $n' \in S_n$, $n' \sqsubseteq e$.
- $\chi = \phi * \psi$: since $m \models \phi * \psi$, there exists $S_m \in J(m)$ such that for any $m' \in S_m$, there exist $a_{m'}, b_{m'}$ such that $m' \sqsubseteq a_{m'} \cdot b_{m'}$, $a_{m'} \vdash \phi$ and $b_{m'} \vdash \psi$. By the stability axiom, there exists $S_n \in J(n)$ such that for any $n' \in S_n$, $n' \sqsubseteq m'$ for some $m' \in S_m$, from which $n' \sqsubseteq a_{m'} \cdot b_{m'}$ follows. Therefore, for any $n' \in S_n$, there exist $a_{n'}, b_{n'}$ such that $n \sqsubseteq a_{n'} \cdot b_{n'}$, $a_{n'} \models \phi$ and $b_{n'} \models \psi$.

□

LEMMA 5.13 *Let* $[\![-]\!]$ *be a GTI,* i.e., *an interpretation which makes* (K) *and* (Sh) *hold for atomic propositions, then* (Sh) *holds for the interpretation* $[\![\chi]\!]$ *of any* **BI** *proposition* χ *containing just propositional letters interpreted by* $[\![-]\!]$.

PROOF For any $m \in M$ and $S \in J(m)$ such that $m' \models \chi$ for all $m' \in S$, we must show that $m \models \chi$. We proceed by induction on the structure of χ.

- $\chi = \mathrm{p}$: this case follows from the assumptions about $[\![-]\!]$.
- $\chi = \top$: for any $n \in M$ including the case that $n = m$, $n \models \chi$.
- $\chi = \phi \wedge \psi$: for any $m' \in S$, $m' \models \phi$ and $m' \models \psi$. By the induction hypothesis, $m \models \phi$ and $m \models \psi$.

- $\chi = \phi \to \psi$: for any $n \sqsubseteq m$ such that $n \models \phi$, by the stability axiom, there exists $S_n \in J(n)$ such that for any $n' \in S_n$, $n' \sqsubseteq m'$ for some $m' \in S$. Also, $n' \sqsubseteq n$ by the sieve condition on S_n. By (K), as stated in Lemma 5.12, $n' \models \phi \to \psi$ and $n' \models \phi$, which implies $n' \models \psi$. By the induction hypothesis, $n \models \psi$.

- $\chi = \phi \vee \psi$: for any $m' \in S$, there exists $S_{m'} \in J(m')$ such that for any $u \in S_{m'}$, $u \models \phi$ or $u \models \psi$. Let $S_m = \bigcup_{m' \in S} S_{m'}$. Then, $S_m \in J(m)$ because of the transitivity axiom. Moreover, for any $u \in S_m$, $u \models \phi$ or $u \models \psi$. Therefore, $m \models \phi \vee \psi$.

- $\chi = \bot$: for any $m' \in S$, $m' \models \bot$ and so $\emptyset \in J(m')$. Since $\emptyset = \bigcup_{m' \in S} \emptyset$ is in $J(m)$ by the transitivity axiom, $m \models \bot$.

- $\chi = I$: for any $m' \in S$, there exists $S_{m'} \in J(m')$ such that $u \sqsubseteq e$ for any $u \in S_{m'}$. Let $S_m = \bigcup_{m' \in S} S_{m'}$. Then $S_m \in J(m)$ by the transitivity axiom. Moreover, for any $u \in S_m$, $u \sqsubseteq e$. Therefore, $m \models I$.

- $\chi = \phi * \psi$: for any $m' \in S$, there exists $S_{m'} \in J(m')$ such that for any $u \in S_{m'}$, there exist a_u, b_u such that $u \sqsubseteq a_u \cdot b_u$, $a_u \models A$ and $b_u \models B$. Let $S_m = \bigcup_{m' \in S} S_{m'}$. Then, by the transitivity axiom, $S_m \in J(m)$. Moreover, for any $u \in S_m$, there exist a_u, b_u such that $u \sqsubseteq a_u \cdot b_u$, $a_u \models \phi$ and $b_u \models \psi$. Therefore, $m \models \phi * \psi$.

- $\chi = \phi \mathrel{-\!\!*} \psi$: for any n such that $n \models \phi$, let $S_{n \cdot m} = \{n \cdot m' \mid m' \in S\}$. Then by the continuity axiom, $S_{n \cdot m} \in J(n \cdot m)$. For any $m' \in S$, since $m' \models \phi \mathrel{-\!\!*} \psi$, $n \cdot m' \models \psi$. That is, for any $u \in S_{n \cdot m}$, $u \models \psi$. By the induction hypothesis, $n \cdot m \models \psi$.

□

Before proceeding to completeness, we note that the class of models in GTMs includes the models based in preordered commutative monoids, introduced in Chapter 1 and developed in Chapter 4, in the following sense:

PROPOSITION 5.14 *For any preordered commutative monoid* $(M, \cdot, e, \sqsubseteq)$, *let* $J(m) = \{\{m'\} \mid m' = m\}$. *Then*

1 $(M, \cdot, e, \sqsubseteq, J)$ *satisfies all of the axioms in this section;*

2 (K) *implies* (Sh); *and*

3 *For any interpretation which makes (K) hold for atomic propositions, the interpretations of* $\phi \vee \psi, \bot, \phi * \psi, I$ *may be simplifed as follows:*

$$
\begin{array}{rll}
m \models \bot & \text{never} & \\
m \models \phi \vee \psi & \text{iff} & m \models \phi \textit{ or } m \models \psi \\
m \models \phi * \psi & \text{iff} & \textit{there exists } n_\phi, n_\psi \textit{ such that} \\
 & & m \sqsubseteq n_\phi \cdot n_\psi,\ n_\phi \models \phi \textit{ and } n_\psi \models \psi \\
m \models I & \text{iff} & m \sqsubseteq e.
\end{array}
$$

□

BI is complete with respect to this class of models. The proof of completeness follows the usual pattern: We construct a term model which has the property that $\models$ in it corresponds exactly to $\vdash$.

PROPOSITION 5.15 (COMPLETENESS) *For any two* **BI** *propositions* ϕ *and* ψ, *if* $\phi \models \psi$, *then* $\phi \vdash \psi$.

PROOF The proof proceeds in a way which is similar to that for the completeness of $(\bot, \vee)$-free fragments, which may be seen, essentially, as constructing a complete model and using Yoneda embedding. Here, in contrast to the term model described for sheaves, disjunction is handled via the Grothendieck topology, J.

We define a GTM, a topology J and a GTI (*i.e.*, a model existence argument again) as follows:

- M is the set of equivalence classes of propositions ϕ, written $[\phi]$, with respect to the relation given by provability (it is not essential to work with equivalence classes here but to we do so in order to emphasize that our semantics is able to handle equality correctly; *cf.* [van Dalen, 1983]);

- $[\phi] \sqsubseteq [\psi]$ iff $\phi \vdash \psi$. It may be easily shown that the choices of ϕ and ψ from the equivalence classes $[\phi]$ and $[\psi]$, respectively, don't matter;

- $[\phi] \cdot [\psi] = [\phi * \psi]$. Also, it may be easily shown that the choices of ϕ and ψ from the equivalence classes $[\phi]$ and $[\psi]$, respectively, don't matter;

- $e = [I]$;

- $J([\phi])$ is the collection of all finite (possibly empty) families $\{[\phi_1], \dots, [\phi_n]\}$ such that $[\phi_i] \sqsubseteq [\phi]$ for all i and $[\phi] \sqsubseteq [\phi_1 \vee \ldots \vee \phi_n]$. Here again, the choices of ϕs don't matter.

We claim that the above entities do indeed satisfy all of the conditions required for a model. It is straightforward to show that $(M, \sqsubseteq, \cdot, e)$ is a preordered commutative monoid and that J satisfies the sieve and maximality axioms. We deal with the other three conditions.

- *Stability*: for any $[\phi], [\psi] \in M$ and $\{[\phi_l]\}_{l\in L} \in J([\phi])$ such that $[\psi] \sqsubseteq [\phi]$, let's consider the family $\{[\phi_l \wedge \psi]\}_{l\in L}$. Since for any $l \in L$, $[\phi_l \wedge \psi] \sqsubseteq [\psi]$ and $[\psi] \sqsubseteq [\bigvee_{l\in L}(\phi_l \wedge \psi)]$, the family $\{[\phi_l \wedge \psi]\}_{l\in L}$ belongs to $J([\psi])$. Moreover, $[\phi_l \wedge \psi] \sqsubseteq [\phi_l]$ for all $l \in L$, from which the other requirement for the stability axiom follows.

- *Transitivity*: for any $[\phi] \in M$, $\{[\phi_l]\}_{l\in L} \in J([\phi])$ and $\{\{[\phi_l^k]\}_{k\in K_l} \in J([\phi_l])\}_{l\in L}$, let $S = \{[\phi_l^k]\}_{l\in L, k\in K_l}$. From the definition of J, for any $l \in L$ and $k \in K_l$, $[\phi_l^k] \sqsubseteq [\phi_l] \sqsubseteq [\phi]$. Again, from the definition of J, $[\phi] \sqsubseteq [\bigvee_{l\in L} \phi_l] \sqsubseteq [\bigvee_{l\in L} \bigvee_{k\in K_l} \phi_l^k]$, which implies $[\phi] \sqsubseteq [\bigvee_{l\in L, k\in K_l} \phi_l^k]$. Therefore, S is in $J([\phi])$.

- *Continuity*: for any $[\phi], [\psi] \in M$ and $\{[\phi_l]\}_{l\in L} \in J([\phi])$, consider the family $\{[\phi_l * \psi]\}_{l\in L}$. Then $[\phi_l * \psi] \sqsubseteq [\phi * \psi]$ for any $l \in L$ and $[\phi * \psi] \sqsubseteq [(\bigvee_{l\in L} \phi_l) * \psi] = [\bigvee_{l\in L}(\phi_l * \psi)]$.

Let the interpretation $[\![-]\!]$ of atomic propositions be given by $[\![\mathrm{p}]\!] = \{[\phi] \mid \phi \vdash \mathrm{p}\}$. Notice that $[\![-]\!]$ satisfies (K) and (Sh). The resulting model has the following property:

For any two propositions ϕ_0 and ψ_0, $[\phi_0] \models \psi_0$ iff $\phi_0 \vdash \psi_0$.

Before considering why the above property holds, notice that the completeness result follows from it in the usual way. We show the above property by induction on the structure of ψ_0.

- $\psi_0 = \mathrm{p}$: this case follows from the definition of $[\![-]\!]$.

- $\psi_0 = \top$: both $[\phi_0] \models \top$ and $\phi_0 \vdash \top$ always hold.

- $\psi_0 = \phi \wedge \psi$:
  $[\phi_0] \models \phi \wedge \psi$ iff $[\phi_0] \models \phi$ and $[\phi_0] \models \psi$ iff (by the induction hypothesis) $\phi_0 \vdash \phi$ and $\phi_0 \vdash \psi$ iff $\phi_0 \vdash \phi \wedge \psi$.

- $\psi_0 = \phi \to \psi$:

  if: for any $[\phi_1]$ such that $[\phi_1] \sqsubseteq [\phi_0]$ and $[\phi_1] \models \phi$, $\phi_1 \vdash \phi$ by the induction hypothesis, From the definition of $\sqsubseteq$, $\phi_1 \vdash \phi_0$. Therefore, $\phi_1 \vdash \phi \to \psi$ and $\phi_1 \vdash \psi$. Again, by the induction hypothesis, $[\phi_1] \models \psi$;

  only if: since $\phi_0 \wedge \phi \vdash \phi$, $[\phi_0 \wedge \phi] \models \phi$ by the induction hypothesis. Since $[\phi_0 \wedge \phi] \sqsubseteq [\phi_0]$, $[\phi_0 \wedge \phi] \models \psi$. Again, by the induction hypothesis, $\phi_0 \wedge \phi \vdash \psi$. Therefore, $\phi_0 \vdash \phi \to \psi$.

- $\psi_0 = \phi \vee \psi$:

if: consider $S = \{[\phi_0 \wedge \phi], [\phi_0 \wedge \psi]\}$. Then, $[\phi_0 \wedge \phi] \sqsubseteq [\phi_0]$ and $[\phi_0 \wedge \psi] \sqsubseteq [\phi_0]$ and $[\phi_0] \sqsubseteq [\phi_0 \wedge (\phi \vee \psi)] = [(\phi_0 \wedge \phi) \vee (\phi_0 \wedge \psi)]$. Therefore, $S \in J([\phi_0])$. Moreover, by the induction hypothesis, $[\phi_0 \wedge \phi] \models \phi$ and $[\phi_0 \wedge \psi] \models \psi$. Thus, $[\phi_0] \models \phi \vee \psi$;

only if: since $[\phi_0] \models \phi \vee \psi$, there exist $S \in J([\phi_0])$ such that for any $[\phi'] \in S$, $[\phi'] \models \phi$ or $[\phi'] \models \psi$. By the induction hypothesis, for any $\phi' \in S$, $\phi' \vdash \phi$ or $\phi' \vdash \psi$, which implies $\phi' \vdash \phi \vee \psi$. $\bigvee_{\phi' \in S} \phi' \vdash \phi \vee \psi$ follows from this. Since $[\phi_0] \sqsubseteq [\bigvee_{\phi' \in S} \phi']$, $\phi_0 \vdash \phi \vee \psi$.

- $\psi_0 = \bot$: $[\phi_0] \models \bot$ iff $\emptyset \in J([\phi_0])$ iff $[\phi_0] \sqsubseteq [\bot]$ iff $\phi_0 \vdash \bot$. This case is the counterpart to the $\emptyset \models \bot$ case in the sheaf-theoretic semantics previously discussed.

- $\psi_0 = I$:

if: $\{[\phi_0]\} \in J([\phi_0])$ and $[\phi_0] \sqsubseteq e = [I]$ because $\phi_0 \vdash I$. Therefore, $[\phi_0] \models I$;

only if: since $[\phi_0] \models I$, there exists $\{[\phi_l]\}_{l \in L} \in J([\phi_0])$ such that $[\phi_l] \sqsubseteq e = [I]$ for any $l \in L$, which implies $\bigvee_{l \in L} \phi_l \vdash I$. Since $[\phi_0] \sqsubseteq [\bigvee_{l \in L} \phi_l]$, $\phi_0 \vdash \bigvee_{l \in L} \phi_l$. Therefore, $\phi_0 \vdash I$.

- $\psi_0 = \phi * \psi$:

if: $\{[\phi_0]\} \in J([\phi_0])$ and $[\phi_0] \sqsubseteq [\phi] \cdot [\psi]$. Moreover, by the induction hypothesis, $[\phi] \vdash \phi$ and $[\psi] \models \psi$. Therefore, $[\phi_0] \models \phi * \psi$;

only if: since $[\phi_0] \models \phi * \psi$, there exists $\{[\phi_l]\}_{l \in L} \in J([\phi_0])$ such that for any $l \in L$, there exist $[\sigma_l], [\tau_l]$ such that $[\phi_l] \sqsubseteq [\sigma_l] \cdot [\tau_l]$, $[\sigma_l] \models \phi$ and $[\tau_l] \models \psi$. By the induction hypothesis, $\sigma_l \vdash \phi$ and $\tau_l \vdash \psi$ for any $l \in L$. For any $l \in L$, since $[\phi_l] \sqsubseteq [\sigma_l * \tau_l]$, $\phi_l \vdash \phi * \psi$. Since $\phi_0 \vdash \bigvee_{l \in L} \phi_l$, $\phi_0 \vdash \phi * \psi$.

- $\psi_0 = \phi \mathrel{-\!\!*} \psi$:

if: for any $[\phi_1]$ such that $[\phi_1] \models \phi$, by induction hypothesis, $\phi_1 \vdash \phi$. Therefore, $\phi_0 * \phi_1 \vdash \psi$. Again, by the induction hypothesis, $[\phi_0 * \phi_1] \models \psi$. Equivalently, $[\phi_0] \cdot [\phi_1] \models \psi$;

only if: by the induction hypothesis, $[\phi] \models \phi$. Since $[\phi_0] \models \phi \mathrel{-\!\!*} \psi$, $[\phi_0] \cdot [\phi] \models \psi$. By the induction hypothesis again, $\phi_0 * \phi \vdash \psi$. Therefore, $\phi_0 \vdash \phi \mathrel{-\!\!*} \psi$.

Completeness now follows in the usual way. □

For any GTM, $(M, \sqsubseteq, \cdot, e, J)$, the semantics given by forcing relation induces, via (K) and (Sh), a functor $\mathbf{Set}^{M^{op}}$ which is a subobject of 1 as well as a sheaf over the Grothendieck topology J. Therefore, the interpretations of all connectives may be considered as operations on sub-objects of 1 in the category of sheaves over J, *i.e.*, operations in $\mathbf{Sub}_{\mathbf{Sh}(\mathbf{Set}^{M^{op}},J)}(1)$. The soundness of models in Grothendieck sheaves follows from the following observation:

LEMMA 5.16 (BI-CDCCS) *The interpretation of* **BI***'s connectives in* $\mathbf{Sub}_{\mathbf{Sh}(\mathbf{Set}^{M^{op}},J)}(1)$ *induces a* **BI***-algebra.* □

PROPOSITION 5.17 (SOUNDNESS) *For any* **BI** *propositions* ϕ *and* ψ*, if* $\phi \vdash \psi$*, then* $\phi \models \psi$. □

PROOF OF LEMMA 5.16 It is easy to see that $\top$, $\wedge$ and $\rightarrow$ induce a Heyting algebra. We must show that I, $*$ and $-\!\!*$ induce a symmetric monoidal closed structure and that $\bot$, $\vee$ induce finite coproducts. Since any two objects in $\mathbf{Sub}_{\mathbf{Sh}(\mathbf{Set}^{M^{op}},J)}(1)$ may have at most one morphism, it is sufficient to show that the following rules hold:

- $\vee$ is a binary coproduct:

$$\frac{}{\psi \models \psi \vee \chi}\,1 \qquad \frac{}{\chi \models \psi \vee \chi}\,2$$

$$\frac{\phi \models \chi \quad \psi \models \chi}{\phi \vee \psi \models \chi}\,3.$$

1 For any $m \in M$ such that $m \models \psi$, since $\{m\} \in J(m)$, $m \models \psi \vee \chi$.

2 For any $m \in M$ such that $m \models \chi$, since $\{m\} \in J(m)$, $m \models \psi \vee \chi$.

3 For any $m \in M$ such that $m \models \phi \vee \psi$, by the interpretation of $\vee$, there exists $S \in J(m)$ such that $m' \models \phi$ or $m' \models \psi$ for all $m' \in S$. Since $\phi \models \chi$ and $\psi \models \chi$, $m' \models \chi$ for all $m' \in S$. Since (Sh) holds for χ, $m \models \chi$.

- $\bot$ is an initial object:

$$\frac{}{\bot \models \phi}\,1.$$

1 For any $m \in M$ such that $m \models \bot$, by the interpretation of $\bot$, $\emptyset \in J(m)$. Since (Sh) holds for ϕ, $m \models \phi$.

- $(I, *)$ induces a symmetric monoidal structure:

$$\frac{}{(\phi * \psi) * \chi \models \phi * (\psi * \chi)}\,1 \qquad \frac{}{\phi * (\psi * \chi) \models (\phi * \psi) * \chi}\,2$$

$$\frac{}{\phi * I \models \phi}\,3 \qquad \frac{}{\phi \models \phi * I}\,4$$

$$\frac{}{\phi * \psi \models \psi * \phi}\,5 \qquad \frac{\phi \models \psi \quad \chi \models \rho}{\phi * \chi \models \psi * \rho}\,6.$$

1 The proof of this case follows from the following lemma:

> LEMMA For any $m \in M$ and **BI** propositions ϕ_0, ψ_0 and χ_0, $m \models (\phi_0 * \psi_0) * \chi_0$ iff there exists $S \in J(m)$ such that for any $m' \in S$, there exists $a_{m'}$, $b_{m'}$ and $c_{m'}$ in M such that $m' \sqsubseteq (a_{m'} \cdot b_{m'}) \cdot c_{m'}$, $a_{m'} \models \phi_0$, $b_{m'} \models \psi_0$ and $c_{m'} \models \chi_0$.

Suppose the lemma above holds. Then, by the associativity and commutativity of $\cdot$, $m \models (\phi * \psi) * \chi$ iff $m \models (\psi * \chi) * \phi$. As will be shown in 5, this is equivalent to $m \models \phi * (\psi * \chi)$. The proof of the above lemma proceeds using axioms of a Grothendieck topology, as follows:

if: for any $m' \in S$, since $\{a_{m'} \cdot b_{m'}\} \in J(a_{m'} \cdot b_{m'})$ by the maximality axiom, $a_{m'} \cdot b_{m'} \models \phi_0 * \psi_0$. Therefore, $m \models (\phi_0 * \psi_0) * \chi_0$.

only if: since $m \models (\phi_0 * \psi_0) * \chi_0$, there exists $S \in J(m)$ such that for any $m' \in S$, there exist $n_{m'}$ and $c_{m'}$ such that $m' \sqsubseteq n_{m'} \cdot c_{m'}$, $n_{m'} \models \phi_0 * \psi_0$ and $c_{m'} \models \chi_0$. We show that for any m' in S, there exists $S_{m'}$ such that, for any $d \in S_{m'}$, there exist $a_{m'}, b_{m'}$ satisfying $d \sqsubseteq (a_{m'} \cdot b_{m'}) \cdot c_{m'}$, $a_{m'} \models \phi$ and $b_{m'} \models \psi$. Then, the conclusion follows from $S_m = \bigcup_{m' \in S} S_{m'}$, which is in $J(m)$ by the transitivity axiom. Choose an m' in S. Since $n_{m'} \models \phi_0 * \psi_0$, there exists $S_{n_{m'}}$ such that for any $u \in S_{n_{m'}}$, there exist a_u and b_u satisfying $u \sqsubseteq a_u \cdot b_u$, $a_u \models \phi_0$ and $b_u \models \psi_0$. Let $S_{n_{m'} \cdot c_{m'}} = \{u \cdot c_{m'} \mid u \in S_{n_{m'}}\}$. Then, by the continuity of $\cdot$, $S_{n_{m'} \cdot c_{m'}} \in J(n_{m'} \cdot c_{m'})$. Since $m' \sqsubseteq n_{m'} \cdot c_{m'}$, by the stability axiom, there exists $S_{m'} \in J(m')$ such that for any $d \in S_{m'}$, $d \sqsubseteq u \cdot c_{m'}$ for some $u \in S_{n_{m'}}$, which, by the monotonicity of $\cdot$, implies that $d \sqsubseteq (a_u \cdot b_u) \cdot c_{m'}$.

2 This case is handled while proving 1.

3 For any $m \in M$ such that $m \models \phi * I$, there exists $S \in J(m)$ such that, for any $m' \in S$, there exist $a_{m'}, b_{m'}$ in M such that $m' \sqsubseteq a_{m'} \cdot b_{m'}$, $a_{m'} \models \phi$ and $b_{m'} \models I$. By the interpretation of I, for any $m' \in S$, there exists $S_{b_{m'}} \in J(b_{m'})$ such that for

any $u \in S_{b_{m'}}$, $u \sqsubseteq e$. By the continuity of $\cdot$, for any $m' \in S$, $\{a_{m'} \cdot u \mid u \in S_{b_{m'}}\} \in J(a_{m'} \cdot b_{m'})$. For any $m' \in S$ and $u \in S_{b_{m'}}$, since $a_{m'} \cdot u \sqsubseteq a_{m'} \cdot e = a_{m'}$ and $a_{m'} \models \phi$, by (K), $a_{m'} \cdot u \models \phi$. Therefore, by (Sh), $a_{m'} \cdot b_{m'} \models \phi$ and since $m' \sqsubseteq a_{m'} \cdot b_{m'}$, (K) implies that $m' \models \phi$, for all $m' \in S$.

4 For any $m \in M$ such that $m \models \phi$, since $m = m \cdot e$, $\{m \cdot e\} \in J(m)$. Since $\{e\} \in J(e)$ and $e \sqsubseteq e$, $e \models I$. Therefore, $m \models \phi * I$.

5 For any $m \in M$ such that $m \models \phi * \psi$, there exists $S \in J(m)$ such that, for any $m' \in S$, there exist $a_{m'}, b_{m'}$ in M such that $a_{m'} \models \phi$, $b_{m'} \models \psi$ and $m' \sqsubseteq a_{m'} \cdot b_{m'}$. Since $\cdot$ is commutative, for any $m' \in S$, $m' \sqsubseteq b_{m'} \cdot a_{m'}$. Therefore, $m \models \phi * \psi$.

6 For any $m \in M$ such that $m \models \phi * \chi$, there is $S \in J(m)$ such that, for any $m' \in S$, there exist $a_{m'}, c_{m'}$ in M such that $a_{m'} \models \phi$, $c_{m'} \models \chi$ and $m' \sqsubseteq a_{m'} \cdot c_{m'}$. Since $\phi \models \psi$ and $\chi \models \rho$, for any $m' \in S$, $a_{m'} \models \psi$ and $c_{m'} \models \rho$. Therefore, $m \models \psi * \rho$.

- $(*, \mathrel{-\!\!*})$ induce a residuated (*i.e.*, closed) structure:

$$\frac{\phi * \psi \models \chi}{\phi \models \psi \mathrel{-\!\!*} \chi}\,1 \qquad \frac{\phi \models \psi \mathrel{-\!\!*} \chi}{\phi * \psi \models \chi}\,2.$$

1 For any $m, n \in M$ such that $m \models \phi$ and $n \models \psi$, by the maximality axiom, $\{m \cdot n\}$ is in $J(m \cdot n)$, from which it follows that $m \cdot n \models \phi * \psi$. Since $\phi * \psi \models \chi$, $m \cdot n \models \chi$.

2 For any $m \in M$ such that $m \models \phi * \psi$, by the interpretation of $*$, there exists $S \in J(m)$ such that for any $m' \in S$, there exist $a_{m'}$ and $b_{m'}$ in S such that $m' \sqsubseteq a_{m'} \cdot b_{m'}$, $a_{m'} \models \phi$ and $b_{m'} \models \psi$. Since $\phi \models \psi \mathrel{-\!\!*} \chi$, for any $m' \in S$, $a_{m'} \models \psi \mathrel{-\!\!*} \chi$, from which it follows that $a_{m'} \cdot b_{m'} \models \chi$. By (K), $m' \models \chi$ for all $m' \in S$. By (Sh), $m \models \chi$.

□

We conclude with a simple example, which is a specific counter-model to the entailment

$$((\mathrm{p} \mathrel{-\!\!*} \bot) \to \bot) \wedge ((\mathrm{q} \mathrel{-\!\!*} \bot) \to \bot) \models (\mathrm{p} * \mathrm{q} \mathrel{-\!\!*} \bot) \to \bot,$$

used in Proposition 4.8. We define a preordered monoid $\mathcal{M} = (M, \cdot, e, \sqsubseteq)$, where

- the carrier set $M = \{e, a, \bot\}$,
- the order is $e \sqsupseteq \bot \sqsubseteq a$,

- the multiplication is

$\cdot$	e	a	$\bot$
e	e	a	$\bot$
a	a	$\bot$	$\bot$
$\bot$	$\bot$	$\bot$	$\bot$

- and the Grothendieck topology is

$$J(\bot) = \{\{\bot\}, \emptyset\}, J(e) = \{\{e\}\}, J(a) = \{\{a\}\}.$$

We define an interpretation and forcing relation as follows:

- $m \models \mathrm{p}$ iff $m = a$ or $m = \bot$;
- $m \models \mathrm{q}$ iff $m = a$ or $m = \bot$.
- $m \models \bot$ iff $m = \bot$.

Now, $e \models (\mathrm{p} \mathrel{-\!\!*} \bot) \to \bot$ iff for all $n \sqsubseteq e$ such that $e \neq \bot$ there is an l such that $l \models \mathrm{p}$ and $n \cdot l \neq \bot$ iff there exists l such that $l \models \mathrm{p}$ and $l \neq \bot$. Since a is such an l, we have $e \models (\mathrm{p} \mathrel{-\!\!*} \bot) \to \bot$. However, $e \models (\mathrm{p} * \mathrm{q} \mathrel{-\!\!*} \bot) \to \bot$ iff for any $n \sqsubseteq e$ such that $n \neq \bot$ there is an l such that $l \models \mathrm{p} * \mathrm{q}$ and $l \cdot n \neq \bot$ iff there are l, l' such that $l \models \mathrm{p}$, $l' \models \mathrm{q}$ and $l \cdot l' \neq \bot$ which cannot be so because, for any l and l', if $l \models \mathrm{p}$ and $l' \models \mathrm{q}$, then $l \cdot l' = \bot$. Therefore, $e \not\models (\mathrm{p} * \mathrm{q} \mathrel{-\!\!*} \bot) \to \bot$

Therefore, $e \models ((\mathrm{p} \mathrel{-\!\!*} \bot) \to \bot) \wedge ((\mathrm{q} \mathrel{-\!\!*} \bot) \to \bot)$, but $e \not\models ((\mathrm{p} * \mathrm{q} \mathrel{-\!\!*} \bot) \to \bot)$ in this model.

Chapter 6

PROPOSITIONAL BI AS A SEQUENT CALCULUS

1. A Sequent Calculus

We reformulate **BI**'s proof theory as a sequent calculus [Gentzen, 1934] in which the elimination rule for each connective, #, in a natural deduction system is replaced by a corresponding left rule which introduces # to the left hand side of a sequent. Along with each left rules comes a right rule, which is typically identical to the corresponding introduction rule.

We call the calculus **LBI**. **BI**'s sequent calculus was sketched in [Pym, 1999]. The rules are given in Table 6.1.

Our presentation takes explicit structural rules of Weakening and Contraction and retains a multiplicative presentation of the binary rules for the additives. However, just as for **NBI**, recovery of the familiar additive presentation is straightforward.

LEMMA 6.1 (ADDITIVE RULES) *The following additive versions of the* Ax, $\wedge R$, $\rightarrow L$ *and* $\vee L$ *rules are derivable in* **LBI**:

$$\frac{}{\phi ; \Gamma \vdash \phi}\ Axiom \qquad \frac{\Gamma \vdash \phi \quad \Gamma \vdash \psi}{\Gamma \vdash \phi \wedge \psi}\ \wedge R$$

$$\frac{\Gamma \vdash \phi \quad \Delta(\Gamma ; \psi) \vdash \chi}{\Delta(\Gamma ; \phi \rightarrow \psi) \vdash \chi}\ \rightarrow L \qquad \frac{\Gamma(\phi) \vdash \chi \quad \Gamma(\psi) \vdash \chi}{\Gamma(\phi \vee \psi) \vdash \chi}\ \vee L.$$

PROOF Easy constructions using the structurals. □

2. Cut-elimination

THEOREM 6.2 (CUT-ELIMINATION) *If* **LBI** *proves* $\Gamma \vdash \phi$, *then* **LBI**$\backslash$Cut *proves* $\Gamma \vdash \phi$.

IDENTITIES

$$\frac{}{\phi \vdash \phi}\ \textit{Axiom} \qquad \frac{\Gamma \vdash \phi \quad \Delta(\phi) \vdash \psi}{\Delta(\Gamma) \vdash \psi}\ \textit{Cut}$$

STRUCTURALS

$$\frac{\Gamma(\Delta) \vdash \phi}{\Gamma(\Delta ; \Delta') \vdash \phi}\ W \qquad \frac{\Gamma(\Delta ; \Delta) \vdash \phi}{\Gamma(\Delta) \vdash \phi}\ C$$

$$\frac{\Gamma \vdash \phi}{\Delta \vdash \phi}\ (\Delta \equiv \Gamma) \quad E$$

UNITS

$$\frac{\Gamma(\emptyset_m) \vdash \phi}{\Gamma(I) \vdash \phi}\ I\,L \qquad \frac{}{\emptyset_m \vdash I}\ I\,R$$

$$\frac{}{\Gamma(\bot) \vdash \phi}\ \bot L$$

$$\frac{\Gamma(\emptyset_a) \vdash \phi}{\Gamma(\top) \vdash \phi}\ \top L \qquad \frac{}{\emptyset_a \vdash \top}\ \top R$$

MULTIPLICATIVES

$$\frac{\Gamma \vdash \phi \quad \Delta(\Delta', \psi) \vdash \chi}{\Delta(\Delta', \Gamma, \phi \mathrel{-\!\!*} \psi) \vdash \chi}\ {-\!\!*}\,L \qquad \frac{\Gamma, \phi \vdash \psi}{\Gamma \vdash \phi \mathrel{-\!\!*} \psi}\ {-\!\!*}\,R$$

$$\frac{\Gamma(\phi, \psi) \vdash \chi}{\Gamma(\phi * \psi) \vdash \chi}\ * L \qquad \frac{\Gamma \vdash \phi \quad \Delta \vdash \psi}{\Gamma, \Delta \vdash \phi * \psi}\ * R$$

ADDITIVES

$$\frac{\Gamma \vdash \phi \quad \Delta(\Delta' ; \psi) \vdash \chi}{\Delta(\Delta' ; \Gamma ; \phi \to \psi) \vdash \chi}\ \to L \qquad \frac{\Gamma ; \phi \vdash \psi}{\Gamma \vdash \phi \to \psi}\ \to R$$

$$\frac{\Gamma(\phi_1 ; \phi_2) \vdash \psi}{\Gamma(\phi_1 \wedge \phi_2) \vdash \psi}\ \wedge L \qquad \frac{\Gamma \vdash \phi \quad \Delta \vdash \psi}{\Gamma ; \Delta \vdash \phi \wedge \psi}\ \wedge R$$

$$\frac{\Gamma(\phi) \vdash \chi \quad \Delta(\psi) \vdash \chi}{\Gamma(\phi \vee \psi) ; \Delta(\phi \vee \psi) \vdash \chi}\ \vee L \qquad \frac{\Gamma \vdash \phi_i}{\Gamma \vdash \phi_1 \vee \phi_2}\ (i = 1, 2) \quad \vee R$$

Table 6.1. **LBI** for Propositional **BI**

PROOF SKETCH By induction on the complexity of proofs in **LBI**. We show that each Cut may replaced by one or more Cuts of lower complexity, *i.e.*, in which the Cut formula has fewer connectives. With respect to the multiplicative fragment, this definition of the complexity of a proof depends exactly on the number of inferences. With respect to the additive fragment, note that we must take a little more care with Contraction. The additive fragment of Cut interacts with Contraction as follows: The derivation

$$\dfrac{\dfrac{\Gamma;\phi;\phi\vdash\psi}{\Gamma;\phi\vdash\psi}\,\mathit{Contraction}\qquad \Gamma'\vdash\phi}{\Gamma;\Gamma'\vdash\psi}\,\mathit{Cut}$$

cannot be reduced to one in which the newer Cuts are of lower complexity unless we work not with Cut but rather with its derivable generalization, Multi-Cut,

$$\dfrac{\Gamma(\phi^n)\vdash\psi\quad \Gamma'\vdash\phi}{\Gamma(\Gamma')\vdash\psi}\;\mathit{Multi\text{-}Cut},$$

where ϕ^n denotes n additively combined copies of ϕ, which may be considered to encode the necessary Contractions within the Cut. The problem does not arise for the multiplicative fragment. See [Troelstra and Schwichtenberg, 1996] for a discussion of this point. The remainder of the argument is sketched below.

In the base case, a Cut rule is replaced by an Axiom, *i.e.*, the basic identity rule. The argument follows a familiar pattern and is not problematic; see, for example, [Trolestra, 1992]. Accordingly, we demonstrate just a few cases.

$\wedge$: The proof figure

$$\dfrac{\dfrac{\Gamma(\phi_1;\phi_2)\vdash\chi}{\Gamma(\phi_1\wedge\phi_2)\vdash\chi}\,\wedge L\qquad \dfrac{\Gamma'\vdash\phi_1\quad\Gamma''\vdash\phi_2}{\Gamma';\Gamma''\vdash\phi_1\wedge\phi_2}\,\wedge R}{\Gamma(\Gamma';\Gamma'')\vdash\chi}\,\mathit{Cut}$$

reduces to the proof figure

$$\dfrac{\dfrac{\Gamma(\phi_1;\phi_2)\vdash\chi\quad\Gamma'\vdash\phi_1}{\Gamma(\Gamma';\phi_2)\vdash\chi}\,\mathit{Cut}\qquad \Gamma''\vdash\phi_2}{\Gamma(\Gamma';\Gamma'')\vdash\chi}\,\mathit{Cut}$$

in which the Cut on $\phi_1\wedge\phi_2$ has been replaced by simpler Cuts on the ϕ_is. Note that had we worked with the form of $\wedge$R given in

Lemma 6.1, then we should have required a Contraction below these two Cuts. Such a Contraction would not contribute to the complexity of the figure.

$*$: The proof figure

$$\dfrac{\dfrac{\Gamma(\phi_1,\phi_2) \vdash \chi}{\Gamma(\phi_1 * \phi_2) \vdash \chi}\,{*L} \qquad \dfrac{\Delta_1 \vdash \phi_1 \quad \Delta_2 \vdash \phi_2}{\Delta_1,\Delta_2 \vdash \phi_1 * \phi_2}\,{*R}}{\Gamma(\Delta_1,\Delta_2) \vdash \chi}\,Cut$$

reduces to the proof figure

$$\dfrac{\dfrac{\Gamma(\phi_1,\phi_2) \vdash \chi \quad \Delta_1 \vdash \phi_1}{\Gamma(\Delta_1,\phi_2) \vdash \chi}\,Cut \qquad \Delta_2 \vdash \phi_2}{\Gamma(\Delta_1,\Delta_2) \vdash \chi}\,Cut$$

in which the Cut on the proposition $\phi_1 * \phi_2$ has been replaced by the two simpler Cuts on ϕ_1 and ϕ_2.

$\rightarrow$: The proof figure

$$\dfrac{\dfrac{\Gamma \vdash \phi \quad \Delta(\Delta';\psi) \vdash \chi}{\Delta(\Delta';\Gamma;\phi \rightarrow \psi) \vdash \chi}\rightarrow L \qquad \dfrac{\Theta;\phi \vdash \psi}{\Theta \vdash \phi \rightarrow \psi}\rightarrow R}{\Delta(\Delta';\Gamma;\Theta) \vdash \chi}\,Cut$$

reduces to the proof figure

$$\dfrac{\dfrac{\Theta;\phi \vdash \psi \quad \Delta(\Delta';\psi) \vdash \chi}{\Delta(\Delta';\Theta;\phi) \vdash \chi}\,Cut \qquad \Gamma \vdash \phi}{\Delta(\Delta';\Gamma;\Theta) \vdash \chi}\,Cut.$$

Here we suppress an Exchange, which does not contribute to the complexity of the figure.

$-\!\!*$: The proof figure

$$\dfrac{\dfrac{\Gamma \vdash \phi \quad \Delta(\Delta',\psi) \vdash \chi}{\Delta(\Delta',\Gamma,\phi -\!\!* \psi) \vdash \chi}-\!\!*\,L \qquad \dfrac{\Theta,\phi \vdash \psi}{\Theta \vdash \phi -\!\!* \psi}-\!\!*\,R}{\Delta(\Delta',\Gamma,\Theta) \vdash \chi}\,Cut$$

reduces to the proof figure

$$\dfrac{\dfrac{\Theta,\phi \vdash \psi \quad \Delta(\Delta',\psi) \vdash \chi}{\Delta(\Delta',\Theta,\phi) \vdash \chi}\,Cut \qquad \Gamma \vdash \phi}{\Delta(\Delta',\Gamma,\Theta) \vdash \chi}\,Cut.$$

Again, we harmlessly suppress an Exchange.

The other cases are similar. □

We conclude with a conjecture about a corollary of Cut-elimination: That propositional **BI** is decidable and that it has the finite model property . However, we defer an analysis of decidability in **BI**, together with its relationship with the decidability properties of propositional classical, intuitionistic and other substructural logics, to another occasion. See [Galmiche and Méry, 2001a, Galmiche and Méry, 2001b, Galmiche et al., 2001] for more discussion of these topics.

3. Equivalence

THEOREM 6.3 (EQUIVALENCE OF **NBI** AND **LBI**) *$\Gamma \vdash \phi$ is provable in* **NBI** *if and only if $\Gamma \vdash \phi$ is provable in* **LBI**.

PROOF We begin by defining mappings from **NBI** to **LBI** and *vice versa*.

Firstly, we consider a map $\mathsf{L} : \mathbf{NBI} \longrightarrow \mathbf{LBI}$. The introduction rules of **NBI** map directly to the right rules of **LBI** and the structural rules map to themselves. Turning to the elimination rules, we note that the basic pattern is that the elimination rule for a given connective, #, maps to the left rule for #, together with a cut on the principal formula of the elimination. For example, a proof in **NBI** ending with a $*$E,

$$\frac{\begin{matrix}\Phi_1 \\ \Gamma \vdash \phi * \psi\end{matrix} \qquad \begin{matrix}\Phi_2 \\ \Delta(\phi, \psi) \vdash \chi\end{matrix}}{\Delta(\Gamma) \vdash \chi} *E,$$

maps under L to the proof

$$\frac{\begin{matrix}\mathsf{L}(\Phi_1) \\ \Gamma \vdash \phi * \psi\end{matrix} \qquad \dfrac{\begin{matrix}\mathsf{L}(\Phi_2) \\ \Delta(\phi, \psi) \vdash \chi\end{matrix}}{\Delta(\phi * \psi) \vdash \chi} *L}{\Delta(\Gamma) \vdash \chi} \; Cut.$$

The other cases are similar, with appropriate uses of ";" for the additives.

Secondly, we consider a map $\mathsf{N} : \mathbf{LBI} \longrightarrow \mathbf{NBI}$. The right rules of **LBI** map directly to the introduction rules of **NBI** and, as for L, the structural rules maps to themselves. (Note that the cut rule is admissible in **NBI**.) To see how N works, consider that a proof in **LBI** ending with a $*$L,

$$\frac{\begin{matrix}\Phi \\ \Gamma(\phi, \psi) \vdash \chi\end{matrix}}{\Gamma(\phi * \psi) \vdash \chi} *L,$$

maps under N to the proof

$$\dfrac{\dfrac{}{\phi * \psi \vdash \phi * \psi}\,\mathit{Axiom} \qquad \begin{array}{c}\mathsf{N}(\Phi)\\ \Gamma(\phi, \psi) \vdash \chi\end{array}}{\Gamma(\phi * \psi) \vdash \chi}\, {*E}.$$

This case illustrates the general pattern, but the case for the implication left rules is, in a sense, pathological, requiring a Cut. For example, a proof in **LBI** ending with a (simplified form of) $\mathrel{-\!\!*}$ L,

$$\dfrac{\begin{array}{c}\Phi_1\\ \Gamma \vdash \phi\end{array} \qquad \begin{array}{c}\Phi_2\\ \Delta, \psi \vdash \chi\end{array}}{\Gamma, \Delta, \phi \mathrel{-\!\!*} \psi \vdash \chi}\, {*L},$$

maps under N to the proof

$$\dfrac{\dfrac{\dfrac{}{\phi \mathrel{-\!\!*} \psi \vdash \phi \mathrel{-\!\!*} \psi}\,\mathit{Axiom} \qquad \begin{array}{c}\mathsf{N}(\Phi_1)\\ \Gamma \vdash \phi\end{array}}{\Gamma, \phi \mathrel{-\!\!*} \psi \vdash \psi}\, {\mathrel{-\!\!*} E} \qquad \begin{array}{c}\mathsf{N}(\Phi_2)\\ \Delta, \psi \vdash \chi\end{array}}{\Gamma, \Delta, \phi \mathrel{-\!\!*} \psi \vdash \chi}\, \mathit{Cut}.$$

Given these mappings, the proof proceeds by routine inductions on the structure of proofs in **LBI** and **NBI**. □

Before concluding this section, it is worth pausing to remark upon the behaviour $\bot$. Consider the following **NBI** proof:

$$\dfrac{\dfrac{\dfrac{}{\bot \vdash \bot}\,\mathit{Axiom}}{\bot \vdash \phi \mathrel{-\!\!*} \psi}\, {\bot E} \qquad \Gamma \vdash \phi}{\bot, \Gamma \vdash \psi}\, {\mathrel{-\!\!*} E}.$$

It translates under L to the following sequent proof:

$$\dfrac{\dfrac{}{\bot \vdash \phi \mathrel{-\!\!*} \psi}\, {\bot L} \qquad \dfrac{\Gamma \vdash \phi \qquad \dfrac{}{\psi \vdash \psi}\,\mathit{Axiom}}{\Gamma, \phi \mathrel{-\!\!*} \psi \vdash \psi}\, {\mathrel{-\!\!*} L}}{\bot, \Gamma \vdash \psi}\, \mathit{Cut}.$$

Under Cut-elimination, this proof reduces to the axiom $\bot, \Gamma \vdash \psi$. Note that this reduction makes explicit use of the affine character of the logic in the presence of $\bot$ on the left. The soundness of the $\bot L$ rule may readily be verified with respect to **BI**'s topological Kripke models and Grothendieck topological models.

It is a straightforward matter to assign terms of $\alpha\lambda$ to **LBI**'s proofs and then to extend the equivalence of **NBI** and **LBI** from consequences to terms.

4. Other Proof Systems

Recent work by Galmiche and Méry [Galmiche and Méry, 2001a, Galmiche and Méry, 2001b] and by Galmiche, Méry and Pym [Galmiche et al., 2001] has developed systems of *semantic tableaux* [Hodges, 1993] for propositional **BI**.

In [Galmiche and Méry, 2001a, Galmiche and Méry, 2001b], the authors develop the theory of semantic tableaux for propositional **BI** without $\bot$. The main technical difficulty derives from the need to manage the the combination and comparison of worlds which is inherent in the possible-worlds semantics of **BI**. The solution borrows the idea of *labelled deductive systems* from Gabbay [Gabbay, 1996]. The system **TBI** is a system of labelled tableaux rules in which labels are drawn from a preordered commutative monoid of worlds. This algebra of worlds is then used to calculate the constraints which are necessary for a proof to be determined.

In [Galmiche et al., 2001], this technique is extended to handle **BI** with bottom. The technical step required is to move from a preordered commutative monoid of labels to a Grothendieck topology of labels (*q.v.* Chapter 5).

In both settings [Galmiche and Méry, 2001a, Galmiche and Méry, 2001b, Galmiche et al., 2001], suitable soundness and completeness theorems are obtained.

Chapter 7

TOWARDS CLASSICAL PROPOSITIONAL BI

1. Introduction

So far our discussion has been confined to what we may call "intuitionistic **BI**" or, indeed, "minimal **BI**", to which we can easily add the additive (intuitionistic) negation, $\neg\phi = \phi \to \bot$. From the semantic point of view, the key characteristic of a classical system is the strength of negation. Within the single our conclusioned formulation of bunched consequence, we can strengthen the intuitionistic negation, via the addition of RAA but it is a move to a multiple-conclusioned formultion which suggests a form of multiplicative negation.

In this chapter, we give a brief discussion of some classical bunched logics. There are four basic possibilities for combining the intuitionistic and classical additives and multiplicatives:

- Intuitionistic additives and intuitionistic multiplicatives;
- Classical additives and intuitionistic multiplicatives;
- Intuitionistic additives and classical multiplicatives;
- Classical additives and classical multiplicatives.

Here, the distinction between the intuitionistic and classical systems should be considered to be a distinction between different strengths of negation [Prawitz, 1965].

From a computational point view, we have several concrete models which are of interest, based on the resource interpretation of **BI**'s classical additives and intuitionistic multiplicatives. These models are described in Chapter 9.

This chapter concludes with a brief discussion of Troelstra's [Trolestra, 1992] analysis of a classical system which combines additive and multiplicative implications, using neither bunches nor exponentials [Girard, 1987].

2. An Algebraic View

In algebraic terms, the four systems may be described conveniently as follows:

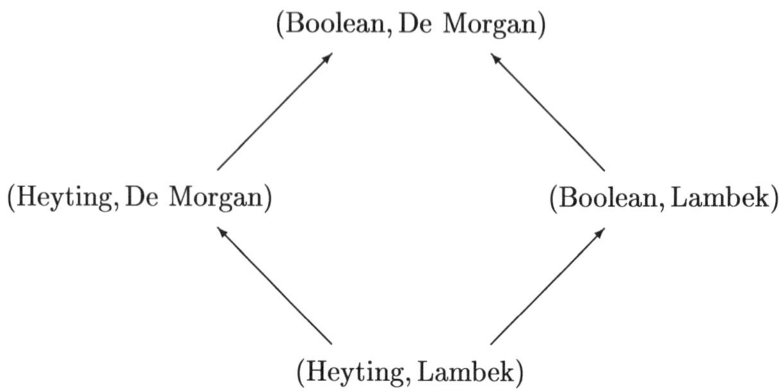

Here each pair describes the logical strength of the (additive, multiplicative) system formed by freely combining the parts. The additive part may be either Boolean (classical) or Heyting (intuitionistic) and the multiplicative part may be either De Morgan (classical) or Lambek (intuitionistic). The diagram is ordered according to the strength of negation. Just as Boolean negation is involutive, *i.e.*, $\neg\neg\phi$ is equivalent to ϕ, so too is De Morgan negation, *i.e.*, $\sim\sim\phi$ is equivalent to ϕ.

Unless stated otherwise, the unmodified "**BI**" refers to (Heyting, Lambek) **BI**.

We begin with the relatively simple system (Boolean, Lambek) in which the additives are treated classically and the multiplicatives continue to be treated intuitionistically. Thus we define "Boolean **BI**" to be the consequence relation generated by the Hilbert-type rules for the **BI**-algebras, here characterized as (Heyting, Lambek), described in Table 3.1 (in Chapter 3) plus *reductio ad absurdum*:

$$\frac{\psi \vdash (\phi \to \bot) \to \bot}{\psi \vdash \phi}\ RAA.$$

An algebraic model for this system is called a *Boolean* **BI**-*algebra*, *i.e.*, a **BI**-algebra in which the Heyting component is in fact Boolean. We give, in Chapter 16, several computational models of Boolean **BI**.

The rule of RAA describes how Boolean negation is related to additive implication. The systems with De Morgan negation, $\sim -$, require an involution which corresponds not to the additive residuation but rather to multiplicative implication. In short, we require that

$$(\phi \mathrel{-\!\!*} \perp\!\!\!\perp) \mathrel{-\!\!*} \perp\!\!\!\perp \;\dashv\vdash\; \phi,$$

where $\perp\!\!\!\perp$, the dual of I is the unit of the evident multiplicative disjunction.

Just as the "Lambek" residuated monoid may be seen as being obtained by collapsing a symmetric monoidal closed category [Mac Lane, 1971, Trolestra, 1992], so the "De Morgan" algebra may be seen as being obtained by collapsing the structure of a *-autonomous category [Barr and Wells, 1995, Barr, 1979].

BI-algebras (for whichever system) are useful as a reference point but the definition itself neither suggests a declarative way of reading formulæ, *i.e.*, meanings for the connectives, nor tells us if there are any interesting **BI**-algebras. However, the study of structures of this type has a long history, with an excellent discussion been given by Lambek in [Lambek, 1993]. Lambek gives a semantics of the (non-commutative) mutliplicatives, including a treatment of De Morgan negation, based on subsets of a monoid $(M, \cdot, e)$. For example, the tensor product is interpreted as follows:[1]

$$\begin{array}{rcl} [\![\phi * \psi]\!] & = & \{p \cdot q \mid p \in [\![\phi]\!], q \in [\![\psi]\!]\} \\ & = & \{r \mid \text{there are } p, q \text{ s.t. } r = p \cdot q,\ p \in [\![\phi]\!] \text{ and } q \in [\![\psi]\!]\} \\ \\ [\![I]\!] & = & \{e\}. \end{array}$$

The interpretation of a classical (De Morgan) negation uses complementation,

$$[\![\sim \phi]\!] \;=\; \{p \mid p \notin [\![\phi]\!]\},$$

and so admits the definition

$$\phi \# \psi = \sim(\sim \phi * \sim \psi).$$

[1]Note that this construction is an instance of the general relationship between Kripke models of a logic and its corresponding algebraic formulation: the carrier set for the algebra is, essentially, the set of all worlds.

Additive connectives are not discussed.

Possible worlds models are another way to begin to address both of the question of A declarative meaning and finding interesting examples. However, the provision of possible worlds models for the classical systems seems to be a delicate matter. We discuss some of the issues in this chapter.

3. A Proof-theoretic View

We can readily describe a multiple-conclusioned sequent calculus, given in Table 7.1, based on the evident notion of bunch for the right-hand side.

"Disjunctive bunches" are defined in just the same way as the "conjunctive bunches" (Chapter 2) we have already considered, *i.e.*, they have two symmetric monoidal operations, ";" and ",", and congruence. However, they are interpreted logically by inference rules which characterize ";" by additive disjunction, $\vee$, and "," by multiplicative disjunction, $\#$. The De Morgan laws hold for $*$, $\#$ and $\sim$.

It is a straightforward matter to check that the calculus given in Table 7.1 admits Cut-elimination. For example, the Cut-reduction for the mutliplicative disjunction, $\#$, goes as follows: The complex Cut on $\phi \# \psi$,

$$\dfrac{\dfrac{\Gamma_1, \phi \vdash \Delta_1 \quad \Gamma_2, \psi \vdash \Delta_2}{\Gamma_1, \Gamma_2, \phi \# \psi \vdash \Delta_1, \Delta_2}\,\#L \qquad \dfrac{\Gamma_3 \vdash \Delta_3(\phi, \psi)}{\Gamma_3 \vdash \Delta_3(\phi \# \psi)}\,\#R}{\Gamma_1, \Gamma_2, \Gamma_3 \vdash \Delta_3(\Delta_1, \Delta_2)}\,Cut,$$

reduces to the simpler Cuts on ψ and ϕ,

$$\dfrac{\Gamma_1, \phi \vdash \Delta_1 \qquad \dfrac{\Gamma_2, \psi \vdash \Delta_2 \quad \Gamma_3 \vdash \Delta_3(\phi, \psi)}{\Gamma_2, \Gamma_3 \vdash \Delta_3(\phi, \Delta_2)}\,Cut}{\Gamma_1, \Gamma_2, \Gamma_3 \vdash \Delta_3(\Delta_1, \Delta_2)}\,Cut.$$

The other cases are, of course, similar.

Now, it may readily be seen that

$$\phi \mathrel{-\!\!*} \psi \dashv\vdash \sim\phi \# \psi .$$

However, the status of the sequent calculus given above is problematic. In the presence of the implication rules, given in Table 7.2, the evident Cut-elimination property fails.

$$\frac{}{\phi \vdash \phi}\ \text{Axiom} \qquad \frac{\Gamma \vdash \Delta(\phi) \quad \Gamma'(\phi) \vdash \Delta'}{\Gamma'(\Gamma) \vdash \Delta(\Delta')}\ \text{Cut}$$

$$\frac{\Gamma(\Gamma') \vdash \Delta}{\Gamma(\Gamma';\Gamma'') \vdash \Delta}\ W\text{L} \qquad \frac{\Gamma \vdash \Delta(\Delta')}{\Gamma \vdash \Delta(\Delta';\Delta'')}\ W\text{R}$$

$$\frac{\Gamma(\Gamma';\Gamma'') \vdash \Delta}{\Gamma(\Gamma') \vdash \Delta}\ (\Gamma' \cong \Gamma'') \quad C\text{L} \qquad \frac{\Gamma \vdash \Delta(\Delta';\Delta'')}{\Gamma \vdash \Delta(\Delta')}\ (\Delta' \cong \Delta'') \quad C\text{R}$$

$$\frac{\Gamma \vdash \Delta}{\Gamma \vdash \perp\!\!\!\perp, \Delta}\ \perp\!\!\!\perp R$$

$$\frac{\Gamma \vdash \Delta}{\Gamma, I \vdash \Delta}\ IL \qquad \frac{}{\emptyset_m \vdash I}\ IR$$

$$\frac{\Gamma \vdash \phi, \Delta}{\Gamma, \sim \phi \vdash \Delta}\ \sim \text{L} \qquad \frac{\Gamma, \phi \vdash \Delta}{\Gamma \vdash \sim \phi, \Delta}\ \sim \text{R}$$

$$\frac{\Gamma(\phi, \psi) \vdash \Delta}{\Gamma(\phi * \psi) \vdash \Delta}\ *\text{L} \qquad \frac{\Gamma \vdash \phi, \Delta \quad \Gamma' \vdash \psi, \Delta'}{\Gamma, \Gamma' \vdash \phi * \psi, \Delta, \Delta'}\ *\text{R}$$

$$\frac{\Gamma, \phi \vdash \Delta \quad \Gamma', \psi \vdash \Delta'}{\Gamma, \Gamma', \phi \# \psi \vdash \Delta, \Delta'}\ \#\,\text{L} \qquad \frac{\Gamma \vdash \Delta(\phi, \psi)}{\Gamma \vdash \Delta(\phi \# \psi)}\ \#\,\text{R}$$

$$\frac{}{\Gamma(\perp) \vdash \Delta}\ \perp L$$

$$\frac{}{\Gamma \vdash \Delta(\top)}\ \top R$$

$$\frac{\Gamma \vdash \phi; \Delta}{\Gamma; \neg\phi \vdash \Delta}\ \neg\text{L} \qquad \frac{\Gamma; \phi \vdash \Delta}{\Gamma \vdash \neg\phi; \Delta}\ \neg\text{R}$$

$$\frac{\Gamma(\phi; \psi) \vdash \Delta}{\Gamma(\phi \wedge \psi) \vdash \Delta}\ \wedge\text{L} \qquad \frac{\Gamma \vdash \phi; \Delta \quad \Gamma' \vdash \psi; \Delta'}{\Gamma; \Gamma' \vdash \Delta(\phi \wedge \psi); \Delta'(\phi \wedge \psi)}\ \wedge\text{R}$$

$$\frac{\Gamma(\phi) \vdash \Delta \quad \Gamma'(\psi) \vdash \Delta'}{\Gamma(\phi \vee \psi); \Gamma'(\phi \vee \psi) \vdash \Delta; \Delta'}\ \vee\text{L} \qquad \frac{\Gamma \vdash \Delta(\phi; \psi)}{\Gamma \vdash \Delta(\phi \vee \psi)}\ \vee\text{R}$$

Table 7.1. Some Sequential Rules for (Boolean, De Morgan) **BI**

$$\frac{\Gamma' \vdash \Delta(\phi) \quad \Gamma(\psi) \vdash \Delta'}{\Gamma(\Gamma', \phi \mathrel{-\!\!*} \psi) \vdash \Delta(\Delta')} \mathrel{-\!\!*} \mathrm{L} \qquad \frac{\Gamma, \phi \vdash \psi, \Delta}{\Gamma \vdash \phi \mathrel{-\!\!*} \psi, \Delta} \mathrel{-\!\!*} \mathrm{R}$$

$$\frac{\Gamma' \vdash \Delta(\phi) \quad \Gamma(\psi) \vdash \Delta'}{\Gamma(\Gamma'; \phi \to \psi) \vdash \Delta(\Delta')} \to \mathrm{L} \qquad \frac{\Gamma; \phi \vdash \psi; \Delta}{\Gamma \vdash \phi \to \psi; \Delta} \to \mathrm{R}$$

Table 7.2. Some Sequential Implicational Rules for (Boolean, De Morgan) **BI**

We should be able to recover Cut-elimination by a restriction of the left rules for the implications to apply only to principal formulæ which occur at the top of a bunch, *i.e.*,

$$\frac{\Gamma' \vdash \Delta, \phi \quad \Gamma, \psi \vdash \Delta'}{\Gamma, \Gamma', \phi \mathrel{-\!\!*} \psi \vdash \Delta, \Delta'} \mathrel{-\!\!*} \mathrm{L} \quad \frac{\Gamma' \vdash \Delta; \phi \quad \Gamma; \psi \vdash \Delta'}{\Gamma; \Gamma'; \phi \to \psi \vdash \Delta; \Delta'} \to \mathrm{L} .$$

However, such a system would be of questionable semantic value.

4. A Forcing Semantics

We sketch a forcing semantics for (Boolean, De Morgan) **BI**. Note that, in contrast to the forcing semantics for (Heyting, Lambek) **BI** given in Chapters 4 and 5 (or, indeed, in Chapter 15), we do not interpret arbitrary propositions in a mathematical structure.[2] Rather, given a collection of worlds M, we take a function $[\![\mathrm{p}]\!] : |\, M \,| \longrightarrow \mathbf{2}$ which determines the truth of atomic propositions at each world, so that

$$m \models \mathrm{p} \quad \text{iff} \quad [\![\mathrm{p}]\!](m) = 1,$$

and a forcing relation, $\models$, subject to the semantic clauses for the additives given in Table 7.3 and the semantic clauses for the multiplicatives given in Table 7.4.

To give a forcing semantics for the Boolean connectives, as in Table 7.3, is straightforward.

However, the classical multiplicatives are more problematic. Given an algebra of worlds having an involution, $-^{\perp\!\!\perp}$, we might expect the following semantics for multiplicative negation:

$$m \models \sim \phi \quad \text{iff} \quad m^{\perp\!\!\perp} \models \phi .$$

[2]It is not immediately clear how to give such a semantics for the classical systems.

$$m \models \phi \wedge \psi \quad \text{iff} \quad m \models \phi \quad \text{and} \quad m \models \psi$$

$$m \models \phi \vee \psi \quad \text{iff} \quad m \models \phi \quad \text{or} \quad m \models \psi$$

$$m \models \neg\phi \quad \text{iff} \quad m \not\models \phi$$

$$m \models \phi \to \psi \quad \text{iff} \quad m \models \neg\phi \quad \text{or} \quad m \models \psi$$

It follows that

$$m \models \phi \vee \psi \quad \text{iff} \quad m \models \neg(\neg\phi \wedge \neg\psi)$$

$$m \models \phi \to \psi \quad \text{iff} \quad m \models \phi \quad \text{implies} \quad m \models \psi$$

Table 7.3. Clauses for Classical Additives

Whilst this definition is dualizing, $m \models \sim\ \sim\phi$ iff $m \models \phi$, it does not interact well with the other connectives. More specifically, it seems that the expected dualities, such as $m \models \phi \mathrel{-\!\!*} \psi$ iff $m \models \sim\psi \mathrel{-\!\!*}\ \sim\phi$, cannot be established.

One solution, essentially following in the relevantists' tradition [Anderson et al., 1992], is to take the following definition of $\sim$:

$$m \models \sim\phi \quad \text{iff} \quad m^{\perp\!\!\perp} \not\models \phi .$$

This negation, the meaning of which is perhaps conceptually unclear, is compatible with the semantics of the other multiplicatives given by Table 7.4, in which $(M, \cdot, e)$ is a commutative monoid with an involution, $-^{\perp\!\!\perp}$, and in which we define $m \# n = (m^{\perp\!\!\perp} \cdot n^{\perp\!\!\perp})^{\perp\!\!\perp}$: The De Morgan laws hold for $*$ and $\#$, and $m \models \phi \mathrel{-\!\!*} \psi$ if and only if $m \models \sim\psi \mathrel{-\!\!*}\ \sim\phi$.

It follows that $m \models \phi \mathrel{-\!\!*} \psi$ iff $m \models \sim\phi \# \psi$.

Kripke models of this kind may be based, for example, on abelian groups of worlds, where $m^{\perp\!\!\perp} = m^{-1}$.

5. Troelstra's Additive Implication

In [Trolestra, 1992], Troelstra discusses a sequent calculus $\mathbf{CLL}_{\mathrm{if}}$ for the following connectives (using Troelstra's notation): $\multimap$ (multiplicative implication), $\leadsto$ (additive implication), $\perp$ (additive false) and $\mathbf{0}$ (multiplicative false). The calculus does not use bunches. Rather, it decomposes the right rule for additive implication into two parts.

$$m \models \phi * \psi \quad \text{iff} \quad \text{for some} \quad n, n' \in M \quad \text{such that} \quad n \cdot n' = m \quad (n \models \phi \quad \text{and} \quad n' \models \psi)$$

$$m \models \phi \# \psi \quad \text{iff} \quad \text{for some} \quad n, n' \in M \quad \text{such that} \quad n^{\perp\!\!\perp} \# n'^{\perp\!\!\perp} = m \quad (n \models \sim \phi \quad \text{and} \quad n' \models \sim \psi)$$

$$m \models \sim \phi \quad \text{iff} \quad m^{\perp\!\!\perp} \not\models \phi$$

$$m \models \phi \mathrel{-\!\!*} \psi \quad \text{iff} \quad \text{for all } n \text{ such that } n \models \phi, \; m \cdot n \models \psi$$

Table 7.4. Clauses for Classical Multiplicatives

$$\frac{}{\phi \vdash \phi}\, Ax \qquad \frac{\Gamma_0 \vdash \phi, \Delta_0 \quad \Gamma_1, \phi \vdash \Delta_1}{\Gamma_0, \Gamma_1 \vdash \Delta_0, \Delta_1}\, Cut$$

$$\frac{}{\Gamma, \bot \vdash \Delta}\, \bot L$$

$$\frac{}{\mathbf{0} \vdash}\, \mathbf{0}L \qquad \frac{\Gamma \vdash \Delta}{\Gamma \vdash \mathbf{0}, \Delta}\, \mathbf{0}R$$

$$\frac{\Gamma_0 \vdash \Delta_0, \phi \quad \Gamma_1, \psi \vdash \Delta_1}{\Gamma_0, \Gamma_1, \phi \multimap \psi \vdash \Delta_0, \Delta_1} \multimap L \qquad \frac{\Gamma, \phi \vdash \psi, \Delta}{\Gamma \vdash \phi \multimap \psi, \Delta} \multimap R$$

$$\frac{\Gamma \vdash \Delta, \phi \quad \Gamma, \psi \vdash \Delta}{\Gamma, \phi \rightsquigarrow \psi \vdash \Delta} \rightsquigarrow L \qquad \frac{\Gamma \vdash \psi, \Delta}{\Gamma \vdash \phi \rightsquigarrow \psi, \Delta} \rightsquigarrow R \qquad \frac{\Gamma, \phi \vdash \psi}{\Gamma \vdash \phi \rightsquigarrow \psi, \Delta} \rightsquigarrow R$$

Table 7.5. The $\mathbf{CLL}_{\mathrm{if}}$ Sequent Calculus

The system $\mathbf{CLL}_{\mathrm{if}}$ is given Table 7.5.

$\mathbf{CLL}_{\mathrm{if}}$ is both cut-free and equivalent, via both additive and multiplicative versions of the familiar classical definitions of the remaining linear connectives, to the usual sequent calculus for classical linear logic

$$\frac{}{\phi \vdash \phi}\ Ax \qquad \frac{\Gamma_0 \vdash \phi, \Delta_0 \quad \Gamma_1, \phi \vdash \Delta_1}{\Gamma_0, \Gamma_1 \vdash \Delta_0, \Delta_1}\ Cut$$

$$\frac{}{\Gamma, \bot \vdash \Delta}\ \bot L$$

$$\frac{\Gamma_0 \vdash \Delta_0, \phi \quad \Gamma_1, \psi \vdash \Delta_1}{\Gamma_0, \Gamma_1, \phi \to \psi \vdash \Delta_0, \Delta_1} \to L \qquad \frac{\Gamma, \phi \vdash \psi, \Delta}{\Gamma \vdash \phi \to \psi, \Delta} \to R$$

$$\frac{\Gamma \vdash \Delta}{\Gamma, \phi \vdash \Delta}\ WL \qquad \frac{\Gamma \vdash \Delta}{\Gamma \vdash \phi, \Delta}\ WR$$

$$\frac{\Gamma, \phi, \phi \vdash \Delta}{\Gamma, \phi \vdash \Delta}\ CL \qquad \frac{\Gamma \vdash \phi, \phi, \Delta}{\Gamma \vdash \phi, \Delta}\ CR$$

Table 7.6. The $\mathbf{CL}_{\text{if}}$ Sequent Calculus

[Girard, 1987, Trolestra, 1992].[3] Semantically, however, the additive implication, $\rightsquigarrow$, is perhaps somewhat deficient: The usual, and highly desirable, residuation (adjunction) between conjunction and implication is broken. So, just as for conjunction and disjunction, linear logic's treatment of additivity, lacking bunches, is problematic.

Another, perhaps more technical, way to understand this situation is via the negations. Both negations are indeed dualizing, *i.e.*,

$$(\phi \multimap \mathbf{0}) \multimap \mathbf{0} \quad \text{iff} \quad \phi \quad \text{iff} \quad (\phi \rightsquigarrow \bot) \rightsquigarrow \bot,$$

but they are *not* distinct, *i.e.*,

$$\phi \multimap \mathbf{0} \quad \text{iff} \quad \phi \rightsquigarrow \bot.$$

Since $\rightsquigarrow$ is not classical implication, it is natural to ask whether we can define a similar classical implication. To this end, consider the calculus $\mathbf{CL}_{\text{if}}$, given in Table 7.6.

$\mathbf{CL}_{\text{if}}$ describes classical implication. Now, consider the calculus $\mathbf{CL}^*_{\text{if}}$, given in Table 7.7, which is obtained from $\mathbf{CL}_{\text{if}}$ by dropping Weakening and Contraction and, correspondingly, adding $\bot R$.

[3] Note that a similar single-conclusioned calculus could be described, thereby facilitating a similar analysis for intuitionistic linear logic.

$$\frac{}{\phi \vdash \phi}\ Ax \qquad \frac{\Gamma_0 \vdash \phi, \Delta_0 \quad \Gamma_1, \phi \vdash \Delta_1}{\Gamma_0, \Gamma_1 \vdash \Delta_0, \Delta_1}\ Cut$$

$$\frac{}{\Gamma, \bot \vdash \Delta}\ \bot L$$

$$\frac{\Gamma_0 \vdash \Delta_0, \phi \quad \Gamma_1, \psi \vdash \Delta_1}{\Gamma_0, \Gamma_1, \phi \to \psi \vdash \Delta_0, \Delta_1} \to L \qquad \frac{\Gamma, \phi \vdash \psi, \Delta}{\Gamma \vdash \phi \to \psi, \Delta} \to R$$

Table 7.7. The $\mathbf{CL}^*_{\mathrm{if}}$ Sequent Calculus

It follows that $\mathbf{CL}^*_{\mathrm{if}}$ is equivalent to $\mathbf{CL}_{\mathrm{if}}$ but does not admit Cut-elimination.

Troelstra explains that the derivability of the structural rules of Weakening and Contraction is related to the failure of Cut-elimination: the derivation of Weakening and Contraction uses Cut together with the fact that $\to$ is simultaneously both additive and multiplicative (also, the $\bot L$ (*ex falso quodlibet*) and $\bot R$ rules). By separating once again the additive and multiplicative rôles,

- $\to$ splits into $\rightsquigarrow$ and $\multimap$,
- $\bot$ splits into additive $\bot$ and $\mathbf{0}$,

and the calculus $\mathbf{CLL}_{\mathrm{if}}$ is recovered.

Chapter 8

BUNCHED LOGICAL RELATIONS

1. Introduction

In [Mitchell and Moggi, 1981], the authors give an elementary definition of Kripke λ-models, *i.e.*, Kripke models of the simply-typed λ-calculus. This calculus, together with conjunctive and disjunctive types, may be viewed as the additive fragment of $\alpha\lambda$. In this section, we present an elementary definition of *Kripke $\alpha\lambda$-models* which generalizes the elementary definition given in [Mitchell and Moggi, 1981].

Kripke λ-models interpret the additive implication and conjunction of $\alpha\lambda$-calculus but are unable to interpret their multiplicative counterparts non-trivially. Here, we present an account of Kripke $\alpha\lambda$-models, starting from the idea of Kripke semantics. We show that Kripke $\alpha\lambda$-models fit, cleanly, via presheaves, into the general semantic framework provided by DCCs and show how the idea of *logical relations* [Statman, 1985b, Plotkin, 1980, Mitchell and Moggi, 1981] may be extended to Kripke $\alpha\lambda$-models.

2. Kripke $\alpha\lambda$-models

In this section, we develop a simple account of Kripke models for $\alpha\lambda$. As usual, our view is inspired by our model of resources, though, of course, the models are mathematically independent of this view. For simplicity and brevity, we restrict our account to the $(\multimap, \rightarrow)$-fragment. It is a straightforward matter to recover $(I, *)$ and $(1, \times)$.

Kripke applicative structures for $\alpha\lambda$ are the elementary functional manifestations of Kripke models, providing a resource interpretation of the functions denoted by $\alpha\lambda$-terms. An application of terms of additive functional type conserves resources, whereas an application of a term of multiplicative functional type combines the resources required to support the function with resources required to support its argument. In the following definition, in addition to the usual transition maps between worlds, we take both additive and multiplicative application maps, $\mathbf{app}_{\rightarrow}^{-,-}$ and $\mathbf{app}_{-*}^{-,-}$:

DEFINITION 8.1 (KRIPKE APPLICATIVE STRUCTURES) *Let*

$$\mathcal{M} = (M, \cdot, e, \sqsubseteq)$$

be a preordered commutaive monoid. A Kripke Applicative Structure, *or KAS, is a triple*

$$\mathcal{A} = (\mathcal{M}, \{\Phi_m^{\psi}\}, App),$$

where App is the following set of functions on the sets Φ_m^{ψ}:

- *For each m, ϕ and ψ,*

$$\mathbf{app}_{\rightarrow}^{\phi,\psi}(m) : \Phi_m^{\phi \rightarrow \psi} \times \Phi_m^{\phi} \longrightarrow \Phi_m^{\psi}$$

- *For each m, n, ϕ and ψ,*

$$\mathbf{app}_{-*}^{\phi,\psi}(m, n) : \Phi_m^{\phi -* \psi} \times \Phi_n^{\phi} \longrightarrow \Phi_{m \cdot n}^{\psi}$$

- *For each m, m' and ϕ,*

$$i_{m \sqsubseteq m'}^{\phi} : \Phi_m^{\phi} \longrightarrow \Phi_{m'}^{\phi}$$

subject to the following conditions:

- *Each*

$$i_{m \sqsubseteq m}^{\phi} : \Phi_m^{\phi} \longrightarrow \Phi_m^{\phi} \tag{8.1}$$

  is the identity;

- *Transitions compose,*

$$i_{m \sqsubseteq m'}^{\phi} ; i_{m' \sqsubseteq m''}^{\phi} = i_{m \sqsubseteq m''}^{\phi}, \tag{8.2}$$

  for all $m \sqsubseteq m' \sqsubseteq m''$;

- *For each $f \in \Phi_m^{\phi \rightarrow \psi}$ and each $a \in \Phi_m^{\phi}$,*

$$i_{m' \sqsubseteq m}^{\psi}(\mathbf{app}_{\rightarrow}^{\phi,\psi}(m)(f, a)) = \mathbf{app}_{\rightarrow}^{\phi,\psi}(m)(i_{m' \sqsubseteq m}^{\phi \rightarrow \psi} f, i_{m' \sqsubseteq m}^{\phi} a),$$

i.e.,

$$\begin{array}{ccc} \Phi^{\phi\to\psi}_{m'} \times \Phi^{\phi}_{m'} & \xrightarrow{\ \mathbf{app}^{\phi,\psi}_{\to}(m')\ } & \Phi^{\psi}_{m'} \\ \uparrow {\scriptstyle i^{\phi\to\psi}_{m'\sqsubseteq m} \times i^{\phi}_{m'\sqsubseteq m}} & & \uparrow {\scriptstyle i^{\psi}_{m'\sqsubseteq m}} \\ \Phi^{\phi\to\psi}_{m} \times \Phi^{\phi}_{m} & \xrightarrow[\ \mathbf{app}^{\phi,\psi}_{\to}(m)\ ]{} & \Phi^{\psi}_{m} \end{array} \tag{8.3}$$

- *For each* $f \in \Phi^{\phi -\!\!* \psi}_{m}$ *and each* $a \in \Phi^{\phi}_{n}$,

$$i^{\psi}_{m'\cdot n'\sqsubseteq m\cdot n}(\mathbf{app}^{\phi,\psi}_{-\!*}(m)(n)(f,a)) = \mathbf{app}^{\phi,\psi}_{-\!*}(m)(n)(i^{\phi -\!* \psi}_{m'\sqsubseteq m}f, i^{\phi}_{n'\sqsubseteq n}a),$$

i.e.,

$$\begin{array}{ccc} \Phi^{\phi -\!* \psi}_{m'} \times \Phi^{\phi}_{n'} & \xrightarrow{\ \mathbf{app}^{\phi,\psi}_{-\!*}(m')(n')\ } & \Phi^{\psi}_{m'\cdot n'} \\ \uparrow {\scriptstyle i^{\phi -\!* \psi}_{m'\sqsubseteq m} \times i^{\phi}_{n'\sqsubseteq n}} & & \uparrow {\scriptstyle i^{\psi}_{m'\cdot n'\sqsubseteq m\cdot n}} \\ \Phi^{\phi -\!* \psi}_{m} \times \Phi^{\phi}_{n} & \xrightarrow[\ \mathbf{app}^{\phi,\psi}_{-\!*}(m)(n)\ ]{} & \Phi^{\psi}_{m\cdot n} \end{array} \tag{8.4}$$

□

We can also require the structures necessary to interpret $(1, \wedge)$ and $(I, *)$. The unit 1 is given by $1_m = \{*\}$ and the unit I is given by $I_m = \mathcal{M}[m, e]$. The interpretation of $\wedge$ uses products in **Set** and the interpretation of $*$ uses Day's construction, in $\mathbf{Set}^{\mathcal{M}^{op}}$, as in Definition 4.1.

DEFINITION 8.2 (ENVIRONMENTS) *An* environment η *for a Kripke Applicative Structure is a partial map from variables and worlds to elements of* Φ *such that if* $\eta x m \in \Phi^{\phi}_{m}$ *and* $m' \sqsubseteq m$, *then* $\eta x m' = i^{\phi}_{m'\sqsubseteq m}(\eta x m)$. *Environments may be understood categorically, in* $\mathbf{Set}^{\mathcal{M}^{op}}$. *Let* $D \in obj([\mathcal{M}^{op}, \mathbf{Set}])$ *be a* domain of individuals. *We define the* interpretation functor with respect to D, $[\![\Gamma]\!]^{D}$, *for the bunch* Γ, *as follows:*

$$\begin{array}{lcllcl} [\![x \in A]\!]^{D} & = & D & & & \\ [\![\emptyset_m]\!]^{D} & = & I & [\![\Gamma, \Delta]\!]^{D} & = & [\![\Gamma]\!]^{D} \otimes [\![\Delta]\!]^{D} \\ [\![\emptyset_a]\!]^{D} & = & 1 & [\![\Gamma; \Delta]\!]^{D} & = & [\![\Gamma]\!]^{D} \times [\![\Delta]\!]^{D} \end{array}$$

where $[\![-]\!]^D$ *is a partial function from bunches to* $obj([\mathcal{M}^{op}, \mathbf{Set}])$. *Here,* $\otimes$ *is Day's tensor product and* $\times$ *is cartesian product of functors. As a result, each* $[\![\Gamma]\!]^D$ *is a functor, so that when* $n \sqsubseteq m$ *and* $\eta \in [\![\Gamma]\!]^D(m)$ *we obtain an element* $[\![\Gamma]\!]^D(n \sqsubseteq m)(\eta) \in [\![\Gamma]\!]^D(n)$. *We say that* $\eta \in [\![\Gamma]\!]^D(m)$ *is an* environment for Γ with respect to D at m *and (abusing notation) write* η_x *to denote the binding for the variable* x *in* Γ *determined by* η. □

If η is an environment and $p \in \Phi^{\phi}_{m}$, we write $\eta[p/x]$ for the environment which is identical to η on all variables other than x and is such that $(\eta[p/x])xm' = i^{\phi}_{m \sqsubseteq m'}p$. We will often write η_m to denote an environment, for whatever variables are determined by the context of the notation, for an "environment η at a world m". Where these associations are evident, we refrain from giving the details.

DEFINITION 8.3 (FORCING) *If* η *is an environment for a KAS* Φ *and if* Γ *is a bunch, then we define* $m \models (\Gamma)[\eta]$, *or* η *satisfies* Γ *at* m, *by induction on the structure of bunches, as follows:*

1 $m \models \emptyset_m[\eta]$ *for all* m *and* η;

2 $m \models \emptyset_a[\eta]$ *iff* $e \sqsubseteq m$, *for all* η;

3 $m \models (x : \phi)[\eta]$ *iff* $\eta x m \in \Phi^{\phi}_{m}$;

4 $m \models (\Gamma, \Delta)[\eta_m]$ *iff there exist* n, n', *where* $m \sqsubseteq n \cdot n'$ *and* $\eta_m = [\eta_n, \eta_{n'}]$, *such that*

$$n \models (\Gamma)[\eta_n] \quad \textit{and} \quad n' \models (\Delta)[\eta_{n'}];$$

5 $m \models (\Gamma; \Delta)[\eta]$ *iff*

$$m \models (\Gamma)[\eta] \quad \textit{and} \quad m \models (\Delta)[\eta].$$

Here, as usual, $[-,-]$ *is Day's pairing operation.* □

We now define the interpretation of $\alpha\lambda$-terms in a KAS. Since we have not required KASs to have combinators, we must impose, in cases (3) and (5) of Definition 8.4, the *environment model condition* [Meyer, 1982], to the effect that the interpretation $[\![-]\!]$ be defined only if suitable elements exist uniquely. In categorical models, this condition is enforced by requiring that the function spaces be given by right adjoints to their respective products.

DEFINITION 8.4 (INTERPRETATION OF TERMS-IN-BUNCHES) *The interpretation in a KAS of terms-in-bunches, with respect to an environment η, is defined by induction on the structure of terms as follows:*

1 *Variables:*

$$[\![\Gamma \vdash x : \phi]\!]\eta m = \begin{cases} \eta x m & \text{if } \Gamma \equiv \Delta\,;\, x : \phi \text{ (some } \Delta) \\ \uparrow & \text{otherwise;} \end{cases}$$

2 *Additive applications:*

$$\begin{aligned} &[\![\Gamma \vdash \mathbf{app}_{\rightarrow}(M, N) : \psi]\!]\eta m \\ &= \\ &\mathbf{app}^{\phi,\psi}_{\rightarrow}(m)([\![\Delta \vdash M : \phi \rightarrow \psi]\!]\eta m, [\![\Delta' \vdash N : \phi]\!]\eta m), \end{aligned}$$

where $\Gamma \equiv \Delta\,;\,\Delta'$;

3 *Additive abstractions:*

$$[\![\Gamma \vdash \alpha x : \phi.M : \phi \rightarrow \psi]\!]\eta m =$$

$$\begin{cases} p \in \Phi^{\phi \rightarrow \psi}_{m} \text{ such that for all } q \in \Phi^{\phi}_{m'} \text{ and } m' \sqsubseteq m, & \\ \mathbf{app}^{\phi,\psi}_{\rightarrow}(m)(i^{\phi \rightarrow \psi}_{m' \sqsubseteq m} p, q) = [\![\Gamma; x : \phi \vdash M : \psi]\!]\eta[q/x]m' & \text{if there is a unique such } p \\ \uparrow & \text{otherwise;} \end{cases}$$

4 *Multiplicative applications:*

$$\begin{aligned} &[\![\Gamma \vdash \mathbf{app}_{-*}(M, N) : \psi]\!]\eta_m m \\ &= \\ &\mathbf{app}^{\phi,\psi}_{-*}(n)(n')([\![\Delta \vdash M : \phi -\!\!* \psi]\!]\eta_n n, [\![\Delta' \vdash N : \phi]\!]\eta_{n'} n'), \end{aligned}$$

where $\Gamma \equiv \Delta\,,\,\Delta'$, $m \sqsubseteq n \cdot n'$ *and* $\eta_m = [\eta_n, \eta_{n'}]$;

5 *Multiplicative abstractions:*

$$[\![\Gamma \vdash \lambda x : \phi.M : \phi -\!\!* \psi]\!]\eta_m m =$$

$$\begin{cases} p \in \Phi^{\phi -* \psi}_{m} \text{ such that} & \\ \text{for all } q \in \Phi^{\phi}_{n} \text{ and all } \eta_n, & \\ \mathbf{app}^{\phi,\psi}_{-*}(m)(n)(p, q) = & \\ [\![\Gamma, x : \phi \vdash M : \psi]\!][\eta_m, \eta_n[q/x]](m \cdot n) & \text{if there is a unique such } p \\ \uparrow & \text{otherwise.} \end{cases}$$

□

Definition 8.4 may be extended to the conjunctions straightforwardly. For example, the units go as follows:

- $[\![\emptyset_m \vdash I : I]\!]\eta m = m \sqsubseteq e$;
- $[\![\emptyset_a \vdash 1 : 1]\!]\eta m = *$.[1]

DEFINITION 8.5 (KRIPKE $\alpha\lambda$-MODELS) *A* Kripke $\alpha\lambda$-model *is a pair*

$$\langle \mathcal{A}, [\![-]\!]\eta \rangle$$

in which $\mathcal{A}$ is a KAS and $[\![-]\!]\eta$ is an interpretation in $\mathcal{A}$ of terms-in-bunches with respect to envionment η. □

Where no confusion is likely, we shall refer to a Kripke $\alpha\lambda$-model, $\mathcal{A}$.

LEMMA 8.6 (TRANSITION) *Let $\mathcal{M}$ be a Kripke $\alpha\lambda$-model in which η satisfies Γ at m. Then, for every well-typed term $\Gamma \vdash M : \phi$ and every $m' \sqsubseteq m$,*

$$[\![\Gamma \vdash M : \phi]\!]\eta m' = i^{\phi}_{m' \sqsubseteq m}([\![\Gamma \vdash M : \phi]\!]\eta m).$$

□

LEMMA 8.7 (SUBSTITUTION) *Let $\mathcal{M}$ be a Kripke $\alpha\lambda$-model in which η satisfies Γ at m.*

1 For all well-typed terms $\Gamma; x : \phi \vdash M : \psi$ and $\Delta \vdash N : \phi$,

$$[\![\Gamma; \Delta \vdash M[N/x] : \psi]\!]\eta m = [\![\Gamma; x : \phi \vdash M : \psi]\!](\eta[[\![\Delta \vdash N : \phi]\!]\eta m/x])m.$$

2 For all well-typed terms $\Gamma, x : \phi \vdash M : \psi$ and $\Delta \vdash N : \phi$,

$$[\![\Gamma, \Delta \vdash M[N/x] : \psi]\!][\eta_m, \eta_n](m \cdot n) \quad = \quad [\![\Gamma, x : \phi \vdash M : \psi]\!]([\eta_m, \eta_x][[\![\Delta \vdash N : \phi]\!]\eta_n n/x])m \cdot n,$$

where $[\eta_m, \eta_x]$ satisfies $\Gamma, x : \phi$ at m and η_n satisfies Δ at n.

□

Lemma 8.7 extends to substitution at arbitrary depth via the evident argument.

DEFINITION 8.8 (FORCING) *We extend satisfaction to well-typed terms and equations as follows:*

[1] Here $*$ denotes the evident unique map.

1 $\eta, m \models (M : \phi)[\Gamma]$ *iff* $m \models (\Gamma)[\eta]$ *and* $[\![\Gamma \vdash M : \phi]\!]\eta m \downarrow$;

2 $\eta, m \models (M = M' : \phi)[\Gamma]$ *iff*

$$m \models (\Gamma)[\eta] \quad \textit{and} \quad [\![\Gamma \vdash M : \phi]\!]\eta m \downarrow = [\![\Gamma \vdash M' : \phi]\!]\eta m \downarrow .$$

The next lemma establishes the familiar structural characterization of satisfaction. Its proof is a straightforward argument by induction on the structure of types and is omitted.

LEMMA 8.9 *The satisfaction of well-typed* $\alpha\lambda$*-terms and equations may be characterized by induction on the structure of types as follows:*

1 $\eta, m \models (M : \mathrm{p})[\Gamma]$ *iff* $[\![\Gamma \vdash M : \mathrm{p}]\!]\eta m \in \Phi^{\mathrm{p}}_{m}$;

2 $\eta, m \models (M : \phi \multimap \psi)[\Gamma]$ *iff, for all* η_m *and and all* n,

$$\eta_n, n \models (N : \phi)[\Delta] \quad \textit{implies} \quad [\eta_m, \eta_n], m \cdot n \models (MN : \psi)[\Gamma, \Delta];$$

3 $\eta, m \models (M : \phi \to \psi)[\Gamma]$ *iff for all* $n \sqsubseteq m$,

$$\eta, n \models (N : \phi)[\Gamma] \quad \textit{implies} \quad \eta, n \models (MN : \psi)[\Gamma].$$

□

If we were to extend our Kripke $\alpha\lambda$-models to include $*$- and $\wedge$-types, we should obtain the following clauses:

- $\eta, m \models (I : I)[\emptyset_m]$ for all $m \sqsubseteq e$;
- $\eta_m, m \models (M : \phi * \psi)[\Theta]$ iff there exist Γ, n, η_n and N, and Δ, n', $\eta_{n'}$ and N', such that $m \sqsubseteq n \cdot n'$, $\eta_m = [\eta_n, \eta_{n'}]$, $\Theta \vdash M = N * N' : \phi * \psi$, $\Theta \equiv \Gamma, \Delta$ and

$$\eta_n, n \models (N : \phi)[\Gamma] \quad \text{and} \quad \eta_{n'}, n' \models (N' : \psi)[\Delta];$$

- $\eta, m \models (1 : 1)[\emptyset_a]$ for all m;
- $\eta, m \models (M : \phi \wedge \psi)[\Gamma]$ iff, for $N = \pi_1(M)$ and $N' = \pi_2(M)$,

$$\eta, m \models (N : \phi)[\Gamma] \quad \text{and} \quad \eta, m \models (N' : \psi)[\Gamma].$$

It should be clear from our definitions that it is now a straightforward matter to establish soundness by induction on the structure of proofs in $\alpha\lambda$. We omit the details of the proof.

PROPOSITION 8.10 (SOUNDNESS) *If* $\Gamma \vdash M = M' : \phi$ *in* $\alpha\lambda$*, then, in any Kripke* $\alpha\lambda$*-model,* $[\![\Gamma \vdash M : \phi]\!] \simeq [\![\Gamma \vdash M' : \phi]\!]$. □

The completeness theorem relies on the construction of a term model. In fact, the model we use is a simpler version of the one we constructed in the proof of model existence for propostional **BI** (Lemma 4.6). We can avoid the need to construct prime bunches because we have avoided $\vee$, $\wedge$ and $*$ (and their units). Accordingly, we just sketch the proof here.

LEMMA 8.11 (MODEL EXISTENCE) *There is a Kripke* $\alpha\lambda$*-model,*

$$\langle \mathcal{M}, [\![-]\!]\eta \rangle,$$

with a world m *such that if* $\Gamma \vdash M = M' : \phi$ *is not provable in* $\alpha\lambda$*, then* $m \models (\Gamma)[\eta]$ *and* $\eta, m \not\models (M = M' : \phi)[\Gamma]$.

PROOF SKETCH The KAS $\mathcal{M}$ is defined as follows:

- Take the preordered commutative monoid $\mathcal{T} = (T, \cdot, e, \sqsubseteq)$ to be the monoid of bunches defined as follows:
  - T is $B/\equiv$, where B is the set of bunches and $\equiv$ is coherent equivalence;
  - $\cdot$ is combination of bunches using the comma, ",";
  - e is $\emptyset_m$; and
  - $\sqsubseteq$ is extension of bunches by semi-colon, ";";
- Each Φ^{Γ}_{ϕ} is defined as follows:

$$\Phi^{\Gamma}_{\phi} = \{\, M \mid M \text{ is a normal } \alpha\lambda\text{-term such that } \Gamma \vdash M : \phi \,\};$$

- *App* is defined as follows:
  - $i^{\phi}_{\Gamma' \sqsubseteq \Gamma}$ is inclusion;
  - The function

$$\mathbf{app}^{\phi,\psi}_{-*}(\Gamma, \Delta) : \Phi^{\phi -* \psi}_{\Gamma} \times \Phi^{\phi}_{\Delta} \longrightarrow \Phi^{\psi}_{\Gamma\,,\,\Delta}$$

  is given by $-*$E and the function $\mathbf{app}^{\phi,\psi}_{\rightarrow}(\Gamma)$ is given similarly by $\rightarrow$E.

The interpretation $[\![-]\!]\eta$ takes each term M to its $\beta\eta$-equivalence class, *etc.*.

The remainder of the argument is a routine structural induction, similar to that sketched in the proof of Lemma 4.6. □

We write $\mathcal{A} \models (M = M' : \phi)[\Gamma]$ if, for every world m in the Kripke $\alpha\lambda$-model $\langle \mathcal{A}, [\![-]\!]\eta \rangle$, we have $\eta, m \models (M = M' : \phi)[\Gamma]$ and write $\Gamma \models M = M' : \phi$ if, for every Kripke $\alpha\lambda$-model $\langle \mathcal{A}, [\![-]\!]\eta \rangle$, $\mathcal{A} \models (M = M' : \phi)[\Gamma]$. We can now have the following completeness result:

PROPOSITION 8.12 (COMPLETENESS)

$$\Gamma \vdash M = M' : \phi \quad \textit{in } \alpha\lambda \textit{ iff} \quad \Gamma \models M = M' : \phi.$$

□

2.1 Kripke $\alpha\lambda$-models and DCCs

It is a straightforward matter to extend our set-theoretic definition of Kripke $\alpha\lambda$-models to include $\wedge$ and $*$ and their respective units. We show that any Kripke $\alpha\lambda$-model $\langle \mathcal{A}, [\![-]\!]\eta \rangle$, so extended with $\wedge$ and $*$, gives rise to a categorical model of **BI**, viewed as $\alpha\lambda$, in a cartesian DCC.

Starting with $\langle \mathcal{A}, [\![-]\!]_{\mathcal{A}}\eta \rangle$, we construct a DCC model, $\langle \mathcal{D}_{\mathcal{A}}, [\![-]\!]_{\mathcal{D}_{\mathcal{A}}} \rangle$. As usual, we regard a preordered commutative monoid $\mathcal{M} = (M, \cdot, e, \sqsubseteq)$ as a category in the usual way, with an arrow from m to n if and only if $n \sqsubseteq m$.

Each proposition ψ determines a functor $\Phi^{\phi} : \mathcal{M}^{op} \longrightarrow \mathit{Set}$:

- $\Phi^{\phi}(m) = \Phi^{\phi}_{m}$;
- $\Phi^{\phi}(n \sqsubseteq m) = i^{\phi}_{m \sqsubseteq n}$.

The functoriality of this definition follows from the definition of KASs.

Mitchell and Moggi [Mitchell and Moggi, 1981], working in the setting of models of the simply-typed λ-calculus in CCCs, have pointed out that even if $\phi \neq \psi$, it may be that $\Phi^{\phi}(m) = \Phi^{\psi}(m)$, with the undesirable consequence that the application functions on distinct types would be identified. Consequently, rather than take the objects of $\mathcal{A}$ to be the functors Φ^{ϕ}, we consider instead the objects to be the propositions ϕ. An arrow from ϕ to ψ is a natural transformation $\Phi^{\phi} \Longrightarrow \Phi^{\psi}$. Conditions (8.3) and (8.4) ensure that the application maps are natural transformations. Definition 8.4 ensures that the denotations of $\alpha\lambda$-terms are natural transformations. The construction is completed by taking the hom set $\mathcal{A}(\phi, \psi)$ to be those natural transformations definable, via $[\![-]\!]$, in $\alpha\lambda$ (*i.e.*, we impose the environment model condition, *q.v.* Definition 8.4, [Meyer, 1982]). A similar discussion, restricted to models of the simply-typed λ-calculus in CCCs but placed in a more general context, may be found in [Hermida, 1993]. It is now a routine matter to show that we have a model of $\alpha\lambda$ in a cartesian DCC. Checking the cartesian closed structure proceeds exactly as in [Lambek and Scott, 1986, Mitchell and Moggi, 1981]. Checking the SMCC structure involves verifying that the definition of $\mathbf{app}^{\phi,\psi}_{-*}(m,n)$ is Day's construction of the right adjoint to $\otimes$ in $\mathbf{Set}^{\mathcal{M}^{op}}$.

Conversely, let $\mathcal{D}$ be a small cartesian DCC and let $\langle \mathcal{D}, [\![-]\!]_{\mathcal{D}} \rangle$ be a model of $\alpha\lambda$ in $\mathcal{D}$. We show that $\mathcal{D}$ determines a Kripke $\alpha\lambda$-model, summarizing the key points in the construction.

- That the cartesian closed structure of $\mathcal{D}$ lifts to the functor category $\mathbf{Set}^{\mathcal{D}^{op}}$ via the Yoneda embedding is familiar from, for example, [Lambek and Scott, 1986]. From Proposition 3.15, we have also that Yoneda preserves SMCC structure. So $\mathbf{Set}^{\mathcal{D}^{op}}$ is a (bi-)cartesian DCC.

- The interpretation $[\![-]\!]_{\mathcal{D}}$ of $\alpha\lambda$-terms lifts directly to an interpretation $[\![-]\!]_{\mathbf{Set}^{\mathcal{D}^{op}}}$ and this interpretation determines a KAS:

  - ϕ is interpreted as a functor $\Phi^{\phi} : \mathcal{D}^{op} \longrightarrow \mathbf{Set}$;
  - $\mathbf{app}^{\phi,\psi}_{\to}(d)$ is interpreted as the evaluation map from $\Phi^{\phi\to\psi}(d) \times \Phi^{\phi}(d)$ to $\Phi^{\psi}(d)$ (and correspondingly on arrows);
  - $\mathbf{app}^{\phi,\psi}_{-*}(d,d')$ is interpreted as the evaluation map from $\Phi^{\phi -* \psi}(d) \otimes \Phi^{\phi}(d')$ to $\Phi^{\psi}(d \otimes d')$ (and correspondingly on arrows).

- Finally, we must recover an interpretation in a presheaf category of the form $\mathbf{Set}^{\mathcal{M}^{op}}$, where $\mathcal{M}$ is a Kripke Monoid. We define a model in $\mathbf{Set}^{\mathcal{M}^{op}}$. The objects of $\mathcal{M}$ are the objects of $\mathcal{D}$ and $n \sqsubseteq m$ in $\mathcal{M}$ just in case $\mathcal{D}[m,n]$ is non-empty. We now force the applicative structure defined on $\mathbf{Set}^{\mathcal{M}^{op}}$ to be equivalent to that in $\mathbf{Set}^{\mathcal{D}^{op}}$, *i.e.*, that $\eta, m \models_{\mathbf{Set}^{\mathcal{M}^{op}}} (M : \phi)[\Gamma]$ iff $[\![\Gamma \vdash M : \phi]\!]_{\mathbf{Set}^{\mathcal{M}^{op}}} \eta m \in \mathcal{D}[[\![\Gamma]\!]_{\mathcal{D}}, [\![\phi]\!]_{\mathcal{D}}]$ is the categorical interpretation of M in the DCC $\mathcal{D}$, as defined in Chapter 3, by induction on the structure of $\alpha\lambda$-terms.[2]

3. Bunched Kripke Logical Relations

In the model theory of the classical λ-calculus, *logical relations* have proved a useful tool. For example, Plotkin [Plotkin, 1980] and Statman [Statman, 1982, Statman, 1985b, Statman, 1985a] have used them to

[2] We note that our analysis here is weaker than the corresponding one, for the simply-typed λ-calculus, obtained in [Mitchell and Moggi, 1981]. Using the Diaconescu cover [Johnstone, 1980], [Mitchell and Moggi, 1981] gives a functorial characterization of the construction of a poset of worlds from an arbitrary CCC and proceeds to establish that the applicative structure so obtained is isomorphic to the one in the arbitrary CCC.

characterize the λ-definable elements of certain models, and they may be used to establish the completeness of $\beta\eta$-conversion for certain models. In [Mitchell and Moggi, 1981], Mitchell and Moggi showed how to generalize Plotkin's *I-relations* [Plotkin, 1980] to Kripke λ-models and used them to analyse extensionality in the λ-calculus.

We give a basic account of *logical relations* for Kripke $\alpha\lambda$-models.[3] For brevity and simplicity, we shall continue to work with just the two function spaces, $-\!\!*$ and $\to$, eliding any consideration of the conjunctions or the disjunction.

DEFINITION 8.13 (BKLRS) *Let*

$$\mathcal{A} = (\mathcal{M}, \{\Phi^{\psi}_{m}\}, App) \quad \text{and} \quad \mathcal{B} = (\mathcal{M}, \{\Psi^{\psi}_{m}\}, App)$$

be Kripke Applicative Structures. A Bunched Kripke Logical Relation *(BKLR) $\mathcal{R}$, over $\mathcal{A}$ and $\mathcal{B}$ is a family of relations*

$$\mathcal{R}^{\phi}_{m} \subseteq \Phi^{\phi}_{m} \times \Psi^{\phi}_{m}$$

satisfying the following conditions:

1 *Monotonicity: if $\mathcal{R}^{\phi}_{m}(p,q)$, then, for every $m' \sqsubseteq m$,*

$$\mathcal{R}^{\psi}_{m'}(i^{\phi}_{m' \sqsubseteq m}(p), i^{\phi}_{m' \sqsubseteq m}(q));$$

2 *Additive comprehension: $\mathcal{R}^{\phi \to \psi}_{m}(f,g)$ iff, for all $m' \sqsubseteq m$ and for all $p, q \in \Phi^{\phi}_{m'}$,*

$$\mathcal{R}^{\phi}_{m}(p,q) \quad \text{implies} \quad \mathcal{R}^{\phi}_{m'}((i^{\phi}_{m' \sqsubseteq m} f)p, (i^{\phi}_{m' \sqsubseteq m} g)q);$$

3 *Multiplicative comprehension: $\mathcal{R}^{\phi -\!* \psi}_{m}(f,g)$ iff, for all n and for all $p, q \in \Phi^{\phi}_{n}$,*

$$\mathcal{R}^{\phi}_{n}(p,q) \quad \text{implies} \quad \mathcal{R}^{\psi}_{m \cdot n}(f(p), g(q)).$$

LEMMA 8.14 (KRIPKE MONOTONICITY) *Let $\mathcal{R}$ be a BKLR over $\mathcal{A}$ and $\mathcal{B}$. For every ϕ and every $m' \sqsubseteq m$,*

$$\mathcal{R}^{\phi}_{m}(p,q) \quad \text{implies} \quad \mathcal{R}^{\phi}_{m'}(i^{\phi}_{m' \sqsubseteq m} p, i^{\phi}_{m' \sqsubseteq m} q)$$

PROOF SKETCH By induction on the structure of ϕ. □

[3] A more general account, for models in arbitrary DCCs, is beyond out present scope. It would seem to require techniques similar to those developed in [Hermida, 1993].

Let $\eta_{\mathcal{A}}$ and $\eta_{\mathcal{B}}$ be environments for $\mathcal{A}$ and $\mathcal{B}$, respectively. We say that $\eta_{\mathcal{A}}$ and $\eta_{\mathcal{B}}$ are *related by $\mathcal{R}$ on Γ at m* if $\mathcal{R}^{\phi}_{m}(\eta_{\mathcal{A}}\,x\,m, \eta_{\mathcal{B}}\,x\,m)$, for all $\Gamma(x : \phi)$.

LEMMA 8.15 (BASIC LEMMA) *Let $\mathcal{R}$ be a BKLR over $\mathcal{A}$ and $\mathcal{B}$ and let $\eta_{\mathcal{A}}$ and $\eta_{\mathcal{B}}$ be environments for $\mathcal{A}$ and $\mathcal{B}$, respectively, which are related by $\mathcal{R}$ on Γ at m. For every $\Gamma \vdash M : \phi$, provable in $\alpha\lambda$, if $[\![\Gamma \vdash M : \phi]\!]\eta_{\mathcal{A}}$ and $[\![\Gamma \vdash M : \phi]\!]\eta_{\mathcal{B}}$ are defined, then*

$$\mathcal{R}^{\phi}_{m}([\![\Gamma \vdash M : \phi]\!]\eta_{\mathcal{A}}\,m, [\![\Gamma \vdash M : \phi]\!]\eta_{\mathcal{B}}\,m).$$

PROOF As for the corresponding result in [Plotkin, 1980], the proof proceeds by induction on the structure of the proof of $\Gamma \vdash M : \phi$. We give illustrative cases, as follows:

- For variables, we are in case $\Delta\,;\, x : \phi \vdash x : \phi$, and the required result is immediate;
- For additive abstraction, suppose that $\Gamma \vdash \alpha x : \phi.M : \phi \to \psi$. By the induction hypothesis, we have that

  $$\mathcal{R}^{\psi}_{m}([\![\Gamma; x : \phi \vdash M : \psi]\!]\eta_{\mathcal{A}}\,m, [\![\Gamma\,;\, x : \phi \vdash M : \psi]\!]\eta_{\mathcal{B}}\,m)$$

  and that

  $$[\![\Gamma \vdash \alpha x : \phi.M : \phi \to \psi]\!]\eta_{\mathcal{A}} m =$$

  $$\begin{cases} \begin{array}{l} p \in \Phi^{\phi\to\psi}_{m} \text{ such that} \\ \text{for all } q \in \Phi^{\phi}_{m'} \text{ and } m' \sqsubseteq m, \\ \mathbf{app}^{\phi;\psi}_{\to}(m)(i^{\phi\to\psi}_{m'\sqsubseteq m}p, q) = \\ [\![\Gamma\,;\, x : \phi \vdash M : \psi]\!]\eta_{\mathcal{A}}[q/x]m' \end{array} & \text{if there is a unique such } p \\ \uparrow & \text{otherwise;} \end{cases}$$

  and

  $$[\![\Gamma \vdash \alpha x : \phi.M : \phi \to \psi]\!]\eta_{\mathcal{B}} m =$$

  $$\begin{cases} \begin{array}{l} p' \in \Phi^{\phi\to\psi}_{m} \text{ such that} \\ \text{for all } q' \in \Phi^{\phi}_{m'} \text{ and } m' \sqsubseteq m, \\ \mathbf{app}^{\phi;\psi}_{\to}(m)(i^{\phi\to\psi}_{m'\sqsubseteq m}p', q') = \\ [\![\Gamma\,;\, x : \phi \vdash M : \psi]\!]\eta_{\mathcal{B}}[q/x]m' \end{array} & \text{if there is a unique such } p' \\ \uparrow & \text{otherwise.} \end{cases}$$

It follows that we have that, if $\mathcal{R}^{\phi}_{m'}(q, q')$, then

$$\mathcal{R}^{\psi}_{m}(\mathbf{app}^{\phi;\psi}_{\rightarrow}(m)(i^{\phi\rightarrow\psi}_{m'\sqsubseteq m}p, q), \mathbf{app}^{\phi;\psi}_{\rightarrow}(m)(i^{\phi\rightarrow\psi}_{m'\sqsubseteq m}p', q'))$$

and so, by additive comprehension, that $\mathcal{R}^{\phi\rightarrow\psi}_{m}(p, p')$, as required.

- For multiplicative abstraction, suppose that $\Gamma \vdash \lambda x : \phi.M : \phi \mathrel{-\!\!*} \psi$. By the induction hypothesis, we have that

$$\mathcal{R}^{\psi}_{m}([\![\Gamma\,,\, x : \phi \vdash M : \psi]\!]\eta_{\mathcal{A}}\, m, [\![\Gamma\,,\, x : \phi \vdash M : \psi]\!]\eta_{\mathcal{B}}\, m)$$

and that

$$[\![\Gamma \vdash \lambda x : \phi.M : \phi \mathrel{-\!\!*} \psi]\!]\eta_{\mathcal{A}} m =$$

$$\begin{cases} p \in \Phi^{\phi \mathrel{-\!*} \psi}_{m} \text{ such that for all and all } \eta_{\mathcal{A}}\ q \in \Phi^{\phi}_{n} \\ \mathbf{app}^{\phi,\psi}_{-\!*}(m)(n)(p, q) = \\ [\![\Gamma\,,\, x : \phi \vdash M : \psi]\!][\eta_{\mathcal{A}}, \eta_{\mathcal{A}}[q/x]](m \cdot n) & \text{if there is a unique such } p \\ \uparrow & \text{otherwise;} \end{cases}$$

and

$$[\![\Gamma \vdash \lambda x : \phi.M : \phi \mathrel{-\!\!*} \psi]\!]\eta_{\mathcal{B}} m -$$

$$\begin{cases} p \in \Phi^{\phi \mathrel{-\!*} \psi}_{m} \text{ such that for all and all } \eta_{\mathcal{B}}\ q \in \Phi^{\phi}_{n} \\ \mathbf{app}^{\phi,\psi}_{-\!*}(m)(n)(p, q) = \\ [\![\Gamma\,,\, x : \phi \vdash M : \psi]\!][\eta_{\mathcal{B}}, \eta_{\mathcal{B}}[q/x]](m \cdot n) & \text{if there is a unique such } p \\ \uparrow & \text{otherwise.} \end{cases}$$

It follows that, if $\mathcal{R}^{\phi}_{n}(q, q')$, then

$$\mathcal{R}^{\psi}_{m \cdot n}(\mathbf{app}^{\phi,\psi}_{-\!*}(m)(n)(p, q), \mathbf{app}^{\phi,\psi}_{-\!*}(m)(n)(p', q'))$$

and so, by multiplicative comprehension, that $\mathcal{R}^{\phi \mathrel{-\!*} \psi}_{m}(p, p')$, as required.

The remaining cases are straightforward. □

These basic results provide the basis for a study of $\alpha\lambda$-definability and extensionality in Kripke $\alpha\lambda$-models.

Chapter 9

THE SHARING INTERPRETATION, I

1. Introduction

We conclude Part I with a chapter on computational interpretations of **BI**, which arise from the following models of computation:

- Proof-search and (propositional) logic programming;
- Interference and non-interference in imperative programming.

The first of these, discussed slightly in [O'Hearn and Pym, 1999], is an example of what is, perhaps, the most immediate and most basic form of computational interpretation of a logical system: the attempt to calculate proofs by treating the rules of the logic as a reductive system. In the propositional setting, the result of the computation is either failure, or success together with the proof which is calculated; there is no answer substitution to be calculated. In the setting of **BI**, we show that the (bunched) structure of the program determines which program clauses have access to which of the atomic assumptions declared in the program.

The second, also discussed in [O'Hearn and Pym, 1999] but developed at more length in [O'Hearn, 1999], is concerned with the sharing and non-sharing of memory by procedures in imperative programming languages of the kind described by Reynolds' Idealized Algol [Reynolds, 1981] or syntactic control of interference, or SCI. The basic idea is that procedures that are combined using multiplicative combination *cannot* share resources with each other and that procedures combined using additive combination *may*, though need not, share resources. The first reference to sharing/non-sharing of variables in this context, though without discussion of SCI or Idealized Algol, is in the work of Ishtiaq and Pym [Ishtiaq and Pym, 1998].

Both the logic and functional/imperative programming interpretations described above are based on (the semantics of) **BI**'s proofs. However, it should be noted that whilst the latter relies on *directly* upon the interpretation of $\alpha\lambda$-terms, the former, for a proper analysis, requires an interpretation not merely of proofs but rather of the larger class of *searches*. Such a semantics is beyond our present scope (see [Galmiche and Pym, 2000, Pym, 2001] for a discussion).

We go on to give three further examples, which are logical (truth-functional) models of propositional **BI** but not of the basic, (Heyting,Lambek) version, which is mainly studied herein, but rather of Boolean **BI**, as described in Chapter 7:

- Petri nets;
- A CCS-like model;
- A pointers model.

Each of these provides support for our interpretation of **BI**'s semantics as an account of resources and their computational properties.

2. Proof-search and (Propositional) Logic Programming

In the introduction to this monograph, we mentioned the idea of *proof-search*, the view of logic in which rule schemata are read not as derivation operators, from premisses to conclusion, but rather as reduction operators, from conclusion to premisses.

The shift in perspective from derivation to reduction has an important computational consequence. We start from a sequent $\Gamma \vdash \phi$, which may or may not be provable, and try to construct a proof of it. So we can view the execution of a search procedure on a given sequent, $\Gamma \vdash \phi$, as a computation which, upon successful termination, will return a proof Φ of $\Gamma \vdash \phi$.

The view of proof-search as the computation of a proof does not on its own, however, constitute an acceptable notion of *programming*. For programming, we require an *operational semantics* which is both independent of the particular program and which provides the programmer with as direct an intuition as possible about the relationship between the logical model of the application and the execution procedure.

In *logic programming*, based on intuitionistic logic, a suitable operational semantics is provided by the idea of *goal-directed* reduction, which may be summarized as follows:

- Fix a program, P;
- Given a complex goal, G, we first reduce G by applying, as a reduction operator, the introduction rule which corresponds to the outermost connective of G. This reduction process is repeated until all remaining goal formulæ are *atomic*. For example, given the goal $G_1 \wedge (G_2 \rightarrow G_3)$, we construct the tree

$$\dfrac{\dfrac{\vdots}{P \mathrel{?\text{-}} G_1} \qquad \dfrac{\dfrac{\vdots}{P; G_2 \mathrel{?\text{-}} G_3}}{P \mathrel{?\text{-}} G_2 \rightarrow G_3} \rightarrow I}{P \mathrel{?\text{-}} G_1 \wedge (G_2 \rightarrow G_3)} \wedge I.$$

  Here, as in the introduction, we write ?- to denote putative consequence. Note that upper rightmost step adds G_2 to the program;
- Given an atomic goal, A, we invoke the program, using a *resolution* step. Suppose the program includes a proposition of the form $G \rightarrow B$, in which B is atomic $B = A$. Then we can immediately proceed to the *sub-goal* G:

$$\frac{P \mathrel{?\text{-}} G}{P \mathrel{?\text{-}} A} \quad (G \rightarrow B \text{ in } \mathcal{P},\ B = A).$$

This operational semantics has several desirable features. Most importantly, perhaps, it is *not very non-deterministic*, thereby reducing the need for backtracking to an acceptable level. More details of this model of computation may be found in [Miller et al., 1991, Pym and Harland, 1994, Miller, 1981, Harland et al., 1996].

Proofs constructed according to this goal-directed strategy are called *uniform proofs* [Miller et al., 1991, Pym and Harland, 1994]. Uniform proofs are not complete for all of **BI** (or indeed for all of intuitionistic logic). For example, consider the following, trivially provable, sequent:

$$G \vee H \vdash G \vee H.$$

To see that there is no uniform proof of this sequent, consider the $\vee I$ rule:

$$\frac{\Gamma \vdash G}{\Gamma \vdash G \vee H} \qquad \frac{\Gamma \vdash H}{\Gamma \vdash G \vee H}.$$

Any attempt to reduce the right hand side of $G \vee H \vdash G \vee H$ first forces a premature choice between G and H leading, for example, to

$$G \vee H \vdash G$$

which is clearly not provable.

However, uniform proofs are complete for the *clausal hereditary Harrop fragment* of intuitionistic logic. The basic idea is to restrict the classes of propositions permitted to occur in each side of a sequent, *program clauses*, P, on the left and *goals*, G, on the right. These two classes (here, for simplicity, we make do with a restricted form) are defined by mutual induction as follows, where A ranges over atomic propositions:

$$\begin{array}{rrcl} \textit{Program clauses} & P & ::= & A \mid P \wedge P \mid G \to A \\[1em] \textit{Goals} & G & ::= & A \mid G \wedge G \mid G \vee G \mid P \to G. \end{array}$$

A sequent is said to be hereditary Harrop if it is of the form $\mathcal{P} \vdash G$, where $\mathcal{P}$ is a finite set of program clauses and G is a goal.

In **BI**, a similar class of hereditary Harrop formulæ supports goal-directed proof (we simplify, for brevity):

$$\begin{array}{rrcl} \textit{Program clauses} & P & ::= & I \mid A \mid P \wedge P \mid G \to A \mid \\ & & & P * P \mid G \mathrel{-\!\!*} A \\[1em] \textit{Goals} & G & ::= & \top \mid I \mid A \mid G \wedge G \mid G \vee G \mid P \to G \mid \\ & & & G * G \mid P \mathrel{-\!\!*} G. \end{array}$$

In intuitionistic logic, each of the reduction operators used in the execution of goal-directed search is additive. In **BI**, however, as in linear logic [Pym and Harland, 1994, Miller, 1981, Harland et al., 1996], we have multiplicatives which introduce a computationally significant difficulty. The typical case is $*R$:

$$\Uparrow \quad \frac{\Gamma_1 \text{ ?- } \phi_1 \quad \Gamma_2 \text{ ?- } \phi_2}{\Gamma \text{ ?- } \phi_1 * \phi_2} \quad \Gamma = \Gamma_1 \, , \, \Gamma_2.$$

Faced with $\Gamma \vdash \phi_1 * \phi_2$, the division of Γ into Γ_1 and Γ_2 must be calculated.

The basic solution, described for linear logic in [Pym and Harland, 1994, Miller, 1981, Harland et al., 1996], is the so-called *input/output* model. It goes as as follows:

- First pass all of Γ to the left hand branch of the proof, leaving the right hand branch undetermined, as follows:

$$\Uparrow \quad \frac{\Gamma \text{ ?- } \phi_1 \quad ? \text{ ?- } \phi_2}{\Gamma \text{ ?- } \phi_1 \otimes \phi_2};$$

- Proceed with the left hand branch until (recursively) it is completed. Then calculate which of the formulæ in Γ have been used to complete the left hand branch and collect them into a finite set, Γ_{left}; The remaining, unused formulæ can now be passed to the right hand branch of the proof:

$$\Uparrow \quad \frac{\dfrac{\vdots}{\Gamma_{\text{left}} \text{ ?- } \phi_1} \quad \Gamma\backslash\Gamma_{\text{left}} \text{ ?- } \phi_2}{\Gamma \text{ ?- } \phi_1 \otimes \phi_2} \quad \Gamma = \Gamma_{\text{left}}, \Gamma\backslash\Gamma_{\text{left}};$$

- Exponentiated formulæ, of the form $!\phi$, are copied to both the left hand and right hand branches.

In **BI**, the problem is made *much* more complex by the mixing of additive and multiplicative structure enforced by the presence of bunches and the basic input/output idea will not work. To see this, consider the following search for a proof of the (provable) sequent

$$\phi, \psi \vdash (\chi \to \psi \wedge \chi) * \phi$$

(note that it is convenient to put the remainder operator, read as "without", in the "current" computation):

$$\frac{\dfrac{\dfrac{}{(\chi; (\phi, \psi))\backslash\phi \text{ ?- } \psi \wedge \chi}\ (\text{not provable})}{(\phi, \psi)\backslash\phi \text{ ?- } \chi \to (\psi \wedge \chi)} \to R \qquad \dfrac{}{\phi\backslash\emptyset_m \text{ ?- } \phi}}{(\phi, \psi)\backslash\emptyset_m \text{ ?- } (\chi \to (\psi \wedge \chi)) * \phi} *R.$$

Consider the left hand branch of the $*R$. In order to get an axiom of the form $\psi \vdash \psi$, we must first remove the χ from the program by performing a Weakening and then perform a subtraction of ϕ, which is required on the right hand branch of the $*R$, so that the result of $\chi; (\phi, \psi)\backslash\phi$ is ψ, *i.e.*, the remainder operator first throws away, via Weakening, the additive bunch surrounding the multiplicative bunch within which ψ, the formula which must be removed, is contained. Now, the remaining bunch is sufficient to form, after an $\wedge R$ reduction, the axiom $\psi \vdash \psi$ but insufficient to form the necessary axiom for χ. Thus we have an incompleteness.

At first sight, thinking of axioms of the form $\Gamma;\alpha \vdash \alpha$, it might seem like we need a subtraction operation which does not perform Weakening. This would solve the problem in the particular case above but it would worsen it in other cases. From being incomplete the system would become unsound. To see this, consider a search for a proof of the (unprovable) sequent $\tau;(\phi,\psi) \vdash (\chi \to (\psi \wedge \tau)) * \phi$. Once the $*R$ rule is applied, after doing $\to R$, the propositions τ and χ are at the same level in $(\chi;\tau;(\phi,\psi))\backslash\phi$ and, if this is taken to be equal to $\chi;\tau;\psi$, both χ and τ are provable.

It may seem that this unsoundness might be fixed by requiring, for example, that the $*R$ rule be applied only on multiplicative bunches. Aside from the unpleasant non-determinism introduced by this solution, it doesn't quite solve the problem either. To see this, it is enough to consider a slight modification of the last search. Consider the unprovable $v,(\tau;(\phi,\psi)) \vdash (\tau \to (\psi \wedge \tau)) * (\phi * v)$. Here, after the application of $*R$ and $\to R$ it would be possible to prove both ψ and τ, which is unsound. It seems clear that the root of the difficulty lies with the interaction between the additives, in particular $\to R$, and the multiplicatives.

The essential idea for the solution, though far from simple in detail, is to introduce a system of stacks which keep a record of which resources have been added to the program as a result of $\to R$ and which manage their interaction with the formation of axioms and with subtraction and passing by continuations. The details are beyond our present scope, and are addressed, in collaboration with Pablo Armelín,, in [Armelín and Pym, 2001]. The essential idea for the operational semantics, with a great deal of technical simplification, is as follows:

- We implement the input/output model in a continuation-passing style;

- We maintain a record of the remainder or unused bunch during a search. For example, the operational $*R$ is

$$\frac{\Gamma\backslash\Delta' \text{ ?- } \phi \quad \Delta'\backslash\Delta \text{ ?- } \psi}{\Gamma\backslash\Delta \text{ ?- } \phi * \psi}.$$

  Here, as we search for a proof, we encounter the need to find a proof of $\phi * \psi$ from Γ with remainder Δ. Applying the rule, we must divide the *available* formulæ from Γ. To this end we introduce a remainder Δ', *i.e.*, Γ_{right}, which we try to calculate by continuing up the left hand branch. Having calculated Δ', we pass $\Delta'\backslash\Delta$ to the right hand branch;

$$
\cfrac{
  \cfrac{
    \cfrac{
      \cfrac{\cfrac{}{\phi \mathrel{?\text{-}} \phi}\,\mathit{Axiom}()}{\phi; (\chi, \psi)\backslash\Delta' \mathrel{?\text{-}} \phi}\,W
      \qquad
      \cfrac{\cfrac{\cfrac{}{\psi \mathrel{?\text{-}} \psi}\,\mathit{Axiom}()}{(\chi, \psi)\backslash\Delta' \mathrel{?\text{-}} \psi}\,\Delta' := \chi}{\phi; (\chi, \psi)\backslash\Delta' \mathrel{?\text{-}} \psi}\,W
    }{\phi; (\chi, \psi)\backslash\Delta' \mathrel{?\text{-}} \phi \wedge \psi}\,\wedge R
  }{(\chi, \psi)\backslash\Delta' \mathrel{?\text{-}} \phi \to (\phi \wedge \psi)}\,\to R
  \qquad
  \cfrac{\cfrac{}{\chi\backslash\Delta \mathrel{?\text{-}} \chi}\,\mathit{Axiom}(\Delta)}{\Delta'\backslash\Delta \mathrel{?\text{-}} \chi}\,\Delta' := \chi
}{(\chi, \psi)\backslash\Delta \mathrel{?\text{-}} (\phi \to (\phi \wedge \psi)) * \chi}\,{*R}
$$

Figure 9.1. A Search Tree

- This mechanism allows the management of the interaction between additive and multiplicative structure. For example, we can formulate the operational $\to R$ rule as

$$\frac{\phi; \Gamma\backslash\Delta \mathrel{?\text{-}} \psi}{\Gamma\backslash\Delta \mathrel{?\text{-}} \phi \to \psi}.$$

  Note that we keep the minor formula ϕ outside of the scope of the calculation of the remainder: ϕ must be used, or deleted via Weakening, on the branch above the occurrence of the $\to R$ rule which introduces it;

- Upon completion of a branch of the search rooted at a multiplicative rule, the remainder bunch is passed as a continuation to the next branch.

We illustrate the procedure with a concrete example. Suppose we are required to calculate a proof of

$$\chi, \psi \vdash (\phi \to (\phi \wedge \psi)) * \chi.$$

We sketch the construction of the search tree given in Figure 9.1, using an informal version of the necessary operational semantics.

We write $\mathit{Axiom}(\Delta)$ to indicate that we can form an axiom sequent provided we pass on the remainder Δ. We calculate assignments to unknown remainders as side-conditions to the formation of axioms. Formally, the interaction between the multiplicatives and the additives, at arbitrary depth within bunches, forces to work not merely with a remainder operator but with a *stack* of bunches to manage the interaction

$$\dfrac{\dfrac{\dfrac{\begin{matrix}\Delta' := \chi\\ \vdots\end{matrix}}{(\phi,\chi,\psi)\backslash\Delta' \mathrel{?\text{-}} \phi * \psi}\,*R}{(\chi,\psi)\backslash\Delta' \mathrel{?\text{-}} \phi \mathrel{-\!\!*} (\phi * \psi)}\,\mathrel{-\!\!*}R \qquad \dfrac{\dfrac{}{\chi\backslash\Delta \mathrel{?\text{-}} \chi}\,Axiom(\Delta)}{\Delta'\backslash\Delta \mathrel{?\text{-}} \chi}\,\Delta' := \chi}{(\chi,\psi)\backslash\Delta \mathrel{?\text{-}} (\phi \mathrel{-\!\!*} (\phi * \psi)) * \chi}\,*R$$

Figure 9.2. A Variation on the Search Tree

between the combination of additive and multiplicative rules and the available formulæ.[1]

- We are faced, at an intermediate stage in a computation (search), with a goal,

$$(\chi\,,\,\psi) \mathrel{?\text{-}} (\phi \to (\phi \wedge \psi)) * \chi\,,$$

  together with an existing remainder, Δ.

- We begin by applying a $*R$ reduction, at which point we must calculate a division of the program. So we introduce a new remainder, Δ', which we must calculate and which will be passed from the left hand branch to the right hand branch. Notice that we must maintain, in the (contingent) right hand branch, the original remainder Δ.

- Proceeding up the left hand branch, we apply an $\to R$ reduction. Strictly speaking, we should introduce a requirement that the ϕ be "consumed", by writing the program as

$$(\phi;\,(\chi\,,\,\psi)\backslash\Delta')\backslash\emptyset_a\,,$$

  but since we have used the additive implication, the ϕ could, if necessary, be removed by weakening at the leaves. This ensures that the ϕ cannot propagate, via a continuation, below the point at which it was introduced to the program. The corresponding case for $\mathrel{-\!\!*} R$ is handled slightly differently, *q.v.* Figure 9.2.

- We conclude by passing the remainder Δ back to the computation within which the branch above arose.

[1] In this example, we have omitted most of the details of the formal operational semantics. In particular, we have omitted all details of the stack of bunches used to manage the calculation of remainders. However, the essential features should be clear. The details may be found in [Armelín and Pym, 2001].

I	a unit goal, requiring $\emptyset_m$ resources
$G * G'$	a pair of goals which cannot share resources
$\top$	a unit goal, requiring $\emptyset_a$ resources
$G \wedge G'$	a pair of goals which may share resources
$G \vee G'$	a choice between two goals, with the same resources available for each choice
$P \mathrel{-\!\!*} G$	a hypothetical goal in which the hypothesis and conclusion cannot share resources
$P \rightarrow G$	a hypothetical goal in which the hypothesis and conclusion may share resources

Figure 9.3. The Sharing Interpretation for Logic Programming Goals

The formal operational semantics of logic programming with **BI**, together with its supporting theory, is described in [Armelín and Pym, 2001].

We can now give, in Figure 9.3, a *sharing interpretation* to our goal-directed operational semantics.

Here resources amount to the program clauses, including both implication clauses and atomic "facts". The clauses themselves may also be given a sharing interpretation, as in Figure 9.4.

We remark that the interpretation of hypothetical goals as modules [Miller, 1981] may be extended to both the sharing or non-sharing cases.

3. Interference in Imperative Programs

In this section, based on [O'Hearn and Pym, 1999, O'Hearn, 1999], we are concerned with imperative programming, based on functional programming notions such as pairs, functions, *etc.*, together with the characteristically imperative notion of *assignment*. Roughly speaking, assignment associates values to variables, or identifiers, which themselves

I	a unit program, yielding $\emptyset_m$ resources
$P * P'$	a pair of clauses which cannot share resources
$\top$	a unit program, yielding $\emptyset_a$ resources
$P \wedge P'$	a pair of clauses which may share resources
$G \mathrel{-\!\!*} A$	a hypothetical clause in which the hypothesis and conclusion cannot share resources
$G \to A$	a hypothetical clause in which the hypothesis and conclusion may share resources

Figure 9.4. The Sharing Interpretation for Logic Programming Clauses

correspond to distinct cells of the computer's memory, *e.g.*,

$$\begin{array}{rcl} x & := & 0; \\ y & := & 1; \end{array}$$

where x and y may be seen as (the addresses of) memory cells:

x	y	...

However, we are also concerned with an additional, and intensional, notion of resource which may be accessed during a computation. The resources are distributed, in the sense that a value or a procedure may have access to some or to none or to all of them.

We describe the interpretation in terms of the kinds of values appropriate to each connective, as follows: The additives $\perp$ and $\vee$ also receive evident interpretations, as empty and a coproduct where sharing is allowed between branches.[2]

[2]A non-sharing disjunction would, of course, amount to the multiplicative disjunction, #, introduced in Chapter 7 but note that our semantics for the logical system with # does not provide for a corresponding λ-calculus.

I	a single element, accessing no resources
$P * Q$	pairs which don't share resources
1	a single element, accessing no resources
$P \times Q$	pairs which may share resources
$P \mathrel{-\!\!*} Q$	procedures that don't share resources with arguments
$P \to Q$	procedures that may share resources with arguments

Figure 9.5. The Sharing Interpretation for Imperative Programming

Notice that this interpretation works as well for the affine variant of **BI**, in which Weakening is permitted for "," and in which $1 \equiv I$. The addition of Weakening is relatively harmless because, just as in linear logic, the control of Contraction is much more significant. It would be interesting, though, to devise a variant interpretation that does not lead to any collapse. In any case, we suggest that the interpretation gives an informal reading of connectives that may serve as a useful guide to provability (in the affine variant).

To illustrate how this informal reading of the connectives works, consider that the example of a proof-term $\lambda x \,.\, \alpha f \,.\, f\, x\, x$, in which a multiplicative assumption is used twice, is illuminated by this interpretation. To see this, consider that the subterm $f\, x$ in the proof-term is of type $E \to F$. By the sharing interpretation, it is allowed to share with its arguments, which is why $(f\, x)\, x$ is reasonable. There is no requirement that an argument to a $\mathrel{-\!\!*}$-typed function be used just once, only that it does not share with other variables in the proof-term. The kind of thing that would be disallowed by the sharing interpretation is a procedure call $(f\, x)\, x$, where f has type $E \mathrel{-\!\!*} E \mathrel{-\!\!*} F$.

This resource reading extends to the other connectives quite straightforwardly as summarized above.

The sharing interpretation in imperative programming is inspired by John Reynolds's work on Syntactic Control of Interference and Idealized

Algol ([Reynolds, 1978, Reynolds, 1981], and also the relevant [O'Hearn et al., 1999]). These are programming languages which use affine and intuitionistic λ-calculus respectively, together with imperative features such as the assignment statement $x := e$. The calculi give the "functional data" of the sharing interpretation, and the computer's store gives the "intensional" component. The sharing interpretation may receive illumination by stating it in these terms.

Contraction in λ-calculus gives rise to the phenomenon of *aliasing* in imperative languages, where two variables denote the same storage cell in the computer memory:

$$(\lambda x\,.\,(\lambda y\,.\,\cdots x := e \cdots y := f \cdots)z)z\,.$$

Here, in intuitionistic λ-calculus, a variable z denoting a storage cell may be passed twice. This has the effect of creating two aliases, x and y, for the same cell, so that assignment to x in the body will change y and *vice versa*. Contraction is what allows z to appear both in $\lambda x\,.(\lambda y\,.\,\cdots x := e \cdots y := f \cdots)z$ and the second z.

In our notation for proof-terms, we use λ for the abstraction corresponding to $-\!\!*$ in **BI**. A consequence of this is that a term of the above form would not be allowed, because $\lambda x\,.(\lambda y\,.\,\cdots x := e \cdots y := f \cdots)z$ would have to have a type of the form $E -\!\!* F$, and it could therefore not be applied to an argument containing z. Generally, whenever we have $\lambda x\,.\,\lambda y\,.\,M$ or even $\alpha x\,.\,\lambda y\,.\,M$ the x and y will not refer to overlapping parts of the store. In contrast, when we have $\alpha x\,.\,M : E \to F$ it is entirely possible for x to share with other variables appearing freely in M. Thus, the use of the two implications of **BI** allows us to control when different variables overlap in their access to the computer's store.

A complete definition of a programming language illustrating these ideas has been given in [O'Hearn, 1999]. Here, as in [O'Hearn and Pym, 1999], for brevity, we describe an example of a DCC, which helps to ground the discussion, and which is a precise counterpart to the informal sharing interpretation of connectives described above.

Let $\mathcal{I}$ be the category of finite sets and injective functions. We think of an object X here as a possible world which identifies a finite set of cells in a computer's memory. These worlds are used to capture the intensional part of the sharing interpretation, that part which refers to resource instead of merely to values.

We give a semantics of the connectives in which each **BI** proposition denotes a functor from $\mathcal{I}$ to **Set**: the valuation $E(X)$ of a functor of world X is thought of as a collection of values of type E, which may access the X-portion of the store. The (object parts of the) functors for implication illustrate the basic elements of the semantics:

$$(E \mathrel{-\!\!*} F)(X) \quad = \quad \mathbf{Set}^{\mathcal{I}}[E(-), F(X + -)]$$

$$(E \rightarrow F)(X) \quad = \quad \mathbf{Set}^{\mathcal{I}}[E(X + -), F(X + -)],$$

where $+$ is the evident functor on $\mathcal{I}$ given by disjoint union in **Set**. Notice how these clauses correspond directly to the sharing interpretation. For $\mathrel{-\!\!*}$, the absence of X in $E(-)$ indicates how a procedure (which lives in world X) and argument must access disjoint sets of cells. For $\rightarrow$, the presence of X in $E(X + -)$ indicates how a procedure may share store with its argument.

We can immediately relate these definitions back to the imperative programming examples above by defining a functor *cell*, which plays the role of the type of storage cells: It is the inclusion functor from $\mathcal{I}$ to **Set**. The value $cellX = X$ of *cell* at world X is the set of cells associated with that world. Now, consider any element $p \in cell \mathrel{-\!\!*} (cell \mathrel{-\!\!*} E)X$. The definition of $\mathrel{-\!\!*}$ says that, for an arbitrary Y and $a \in cellY$, $p[Y]a$ is an element of $(cell \mathrel{-\!\!*} E)(X + Y)$. Unpacking further, this says that, for an arbitrary Z and $b \in cellZ$, $(p[Y]a)[Z]b \in E(X + Y + Z)$. Here, the crucial point is that a is in the X-component and b in the Z-component: so they must be different cells. This shows how the interpretation of $\mathrel{-\!\!*}$ in the functor category reflects exactly the discussion of aliasing above.

Returning to the definition of the DCC, the other additive connectives are interpreted pointwise as is standard for functor categories:

$$\begin{array}{rcl} 0(X) & = & \{\} \\ (E \vee F)(X) & = & E(X) + F(X) \end{array}$$

$$\begin{array}{rcl} 1(X) & = & \{*\} \\ (E \wedge F)(X) & = & E(X) \times F(X). \end{array}$$

To define the multiplicative conjunction, we introduce an auxiliary notion of non-interference. If $(a, b) \in (E \times F)X$ then define $a\%b$ to mean

$$\begin{array}{ll} \exists Y, Z \subseteq X\,. & Y \cap Z = \emptyset \\ & \text{and} \quad a \in range(E(Y \hookrightarrow X) \\ & \text{and} \quad b \in range(F(Z \hookrightarrow X), \end{array}$$

where $\hookrightarrow$ denotes an inclusion function. This definition says that a and b "come from" disjoint possible worlds, so that, intuitively, they access disjoint portions of the store. Then

$$\begin{array}{rcl} I(X) & = & \{*\} \\ (E * F)(X) & = & \{(a, b) \in E(X) \times F(X) \mid a\%b\}. \end{array}$$

For an example with $*$, take a pair $(a, b) \in (cell * cell)X$. We know from the definition of $*$ that we can partition X into disjoint subsets Y and Z where a comes from Y and b comes from Z. This means that a and b must be different storage cells.

We conclude by remarking that this section provides a concrete example of a "spatial" interpretation of **BI**'s multiplicatives.

4. Petri Nets

In this section, we provide a model of **BI** which is based on *Petri nets* (see [Reisig, 1998] for an introduction). Our brief discussion is essentially taken from [Pym et al., 2000], wherein more discussion may be found.

Petri nets provide a basic, concrete, model of computation which fits well with **BI**'s resource interpretation. A central tenet of the theory of Petri nets is that resource is distributed throughout a net, in the form of *tokens* that reside in *places*. A distribution of tokens is called a marking; a net evolves according to local rules which show how to go from one marking to another. As in [Engberg and Winskel, 1993], we consider a basic notion of net which does not have capacities.

Formally, a net $\mathcal{N} = (P, T, pre, post)$ consists of sets P and T of places and transitions and two functions $pre, post : T \to \mathcal{M}$, from transitions to markings, where a marking is a finite multiset of places and $\mathcal{M}$ denotes the set of all markings. A marking may be regarded as a function $M : P \to N$ from places to natural numbers that is zero on all but finitely many places. Addition of markings is given by $(M + N)p = Mp + Np$. We let $[-]$ denote the empty marking.

There are several ways that nets may be used to provide a model of **BI**. One way internalizes the reachability relation on markings, by conflating it with the intuitionistic ordering in the model. If M and N are markings, then define

$$M \Rightarrow N \quad \text{iff} \quad \begin{array}{l} \text{there are } t, M' \text{ such that} \\ M = pre(t) + M' \text{ and } N = post(t) + M'. \end{array}$$

We can then define a preorder on markings by

$$M \sqsubseteq N \quad \text{iff} \quad \begin{array}{l} \text{there are } M_1, \ldots, M_n \text{ such that} \\ M = M_1 \Rightarrow \cdots \Rightarrow M_n = N. \end{array}$$

Then $(\mathcal{M}, [-], +, \sqsubseteq)$ is a preordered commutative monoid and so this gives us an interpretation of all the connectives.

This model is just the Petri net semantics of linear logic described by Engberg and Winskel, with the addition of $\to$. In retrospect, the

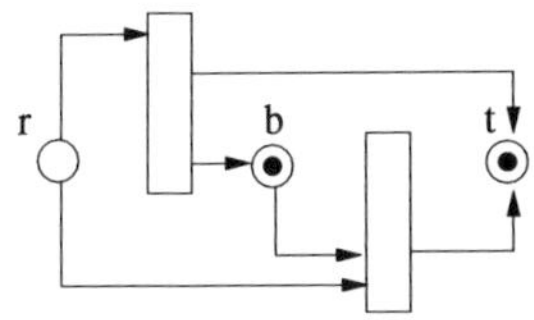

Figure 9.6. Net for a Buffer

omission of $\rightarrow$ seems strange, given that it exists naturally in the model. Admitting it enables some of the discrepancies between model and logic observed by Engberg and Winskel to be avoided. These include the need to state an axiom for distribution of $\vee$ over $\wedge$, which is implied by the more primitive rules for $\rightarrow$, as well as the ability to state negative properties of nets using $\neg\phi = \phi \rightarrow \bot$.

A basic example of the use of **BI**'s connectives is provided by *mutual exclusion*, where we say that two given places cannot be marked at the same time. To see how this works, consider the net in Figure 9.6, which represents processes either producing an item to a buffer or consuming an item from the buffer. The terms r and t denote a ready process and a terminated process, respectively, and b represents a buffer whose tokens are items produced. Then we can express that a process is not both ready and terminated using the proposition $\neg(\mathrm{r} * \mathrm{t} * \top)$. Using $-\!\!*$, we can further express that a process is not both ready and terminated in any marking reachable from a given marking M_0 : $M_0 -\!\!* \neg(\mathrm{r} * \mathrm{t} * \top)$. Note the role of $\top$ in $\mathrm{r} * \mathrm{t} * \top$. It enables the state, at a given time, to be partitioned into three parts where r is true in one, t in another, and where the third part is arbitrary.

In this model, we have internalized the reachability relation of nets. It may be argued, however, that this is an unnatural choice. In fact, in describing the mutual exclusion example we skated over two complications. Firstly, the interpretation of a place as a proposition, say r, cannot be simply a singleton marking: rather it has to be the set of all markings that may reach the singleton marking of interest, from the past. Secondly, if we unwind the formula $\neg(\mathrm{r} * \mathrm{t} * \top)$, then, semantically, there is a complex back-and-forth between future and past; this seems to hinder rather than help understanding. Also, in this semantics, all expressible properties are closed under (backwards) reachability, and this excludes many properties of interest.

An alternative model is given by the monoid $(\mathcal{M}, [-], +, =)$, in which we do not include reachability. This results in a classical semantics which allows properties of nets to be expressed directly. For example,

$$\neg(\mathrm{r} * \mathrm{t} * \top)$$

is true of a marking just when r and t are not both marked. This straightforward use of classical logic is similar to the approach taken by Reisig [Reisig, 1998]. However, our use of multiplicatives is an extension which provides the improved expressiveness we have discussed.

Since we have removed the reachability relation from the semantics of **BI**'s formulæ, we have to incorporate net-dynamics by other means. There are several standard devices for doing this, including modal or temporal logics. For example, a modal operator for transitions may be defined as follows:

$$M \models \langle t \rangle \phi \quad \text{iff} \quad \text{there is an } M' \text{ such that } M = pre(t) + M' \text{ and } post(t) + M' \models \phi.$$

As an example of a specification using this modality,

$$\langle t_1 \rangle \top \;\wedge\; \langle t_2 \rangle \top \;\wedge\; \neg(\langle t_1 \rangle \top \;*\; \langle t_2 \rangle \top)$$

says that t_1 and t_2 are currently enabled, but in conflict, so that one of them, but not both of them, may fire. We can also describe a valid inference rule that expresses the local nature of transitions:

$$\frac{P \to \langle t \rangle P'}{P * Q \to \langle t \rangle (P' * Q)} \,.$$

A construction by Dominique Larchey-Wendling and Didier Galmiche [Galmiche and Larchey-Wendling, 1998, Larchey-Wendling and Galmiche, 2000] is closely related to our Petri nets model of **BI**. Larchey-Wendling and Galmiche give models of intuitionistic linear logic, ILL) [Girard, 1987, Benton et al., 1992] based on a construction of quantales (not distributive as a lattice, as completions of ordered monoids. They obtain a completeness theorem for ILL, without !, and give examples based on both the natural numbers and the rational numbers.

5. CCS-like Models

Process calculi, such as those introduced by Milner [Milner, 1989] and Hoare [Hoare, 1985], provide a model of computation in which the primitive notions are *events* and the *order of occurrence* of events. This view stands in contrast to the model of computation provided by, for example, dynamic and temporal logics , in which the primitives are *states* and the *times* at which they obtain. A connection between the purely operational perspective of process calculi and the logical perspective is provided by *Hennessy-Milner* logic [Milner, 1989], in which the basic semantic judgement is

$$P \models \phi,$$

read as "the process P satisfies the propositional assertion ϕ".

In this section, we provide a model of **BI** based on a CCS-like process calculus. The model is taken from [Pym et al., 2000]. We exploit a notion of *modality* for **BI**, in the usual sense of modal logic rather than that of the exponentials in linear logic. A systematic analysis of modality for **BI** is beyond our present scope.

Process calculi typically contain an operation for the parallel composition of processes which is associative and commutative, with a unit. We consider a small language of processes, as follows:

$$\begin{array}{rclll} P & ::= & 0 & & \text{Null Process} \\ & | & P \mid P & & \text{Parallel Composition} \\ & | & P + P & & \text{Non-deterministic Composition} \\ & | & \alpha.P & & \text{Prefixing} \\ & | & \ldots & & \end{array}$$

Here, α ranges over actions, often including a silent action, names and co-names.

Associated with process terms are a number of equivalence relations $P \equiv P'$, the most significant of which are strong and weak bisimulation. It is also possible to take a very fine equivalence, such as structural congruence. We do that here, simply taking $\equiv$ to be the equivalence relation generated by commutative monoid equations for 0 and $|$, and for 0 and $+$. Since $|$ is commutative and associative with respect to $\equiv$ it immediately gives rise to a monoid model $(\mathcal{P}, 0, |, \equiv)$, where the process terms are possible worlds. Because two processes are related just when they are equivalent, this is a classical model, *i.e.*, a model of Boolean **BI**.

Just as with the second Petri net model, this semantics does not, by itself, account for the *dynamics* of processes. For this, we could add a modality of the kind found in logics of the Hennessy-Milner kind:

$$P \models [a]\phi \quad \text{iff} \quad \text{for all } Q, \text{ if } P \xrightarrow{a} Q, \text{ then } Q \models \phi.$$

where $\xrightarrow{a}$ is a relation which says that "P may do an a and evolve to Q".

This line was followed by Dam in his thesis [Dam, 1990] but with limited success. From our point of view, the most curious point was that he imposed additional conditions on the models, which had the effect of removing the additive implication. This seems unfortunate because the additive connectives naturally exist and a much simpler model is obtained by retaining them. One might speculate that a better logic would obtain if one were to keep the (Boolean) additives. In fact, this is, essentially, the approach followed by Cardelli and Gordon in their logic of ambients [Cardelli and Gordon, 2000].

At this point we have internalized $|$ as a connective, which has a corresponding implication, and the reader might ask: what about $+$? That is, we have two implications in **BI**; why not three, or four ? There is no technical reason why not. Indeed, we can define

$$P \models \phi \circ \psi \quad \text{iff} \quad \text{there are } Q, Q' \text{ such that } Q + Q' \equiv P, \; Q \models \phi \text{ and } Q' \models \psi$$

$$P \models \phi \rightharpoonup \psi \quad \text{iff} \quad \text{for all } Q,\, P \models \phi \text{ implies } P + Q \models \psi.$$

Non-deterministic composition tends to come along with a law $\phi + \phi \equiv \phi$ and we hereby add it to the basic (structural) congruence mentioned above. The result is that $P \models P \circ P$ and we have a model of *fusion* and *relevant implication.* Put together, $+$ and $|$ give rise to a combination of three logics: multiplicative relevant logic (I, $\circ$, $\rightharpoonup$), multiplicative intuitionistic linear logic (I, $*$, $-\!\!*$) and classical logic.

This mixture of logical operations, plus a Hennessy-Milner modality, enables us to state logical axioms which relate $+$ and $|$. For instance, a cousin of the expansion law is the implication

$$\langle x\rangle\top * \langle y\rangle\top \;\to\; \langle x\rangle\langle y\rangle\top \circ \langle y\rangle\langle x\rangle\top.$$

This law is only true for certain process equivalences, such as bisimulation, and actually fails for the structural congruence we have considered here.

6. A Pointers Model

In this section, we present an example model in which "resource" corresponds to "portion of a computer's store". It should be noted that this model is an example of the semantics of **BI**, mentioned in Chapter 5, which is based on partial monoids. Although the theory of the partial monoid semantics (see the discussion by Galmiche, Méry and Pym in [Galmiche et al., 2002]) is not developed herein, our inclusion of it is justified at the end of this section, where we show that it may be cast a

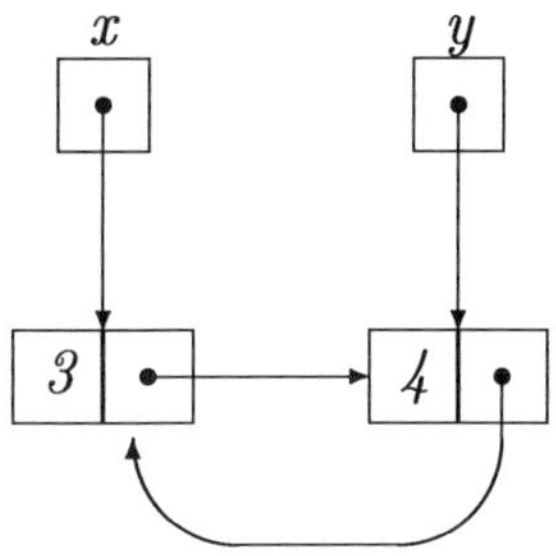

Figure 9.7. Pointers and Aliases

Grothendieck sheaf-theoretic model. This example is taken from [Pym et al., 2000] in which more details may be found. A still more detailed discussion may be found in [Ishtiaq and O'Hearn, 2001]. A related example, concerning reference types in the programming language ML, is provided in Chapter 16.

The store is made up of cons cells, which may have basic data (such as integers) in their components, or pointers to other cons cells. The model presented in this section is from work on using **BI** to reason about pointers [Ishtiaq and O'Hearn, 2001], which builds on work of Reynolds [Reynolds, 1978, Reynolds, 1981, Reynolds, 2000]. A related example, presented from the point of view of a dependently-typed λ-calculus which is intimately related to **BI**, may be found in [Ishtiaq and Pym, 1998].

The inclusion of pointers brings out several issues, most importantly *sharing*: data structures are often constructed so that there are two or more pointers to the same cell, as happens when considering graphs or circular or doubly-linked lists. When this happens, there are multiple ways to refer to the same cell or, in short, there is *aliasing*. For example, if we use the notations $x.1$ and $x.2$ to refer to the first and second components of a cons cell then x, $y.2$ and $x.2.2$ are all aliases in the situation represented by the following box-and-pointer diagram in Figure 9.7.

Since aliasing is extremely delicate, we take more time over this model than for the others. Sharing was not present in the Petri net, process or cost models: they only exhibit combining.

Formally, the worlds in this model are heaps $h \in H$, which are thought of as collections of cons cells in storage. We define

$$\begin{array}{rcl} Val & = & Int \cup \{nil\} \cup Loc \\ H & = & Loc \rightharpoonup_{fin} Val \times Val\,. \end{array}$$

Here, $Loc = \{\ell, ...\}$ is an infinite set of locations and $\rightharpoonup_{fin}$ is for finite partial functions. Each cell in memory is identified by a location and when $h(\ell) = (a, b)$ this represents a situation in which ℓ has a in its first component and b in its second. When $h(\ell)$ is undefined this represents a situation where there is no cell in the heap corresponding to ℓ.

We use a combining operation on heaps that is partial:

> $h \cdot h'$ denotes the union of disjoint heaps (*i.e.*, the union of functions with disjoint domains); e is the empty heap. When the domains of h and h' overlap, $h \cdot h'$ is undefined.

The order we consider at this point is discrete, being given by the equality relation on H. Consequently, the additive part of the logic is classical. (An alternative, intuitionistic, model is also of interest: it works by taking the relation $h \sqsubseteq h'$ between worlds to be a graph-superset of partial functions [Reynolds, 2000].)

The adjustments we must make are to the semantics of multiplicatives; we include definedness conditions:

$$h \models \phi * \psi \quad \text{iff} \quad \text{there are } h_0,\, h_1 \text{ such that} (h_0 \cdot h_1)\downarrow \text{ and } h_0 \cdot h_1 = h \text{ and } h_0 \models \phi \text{ and } h_1 \models \psi$$

$$h \models \phi \mathrel{-\!\!*} \psi \quad \text{iff} \quad \text{for all } h', \text{ if } (h \cdot h')\downarrow \text{ and } h' \models \phi \text{ then } h \cdot h' \models \psi$$

The clauses for the other connectives remain as in the elementary monoid semantics. This gives us a Boolean **BI**-algebra, where the Boolean algebra part is just the set of subsets of H.

In order to describe atomic propositions, we assume a function $s : Var \to Val$ where $Var = \{x, y, ...\}$ is a set of variables. Since we consider s to be given once and for all, we are technically remaining in a propositional set up; this obviously paves the way, however, to a consideration of quantifiers.

The basic proposition is the points-to relation, which has the form $x \mapsto (E, F)$, where E and F range over variables, integers and *nil*:

$$h \models x \to (E, F) \quad \text{iff} \quad \{sx\} = dom(h) \text{ and } h(sx) = \langle [\![E]\!]s, [\![F]\!]s \rangle,$$

where $[\![E]\!]s$ gives the value of E in s.

As a first example in this model, the formula $(x \mapsto 3, y) * (y \mapsto 4, x)$ corresponds to the box-and-pointer diagram pictured earlier. To relate this picture to the formal definition, if the formula is true at a heap h, then we must have that sx and sy are locations, by the definition of $\mapsto$, and that they are distinct, by the definition of $*$. For, $*$ splits h into two subheaps, one where sx is the only defined location and the other where sy is defined. Notice the importance of dangling pointers here: the picture corresponding to the left conjunct is

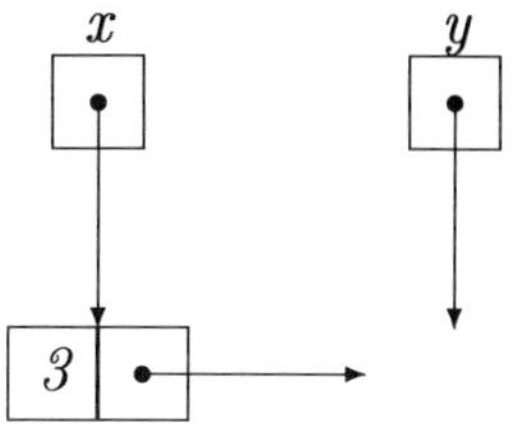

while that for the right is

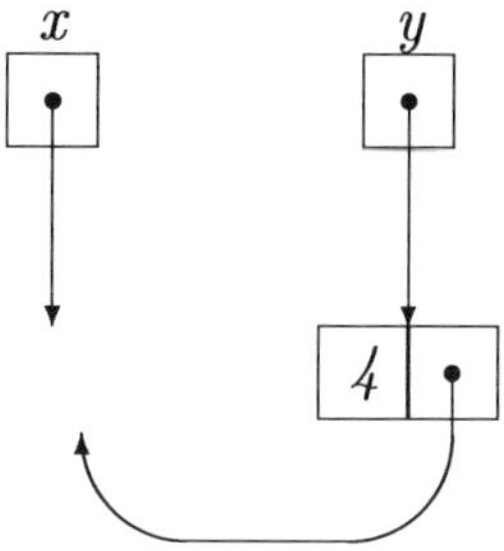

Notice that in each subheap we have a dangling pointer which is a location not in the domain of the heap.

While $*$ is about separation, the implication $-\!*$ may be used to describe new, or fresh, pieces of memory. These two connectives interact in an interesting way: The formula

$$(x \mapsto 3, 5) * ((x \mapsto 7, 5) -\!* P).$$

says that $(x \mapsto 3, 5)$ is true in the current heap but also that if we update the first component to 7 then P will be true. To see why, first note that the semantics of $*$ splits the heap, say,

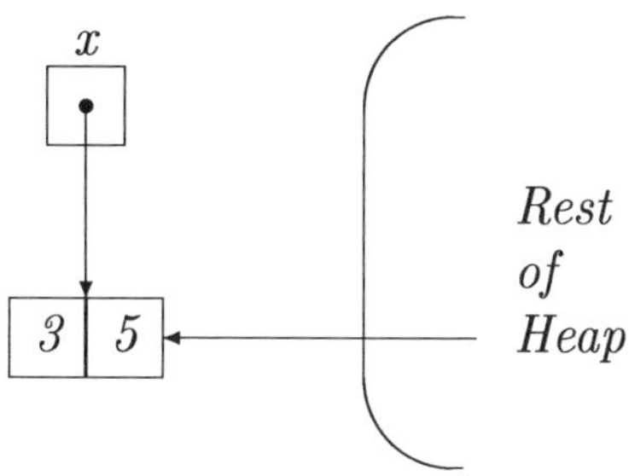

into two portions, one heap in which $(x \mapsto 3, 5)$ and a another heap in which the location denoted by x is dangling:

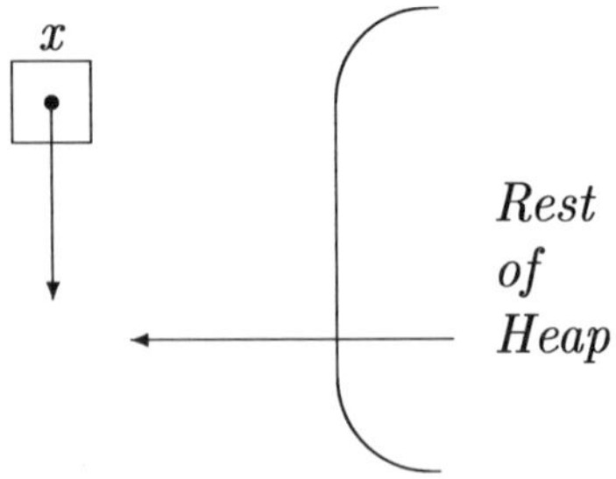

We have included here a dangling pointer out of the rest of the heap in order to emphasize that the location might be referenced from within a heap cell, as well as from x. Because the association $(x \mapsto 3, 5)$ has been, in a sense, retracted by deleting the association from the heap in the right conjunct, this frees $-\!\!*$ to extend the second heap with a different cons cell. The semantics of $-\!\!*$ then ensures that P must be true when this second heap is extended with a new binding of location to contents that makes $(x \mapsto 7, 5)$ true.

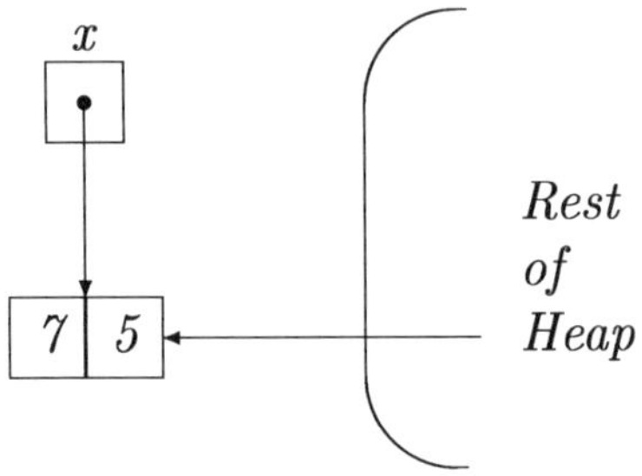

The intuitive description, in terms of updating, follows from several steps in the semantics, which amount to "update as deletion followed by extension". This idea may be used to formulate a program logic axiom for statements $x.1 := E$ and $x.2 := E$ that alter the first or second component of a cons cell in the heap. Here is the Hoare-triple axiom for $x.1$, together with an axiom for an operation for allocating a cons cell and an axiom for an operation which de-allocates a cell:[3]

$$\{\exists w, z.\,(x \mapsto w, z) * ((x \mapsto y, z) -\!\!* P)\}$$
$$x.1 := y$$
$$\{P\}$$

$$\{\forall x.\,(x \mapsto y, z) -\!\!* P\}$$
$$x := cons(y, z)$$
$$\{P\}$$

[3]The axioms require the evident extension of the model to include ***additive*** quantifiers, *q.v.* Part II. We include this example in Part I because it makes ***essential*** use of properties of **BI** only at the propositional level.

$\{P * \exists y, z. (E \mapsto y, z)\}$
dispose(x)
$\{P\}$

The *Dispose* axiom is particularly interesting. It says that that you simply shouldn't depend on what contents the disposed location might or might not have in the post-condition. More details are given in [Pym et al., 2000], from which this example has been taken. More still, with a detailed formulation of **BI**'s "pointer logic", may be found in [Ishtiaq and O'Hearn, 2001].

The examples so far have involved the multiplicative connectives only. We have interpreted the others in a classical fashion. For example, if z denotes a cell, then the formula $\neg(\exists x.\exists y.(x \mapsto y, z) * \top)$ simply says that there is no cell in the heap which has a pointer to z in its second component. More importantly, if we want to describe a logical property of a given collection of cells, such as "this collection represents a sorted list", then we are at liberty use classical logic in the usual way. The multiplicative connectives allow such statements to be put together in a way that guarantees independence properties with respect to the store.

Earlier, we referred to the idea that a judgement $m \models \phi$ may be understood in a "local" way, in which the resources to which ϕ refers are confined to those in m. To see how this works in the pointer model, consider the formula $(x \mapsto y, z) * (z \mapsto w, nil)$. It may be that y and w are pointers but the truth of this statement is judged in a way that ignores what they point to. The reason is that the semantics of $*$ and $\mapsto$ together ensures that the truth of the formula may be established only at a heap in which there are precisely two defined locations:

$$[x \mapsto (y, z), z \mapsto (w, nil)].$$

This local nature of specifications is used in [Ishtiaq and O'Hearn, 2001] to formulate an axiom for inferring certain "frame axioms", which describe invariants of the heap outside the area referred to by a specification.

We remark, echoing Example 3.9, that the semantic structure of this model is incompatible with the formal system of linear logic. To see this, consider that $\phi \multimap \psi \models \phi \to \psi$ always holds in linear logic, using the decomposition $\phi \to \psi \;\; = \,!\phi \multimap \psi$ and the rule of Dereliction for $!$. However, here,

$$(x \mapsto 1, 2) \mathrel{-\!\!*} \bot \not\models (x \mapsto 1, 2) \to \bot\,.$$

because the antecedent may hold in a heap where $x \mapsto 1, 2$ while the consequent cannot. This shows that there may be no ! which decomposes $\phi \to \psi$ into $!\phi \multimap \psi$ in this model.

We conclude this section by noting that the pointers model may be understood as a Grothendieck sheaf model, as described in Chapter 5. The argument is straightforward and is given in [Pym et al., 2000]. We include it here for convenience.

Let $H_\perp$ be the set of heaps, extended with a new least element, $\perp$. We can define an operation $\cdot$ in which $h \cdot h'$ is the union of $h, h' \in H$ if they are disjoint and $\perp$ otherwise. Also, $\cdot$ is strict in both arguments and the unit is again the empty heap. The ordering we take is the flat one, in which $\perp$ is least and all other elements are incomparable.

We can define a Grothendieck topology on $H_\perp$, by setting

$$\begin{array}{rcll} J(\perp) & = & \{\{\perp\}, \emptyset\} & \\ J(m) & = & \{\{m\}\} & \text{if } m \neq \perp. \end{array}$$

The points-to relation is extended so that $\perp$ always forces it. Notice that since $J(\perp)$ contains $\emptyset$, it follows from the semantic clauses that $\perp \models \phi$ always holds.

The connection between the pointer model and this sheaf presentation may then be stated as follows:

For every $h \in H$, $h \models \phi$ in the sheaf model just given iff $h \models \phi$ in the pointer model.

This does not mention $\perp$ but, because of the way it is treated in the topology, the two models do indeed agree on logical consequence:

$\psi \models \phi$ in the sheaf model just given iff $\psi \models \phi$ in the pointer model.

Finally, the pointer model of Reynolds [Reynolds, 2000] may also be seen as a Grothendieck sheaf model. The underlying set of worlds is $H_\perp$, as above, but this time the ordering on worlds is the one in which $h \sqsubseteq h'$ if the graph of h is a superset of the graph of h'. This is an intuitionistic model, whereas the previous one provides a model of Boolean **BI**.

II

PREDICATE BI

Chapter 10

INTRODUCTION TO PART II

Some of the content of Part II has appeared in [O'Hearn and Pym, 1999, Pym, 1999, Ishtiaq and Pym, 1998, Ishtiaq and Pym, 1999, Ishtiaq and Pym, 2000, Ishtiaq and Pym, 2001, Armelín and Pym, 2001]. References are given in the text as appropriate.

—DJP

1. A Proof-theoretic Introduction to Predicate BI

Just as the propositional assumptions in a sequent may be structured as bunches, supporting the distinction between additive and multiplicative connectives, so the collection of (first-order) variables over which such a sequent is formed may be structured as *bunches of variables.* With variables structured as bunches, each connective may be formulated either with additive maintenance of variables or with multiplicative maintenance of variables. So, just as explicit control of the structural rules at the *propositional* level decomposes conjunction and implication into additive and multiplicative parts, so explicit control of the structural rules in the formation of terms further decomposes each *predicate* connective into additive and multiplicative parts. We describe this idea here using a single-sorted version of predicate **BI**. We suppress the details of function symbols and the formation of terms. For simplicity in this introduction, we work with bunches of variables without types.

BUNCHES OF VARIABLES
x ranges over variables, and X over bunches of variables:

$$X \ ::= \ x \mid X;X \mid X,X \mid \emptyset_a \mid \emptyset_m$$

Bunches of variables are subject to the linearity restriction: any variable appears at most once in a bunch. We let $X \cong X'$ denote isomorphism of labelled trees.

JUDGEMENTS

We consider terms- and propositions-in-context, with a syntax of the form

$$X \vdash t : \mathbf{Term} \quad \text{and} \quad X \vdash \phi : \mathbf{Prop}$$

asserting that a term or predicate is well-formed in context X [Martin-Löf, 1996]. Constants and predicate letters may be considered to be given by schematic judgements and, as in the bunched logic itself, Contraction and Weakening are allowed for ";" but not for ",". We omit a formal definition and move on to consider the quantifiers.

Logical judgements have the form

$$(X)\Gamma \vdash \phi$$

asserting that ϕ is a consequence of Γ, where the terms and atomic predicates in the sequent $\Gamma \vdash \phi$ are well-formed in X.

Given this formulation of predication, we can see that for each propositional connective we must choose whether to adopt additive or multiplicative maintenance of variables. For example, the binary rule for introducing $*$ may be formulated either as

$$\frac{(X)\Gamma \vdash \phi \quad (Y)\Delta \vdash \psi}{(X, Y)\Gamma, \Delta \vdash \phi * \psi} \; *I(m),$$

in which the two premisses use distinct bunches of variables, combined multiplicatively in the conclusion, or as

$$\frac{(X)\Gamma \vdash \phi \quad (Y)\Delta \vdash \psi}{(X; Y)\Gamma, \Delta \vdash \phi * \psi} \; *I(a),$$

using ";" to combine bunches of variables. It follows, via Contraction

$$\frac{Y(X; X')\Gamma \vdash \phi}{Y(X)\Gamma[X/X'] \vdash \phi[X/X']} \; (X \cong X') \quad \text{Contraction},$$

that the following, familiar, form of $*I(a)$ is sound:

$$\frac{(X)\Gamma \vdash \phi \quad (X)\Delta \vdash \psi}{(X)\Gamma, \Delta \vdash \phi * \psi} \; *I(a).$$

This latter form is the one taken in linear logic's introduction rule for tensor product, $\otimes$ [Trolestra, 1992, Girard, 1987]. Here we develop the version in which multiplicative (additive) variable maintenance is paired multiplicative (additive) propositional connectives.

The predicate version of **BI** has the familiar additive, or intuitionistic, or extensional, quantifiers $\forall$ and $\exists$. It also has multiplicative, or intensional, quantifiers, obtained by observing structural restrictions on the level of terms as well as propositions. The universal introduction rules have the usual restriction that x must not occur in Γ; similarly, the existential elimination rules have the restriction that x must not occur in Δ or ψ. However, in **BI**, these restrictions must be extended to those variables which are constrained by the formation of axioms to be equal to such an x. $X \vdash \Gamma : \mathbf{Prop}$ has its evident meaning.

MULTIPLICATIVES

$$\frac{(X,Y)\;\Gamma \vdash \phi[t/x] \quad Y \vdash t : \mathbf{Term} \quad X \vdash \Gamma : \mathbf{Prop}}{(X)\;\Gamma \vdash \exists_{\mathbf{new}} x . \phi}\;\exists_{\mathbf{new}} I$$

$$\frac{(X)\;\Gamma \vdash \exists_{\mathbf{new}} x . \phi \quad (Y,x)\;\Delta, \phi \vdash \psi}{(X,Y)\;\Gamma, \Delta \vdash \psi}\;\exists_{\mathbf{new}} E$$

$$\frac{(X,x)\Gamma \vdash \phi}{(X)\Gamma \vdash \forall_{\mathbf{new}} x \,.\, \phi}\;\forall_{\mathbf{new}} I$$

$$\frac{(X)\;\Gamma \vdash \forall_{\mathbf{new}} x \,.\, \phi \quad Y \vdash t : \mathbf{Term}}{(X,Y)\;\Gamma \vdash \phi[t/x]}\;\forall_{\mathbf{now}} E$$

ADDITIVES

$$\frac{(X;Y)\;\Gamma \vdash \phi[t/x] \quad Y \vdash t : \mathbf{Term}}{(X)\;\Gamma \vdash \exists x . \phi}\;\exists I$$

$$\frac{(X)\;\Gamma \vdash \exists x . \phi \quad (X;x)\;\Delta, \phi \vdash \psi}{(X)\;\Gamma; \Delta \vdash \psi}\;\exists E$$

$$\frac{(X;x)\;\Gamma \vdash \phi}{(X)\;\Gamma \vdash \forall x \,.\, \phi}\;\forall I \qquad \frac{(X)\;\Gamma \vdash \forall x \,.\, \phi \quad Y \vdash t : \mathbf{Term}}{(X;Y)\;\Gamma \vdash \phi[t/x]}\;\forall E$$

The idea of the introduction rule for $\forall_{\mathbf{new}}$ is that we can infer $\forall_{\mathbf{new}} x \,.\, \phi$ in the usual way for universal quantification, except that the variable x must be in multiplicative combination with all of the other variables. In the elimination rule we must be careful not to substitute an arbitrary term for x, but only one that is respective of the multiplicative relationship between x and other variables in X. In particular, t cannot contain any of these other variables appearing in X; this requirement is implemented by the linearity restriction.

Formally, this set-up will not quite do. As we shall see in Chapter 11, we must also indicate whether each variable in a bunch is to be bound additively or multiplicatively. To this end our definition of a bunch of variables permits two forms of declaration, $x!A$ for additively binding variables and $x : A$ for multiplicatively binding variables.

The form of the quantifier rules is intimately related to the form of axiom sequents in predicate **BI**. In order to permit the combination of bunches of variables, using either "," or ";", axiom sequents have the form

$$\frac{X_1 \vdash \mathrm{p}(X_1) : \mathbf{Prop} \qquad X_2 \vdash \mathrm{p}(X_2) : \mathbf{Prop}}{(X)\ \mathrm{p}(X_1) \vdash \mathrm{p}(X_2)}\ \mathrm{Axiom}(X_1, X_2),$$

where p is a predicate letter and X is X_1, X_2 or $X_1; X_2$. The side-condition $\mathrm{Axiom}(X_1, X_2)$ is used, in the Substitution and predicate Cut rules, which are not given this Introduction, to ensure that the bunches X_1 and X_2 are treated as being the same in a subsequent derivation. Such side-conditions are also used in the quantifier rules to enforce the maintenance of the usual side-conditions not only for the given x but also for variables constrained to be equal to it via $\mathrm{Axiom}(-, -)$.

A consequence of these observations is that $\forall_{\mathbf{new}}$ should not be read literally as "for all". Rather, we must understand that, in $\forall_{\mathbf{new}} x \,.\, \phi$, the multiplicative relationship between x and other variables must be observed. We read it as "for all x different from the variables in ϕ" or, more concisely, for all *new* x.

It is important, at least for intuitionistic multiplicatives, not to think of $\forall_{\mathbf{new}}$ as an infinite multiplicative conjunction; this idea is sometimes mentioned in analogy with the view of additive quantification as an infinite additive conjunction. Rather, $\forall_{\mathbf{new}}$ is closely allied to the multiplicative implication $-\!\!*$, although a version allied to multiplicative conjunction is possible under restricted circumstances. It would be interesting to try to formulate a dependent function type, along the lines of the $\lambda\Lambda$-calculus, discussed in Chapter 15 and based on [Ishtiaq and Pym, 1998, Ishtiaq and Pym, 2001, Ishtiaq and Pym, 1999, Ishtiaq and Pym, 2000], which generalizes both of them.

For the additive universal quantifier, one might expect the elimination rule to be

$$\frac{(X)\ \Gamma \vdash \forall x \,.\, \phi \qquad X \vdash t : \mathbf{Term}}{(X)\ \Gamma \vdash \phi[t/x]}\ \forall E.$$

Such a rule is admissible if we have Contraction for ";" on the level of terms. Similar considerations apply to the additive existential quantifier.

2. Kripke Semantics for Predicates and Quantifiers

We can extend the Kripke semantics of propositional **BI** to define predicates and quantifiers in **BI**. Just as propositional truth may be defined in a Kripke model, so may predicate truth. Just as a Kripke model supports both additive and multiplicative propositional structure, so a Kripke model supports both the additive and multiplicative formation of predicates.

The key idea is to form predicates not over a list, or set, of variables but rather over a *bunch of variables*, formed just as in the definition of the $\alpha\lambda$-calculus. Syntactically, we consider terms- and propositions-in-context, with a syntax of the form

$$X \vdash t : \mathbf{Term} \quad \text{and} \quad X \vdash \phi : \mathbf{Prop},$$

which assert that a term or proposition is well-formed in context (*i.e.*, bunch of variables) X. It follows that we must work with sequents of the form

$$(X)\Gamma \vdash \phi,$$

asserting that ϕ is a consequence of Γ and in which the terms and predicates in the Γ and ϕ are well-formed according to X.

In order to extend the definition of satisfaction, $\models$, the variables declared in a bunch X must be interpreted at a world $m \in M$, where $\mathcal{M} = (M, e, \cdot, \sqsubseteq)$ is a preordered commutative monoid. Since a bunch may be formed using either multiplicative combination, ",", or additive combination, ";", we must exploit Day's [Day, 1970] construction, discussed in Chapters 1, 4 and 5, of a tensor product on the functor category $\mathbf{Set}^{\mathcal{M}^{op}}$, in which we consider $\mathcal{M}$ as a monoidal category with an arrow from m to n just in case $m \sqsubseteq n$.

In Chapters 4 and 5, we have shown how Day's construction provides a semantics for **BI**'s propositional multiplicative connectives, with the propositional additive connectives interpreted in the usual way [Lambek and Scott, 1986]. Each propositional connective gives rise to a predicate connective. For example, given that we can form propositional multiplicative conjunctions $\phi * \psi$, we can ask for predicate conjunctions $(\phi * \psi)(X)$, where X is the bunch of variables occurring in $\phi * \psi$. The satisfaction relation, $\models$, for predicate **BI** must generalize the one for propositional **BI**, clause by clause. To formulate $\models$, we must interpret the bunch of variables, X, in $\mathbf{Set}^{\mathcal{M}^{op}}$. A variable x is interpreted as a given functor, a domain of individuals, $D = [\![x]\!]^D$ in $obj(\mathbf{Set}^{\mathcal{M}^{op}})$. The interpretation $[\![X]\!]$ of a bunch of variables X is then built using cartesian product, $\times$, of functors and Day's tensor product, $\otimes$, so that an *environment* for X at m is an element u of the set $[\![X]\!]^D(m)$. Given such an interpretation, we can set up a satisfaction relation for predicate **BI** in the following form:

$$(X)u \mid m \models \phi(X).$$

Here, X denotes a bunch of variables, m is a world, ϕ is a proposition, possibly involving quantifiers and predicate letters and u is an environment for X at world m.[1] The whole judgement is read as follows: the proposition ϕ, with variables in the bunch X, is satisfied (or forced) at world m with respect to environment u. The functorial action of environments enables us to formulate an extension of the propositional Kripke Monotonicity, or Hereditary, condition:

$$(X)\ u \mid m \models \phi \quad \text{and} \quad n \sqsubseteq m \quad \text{implies} \quad ([\![X]\!]^D(n \sqsubseteq m)u) \mid n \models \phi.$$

In this Kripke semantics, we can distinguish two versions of the proposition $(\phi * \psi)(X)$, for example, as follows:

- *Additive predication*:

$$\begin{array}{llll} (X) & u \mid m \models (\phi * \psi)(X) \ \text{ iff } & (X) & u \mid n \models \phi(X) \ \text{ and} \\ & & (X) & u \mid n' \models \psi(X), \end{array}$$

  where $m \sqsubseteq n \cdot n'$.

- *Multiplicative predication*:

$$\begin{array}{llll} (X) & u_X \mid m_X \models (\phi * \psi)(X) \ \text{ iff } & (Y) & u_Y \mid m_Y \models \phi(Y) \ \text{ and} \\ & & (Z) & u_Z \mid m_Z \models \psi(Z), \end{array}$$

  where $X = (Y\,,\,Z)$, $u_X = [u_Y, u_Z]$ and $m_X \sqsubseteq m_Y \cdot m_Z$. Note that we have made essential use of the pairing operation, $[-,-]$, for Day's product.

Additive predication is used not only in all the connectives and quantifiers of intuitionistic logic but also in all the connectives and quantifiers of *linear logic*, including the tensor product, or multiplicative conjunction, as described above. Multiplicative predication forces a separation between the variables which occur in each component of the product. In **BI**, we shall adopt multiplicative predication for the multiplicative

[1] We emphasize here that an environment is an element of a set, and not a map in $\mathbf{Set}^{\mathcal{M}^{op}}$.

connectives, $*$ and $-\!\!*$, and additive predication for the additive connectives, $\wedge$, $\vee$ and $\rightarrow$. We shall discuss the *cross cases* such as additive predication with $*$ as described above, in Chapter 11 (see also [Pym, 1999]).

Turning to the semantics of the quantifiers, it follows that, since predication may be treated both additively and multiplicatively, so may the quantifiers. The semantics of the additive quantifiers is the standard one in the functorial account of Kripke models [Lambek and Scott, 1986]:

$$(X)\ u \mid m \models \forall x.\phi \quad \text{iff} \quad \text{for all} \quad n \sqsubseteq m \quad \text{and all} \quad d \in D(n) \quad (X;x)\ ([\![X]\!](n \sqsubseteq m)u, d) \mid n \models \phi$$

$$(X)\ u \mid m \models \exists x.\phi \quad \text{iff} \quad \text{there exists} \quad d \in D(m) \quad (X;x)\ (u,d) \mid m \models \phi.$$

Here, the bunch of variables X is extended by x using the semi colon, ";", and the additive universal uses the preorder structure of $\mathcal{M}$, just as in the Kripke semantics of **IL**. An environment for the bunch $X;x$ is obtained by taking the cartesian pairing (u,d) of environments u for X and d for x .

The multiplicative quantifiers are obtained by working with a bunch of variables X which is extended by x using the comma, ",":

$$(X)\ u \mid m \models \forall_{\mathbf{new}} x.\phi \quad \text{iff} \quad \text{for all } n \text{ and all } d \in D(n),\ (X,x)\ [u,d] \mid n \cdot m \models \phi$$

$$(X)\ u \mid m \models \exists_{\mathbf{new}} x.\phi \quad \text{iff} \quad \text{there exists } n \text{ and } d \in D(n) \text{ such that } (X,x)\ [u,d] \mid n \cdot m \models \phi.$$

Here, we have used the pairing operation $[u,d]$ for Day's tensor product [Day, 1970]: it takes an environment $u \in [\![X]\!](m)$ and an element $d \in D(n)$ and forms the element $[u,d] \in ([\![X]\!] * D)(m \cdot n)$. Notice that we used the monoid structure just as in the semantics of $-\!\!*$.[2]

The difference between $\forall$ and $\forall_{\mathbf{new}}$ is that the former considers elements $d \in D(n)$ only in accessible worlds, where $n \sqsubseteq m$, whereas for the latter we look to a completely separate world n, and then use the monoid operation $\cdot$ to combine it with m. In this sense, $\forall$ is "local", whereas $\forall_{\mathbf{new}}$ is "global". In terms of our resource semantics, the reading of this connective is essentially the same as that for multiplicative implication, as the cost of applying the function $\forall_{\mathbf{new}} x.\phi$, which costs m to an argument which costs n.

[2]From the definition of environment functors, this functor $[\![X]\!] * D$ is in fact equal to $[\![X,x]\!]$, so the definitions are type correct.

The multiplicative existential is different in another respect to its additive counterpart. The additive $\exists$, from intuitionistic logic, is often described as being "local", in that the definition stays at the same world, whereas for $\forall$ you travel to accessible worlds to find elements. The multiplicative $\exists_{\mathbf{new}}$ does not stay at the same world when it looks for an element that exists, but neither does it travel along accessible lines; it hops to an arbitrary world, accessible or not, to find an element, and then considers the resulting formula in an environment formed by multiplicative combination. In this sense, $\exists_{\mathbf{new}}$ is "global" whereas $\exists$ is "local".

Combining local and global operators, such as $\forall$ and $\exists$ and $\forall_{\mathbf{new}}$ and $\exists_{\mathbf{new}}$, with the same logical status within a single system of logic provides a basis for combined local and global reasoning.

In terms of the sharing interpretations, described in Chapters 9 and 16, and in [O'Hearn and Pym, 1999, O'Hearn, 1999, Pym et al., 2000], we would read $\forall_{\mathbf{new}}$ as "for all xs that don't share resources with other variables in ϕ" or, more briefly, for all *new* x. Similar considerations apply to $\exists_{\mathbf{new}}$. We choose the names $\forall_{\mathbf{new}}$ and $\exists_{\mathbf{new}}$ partly because their definitions refer to new, unrelated worlds and partly because of an appealing connection with the local variables described by the **new** construct in Idealized Algol [O'Hearn, 1999]. These ideas, which are quite general and go well beyond Idealized Algol into SCI and logic programming, are discussed in Chapters 9 and 16, and in [O'Hearn and Pym, 1999, O'Hearn, 1999, Pym, 1999, Pym et al., 2000].

Just as in propositional **BI**, the elementary Kripke semantics works well only in the absence of inconsistency, $\perp$, the unit of $\vee$. For an adequate treatment of $\perp$, we must move either to topological Kripke semantics or to a fibred semantics of proofs. We develop the former for predicate **BI**, a straightforward extension of the propositional case, in Chapter 14.

3. Fibred Semantics and Dependent Types

Although we have provided a forcing semantics for predicate **BI**, we have not provided a semantics for predicate **BI**'s proofs. The natural setting for this would seem, as for intuitionistic predicate logic, to be that of fibred categories.

The basic idea is to interpret the first-order terms in a base category, over which is fibred a family of categories in which proofs are interpreted: The fibred structure is used both to manage the occurrences of variables in propositions and proofs and to define the interpretations of quantifiers.

Rather than develop this idea in detail for predicate **BI**, we merely introduce the general ideas, leaving their completion as conjecture, and proceed to provide a more detailed treatment, in the fibred setting, of the semantics of a dependent type theory, $\lambda\Lambda$, which stands in a (slightly weak) propositions-as-types correspondence with a structural variant of a fragment of **BI**. Although $\lambda\Lambda$ may be seen as a bunched system, particularly in its treatment of quantifiers, it also owes much to linear logic and, accordingly, the variant of **BI** to which it corresponds is a structural step, *i.e.*, Dereliction, towards linear logic. The theory of $\lambda\Lambda$ and the RLF logical framework have been introduced in [Ishtiaq and Pym, 1998, Ishtiaq and Pym, 2001, Ishtiaq, 1999, Ishtiaq and Pym, 1999, Ishtiaq and Pym, 2000].

One motivation for studying $\lambda\Lambda$ is its provision of a linear dependent function space. Another, more logical, motivation is its rôle as the linguistic basis for the RLF logical framework, described in Chapter 15. RLF uniformly encodes intuitionistic linear logic and other substructural systems and provides a useful meta-logical view of these logics.

We give a semantics for $\lambda\Lambda$ based on monoid-indexed families of (functorial) Kripke models, thereby giving a "resource" semantics as a refinement, via "phase shifts" of the corresponding models of the corresponding intuitionstic (minimal) type theory [Pym, 2000a], $\lambda\Pi$. In this way, we make an explicit connection with the method used by Hodas and Miller [Hodas and Miller, 1994], emphasizing the sense in which the logical frameworks LF and RLF may be interpreted as logic programming languages (*cf.* Kowalski's early work [Kowalski, 1979]), to give a semantics to the linear logic programming language Lolli. Note, however, that whereas our semantics may be used to give a semantics to disjunctive goals, the semantics in [Hodas and Miller, 1994] does not adequately extend to linear logics additive disjunction, $\oplus$, because linear logic admits the distribution of additive conjunction, $\&$, over $\oplus$.[3] We establish soundness and completeness theorems for this semantics and give a construction of a family of set-theoretic models.

We conclude Chapter 15 with a somewhat speculative discussion of a truly bunched dependent type theory, *i.e.*, one which corresponds to the basic system of predicate **BI**. It seems that such a system presents both proof- and model-theoretic challenges. Thus our study of $\lambda\Lambda$ serves

[3]Recall that

$$\phi\&(\psi \oplus \chi) \multimapboth (\phi\&\psi) \oplus (\phi\&\chi)$$

is not a theorem of linear logic.

mainly to introduce the problems which must be overcome in order to construct such a type system.

4. Computational Interpretations

We conclude Part II with brief discussions of two computational interpretations of **BI**:

- Proof-search and (predicate) logic programming [Armelín and Pym, 2001];
- The representation of ML with reference types in the logical framework RLF [Ishtiaq, 1999, Ishtiaq and Pym, 1998, Ishtiaq and Pym, 2000].

The first extends the view of propositional proof-search and logic programming given in Chapter 9 to predicate **BI**, in which the sharing interpretation may be seen to apply to *answer substitutions.*

The second is an example of a representation of a logic in the RLF logical framework. We show how to represent the operational semantics of ML [Milner et al., 1997] with reference types.

Chapter 11

THE SYNTAX OF PREDICATE BI

1. The Syntax of Predicate BI

The syntax of a sequent in first-order intuitionistic predicate **MILL** may be extended, to indicate the set (or list) X of variables used to form its propositions, as follows:

$$(X)\,\Gamma \vdash \phi .$$

With this formulation, **MILL**'s $-\!\!\circ L$ rule, for example, has the form

$$\frac{(X)\,\Gamma \vdash \phi \quad (X)\,\Delta, \psi \vdash \chi}{(X)\,\Gamma, \Delta, \phi -\!\!\circ \psi \vdash \chi} \; -\!\!\circ L,$$

indicating that the same set (or list) X of variables is used to form each of the premisses and also the conclusion, *i.e.*, that the variables display additive behaviour.

However, just as we can move from antecedents which are lists of propositions to antecedents which are bunches, so can we move from lists of variables to *bunches of variables*. We will assume that variables are many-sorted, with sorts A, B, *etc.*.

TYPES

A ::=		α	atomic types
	|	I	multiplicative unit
	|	$A * A$	multiplicative conjunction
	|	$A \multimap A$	multiplicative implication
	|	$\top$	additive unit
	|	$A \wedge A$	additive conjunction
	|	$A \to A$	additive implication
	|	$\bot$	additive disjunctive unit
	|	$A \vee A$	additive disjunction

BUNCHES OF VARIABLES

x is used to range over variables, and X over bunches of variables:

$$X \ ::= \ x : A \mid x!A \mid X \, , \, X \mid X \, ; \, X \mid \emptyset_m \mid \emptyset_a$$

Bunches of variables are subject to the linearity restriction: any variable appears at most once in a bunch. Bunches are structured as trees, with internal nodes labelled with either "," or ";" and leaves labelled with declarations. We write $x \in A$ to stand for either $x : A$ or $x!A$. Expressions of the form $X(x \in A)$ should be understood to denote either an expression of the form $X(Y, x : A)$ or of the form $X(Y; x!A)$. Bunches may be represented using lists of lists, *etc.* as described in [Read, 1988]. We write $X(Y)$, and refer to Y as a *sub-bunch* of X, for a bunch X in which Y appears as a sub-tree and write $X[Y'/Y]$ for X with Y replaced by Y'. We write $X(-)$ to denote a bunch X which is incomplete and which may be completed by placing a bunch in its hole. We require that "," and ";" be commutative monoids, giving rise to the coherent equivalence, $X \equiv X'$, as follows:

COHERENT EQUIVALENCE: $X \equiv X'$

1 Commutative monoid equations for $\emptyset_a$ and ";".

2 Commutative monoid equations for $\emptyset_m$ and ",".

3 Congruence: if $Y \equiv Y'$ then $X(Y) \equiv X(Y')$.

Note that ";" and "," do not distribute over one another. We use $=$ for syntactic identity of bunches. Congruence or tree isomorphism, $\cong$, is extended to include the distinction between ":" and "!".

We consider terms- and propositions-in-context, with a syntax of the form

$$X \vdash_{\Sigma} t : A \quad \text{and} \quad X \vdash_{\Sigma,\Xi} \phi : \mathbf{Prop}$$

which assert that a term t is well-formed of type A, with respect to a *term signature*, Σ, and that a proposition ϕ is well-formed in the bunch X, with respect to term signature, Σ, and *predicate signature*, Ξ.

The term-language of predicate **BI** is given by the evident first-order fragment of the $\alpha\lambda$-calculus, *i.e., without α- or λ-abstractions*, $\wedge$, $\vee$ or units, extended with term signatures, in which are declared function symbols. The syntax of signatures is given as follows:

$$\begin{array}{rcll} \Sigma & ::= & \emptyset_a & \text{empty signature} \\ & | & c!A & \text{constants (0-ary functions)} \\ & | & f!A & \text{functions (higher type)} \\ & | & \Sigma\,;\,\Sigma & \text{additive combination.} \end{array}$$

Note that here we have made signatures purely additive: Such a restriction is inessential and is made here only for simplicity. To get multiplicative signatures, which would correspond to a semantics in which different constants exist at different worlds, we should add

$$\begin{array}{rcll} \Sigma & ::= & \emptyset_m & \text{empty signature} \\ & | & c : A & \text{constants (0-ary functions)} \\ & | & f : A & \text{functions (higher type)} \\ & | & \Sigma\,,\,\Sigma & \text{multiplicative combination} \end{array}$$

to the clauses given above. We will return to this issue briefly in Chapter 16.

Judgements $X \vdash_{\Sigma} t : A$ are generated by the rules of the $\alpha\lambda$-calculus, introduced in Chapter 2 and given below in Table 11.1 for ease of reference. In Chapter 2, $\alpha\lambda$ is introduced as a representation of proofs in **NBI**, **BI**'s natural deduction system. Here we regard it simply as a calculus of typed functional terms. The equational theory of $\alpha\lambda$ and its meta-theory are given in Chapter 2.

Predicate signatures, Ξ, contain declarations of *predicate letters*:

$$\begin{array}{rcll} \Xi & ::= & \emptyset_a & \text{empty signature} \\ & | & \mathrm{p}!A & \text{predicate letters} \\ & | & \Xi\,;\,\Xi & \text{additive combination.} \end{array}$$

Predicate letters are constant symbols $\mathrm{p} : A$ with arity a bunch X of variables, where, with the obvious abuse of notation, A is the type of X.

IDENTITY AND STRUCTURE

$$\frac{}{x \in A \vdash_\Sigma x : A}\ Axiom \qquad \frac{X \vdash_\Sigma s : A \quad Y(x \in A) \vdash_\Sigma t : B}{Y(X) \vdash_\Sigma t[s/x]B}\ Cut$$

$$\frac{X(Y) \vdash_\Sigma s : A}{X(Y;Y') \vdash_\Sigma s : A}\ W \qquad \frac{X(Y;Y') \vdash_\Sigma s : A}{X(Y) \vdash_\Sigma s[i(Y)/i(Y')] : A}\ (Y' \cong Y)\ \ C$$

$$\frac{X \vdash_\Sigma s : A}{Y \vdash_\Sigma s : A}\ (Y \equiv X)\ \ E$$

UNITS

$$\frac{}{\emptyset_m \vdash I : I}\ I\,I \qquad \frac{X(\emptyset_m) \vdash_\Sigma t : \phi \quad \Delta \vdash_\Sigma s : I}{X(Y) \vdash \mathbf{let}\, I\, \mathbf{be}\, s\, \mathbf{in}\, t : \phi}\ I\,E$$

$$\frac{X \vdash_\Sigma t : \bot}{X \vdash_\Sigma \bot_\phi(t) : \phi}\ \bot E$$

$$\frac{}{\emptyset_a \vdash_\Sigma \top : \top}\ \top I \qquad \frac{X(\emptyset_a) \vdash_\Sigma t : \phi \quad \Delta \vdash s : \top}{X(Y) \vdash_\Sigma t[s/\top] : \phi}\ \top E$$

MULTIPLICATIVES

$$\frac{X, x : A \vdash_\Sigma s : B}{X \vdash_\Sigma \lambda x : A . s : A \multimap B}\ \multimap I \qquad \frac{X \vdash_\Sigma s : A \multimap B \quad Y \vdash_\Sigma t : A}{X, Y \vdash_\Sigma \mathbf{app}_{\multimap}(s,t) : B}\ \multimap E$$

$$\frac{X \vdash_\Sigma s : A \quad Y \vdash_\Sigma t : B}{X, Y \vdash_\Sigma M * N : A * B}\ *I \qquad \frac{X(x : A, y : B) \vdash_\Sigma t : C \quad Y \vdash_\Sigma s : A * B}{X(Y) \vdash_\Sigma \mathbf{let}\,(x,y)\,\mathbf{be}\, s\, \mathbf{in}\, t : C}\ *E$$

ADDITIVES

$$\frac{X ; x : A \vdash_\Sigma s : B}{X \vdash_\Sigma \alpha x : A . s : A \to B}\ \to I \qquad \frac{X \vdash_\Sigma s : A \to B \quad Y \vdash_\Sigma t : A}{X; Y \vdash_\Sigma \mathbf{app}_{\to}(s,t) : B}\ \to E$$

$$\frac{X \vdash_\Sigma s : A \quad Y \vdash_\Sigma t : B}{X; Y \vdash_\Sigma \langle s,t \rangle : A \wedge B}\ \wedge I \qquad \frac{X \vdash_\Sigma s : A \wedge B \quad X; x : A; y : B \vdash t : C}{X \vdash_\Sigma t[\pi_1 s/x, \pi_2 s/y] : C}\ \wedge E$$

$$\frac{X \vdash_\Sigma s : A_i}{X \vdash_\Sigma \mathbf{in}_i(s) : A_1 \vee A_2}\ (i = 1,2)\ \ \vee I$$

$$\frac{X \vdash_\Sigma s : A \vee B \quad Y(x : A) \vdash_\Sigma t : C \quad Y(y : B) \vdash_\Sigma t' : C}{Y(X) \vdash_\Sigma \mathbf{case}\, s\, \mathbf{of}\, \mathbf{in}_1(x) \Rightarrow t\ \mathbf{or}\ \mathbf{in}_2(y) \Rightarrow t' : C}\ \vee E$$

Table 11.1. **NBI** for $\alpha\lambda$

We will write $\mathrm{p}(X)$ to denote that p has arity X. As with signatures, we could add clauses to permit multiplicative predicate signatures, corresponding to a semantics in which different predicate letters are available at different worlds.

Judgements $X \vdash_{\Sigma,\Xi} \phi : \mathbf{Prop}$, "$\phi$ is a well-formed proposition in term signature Σ and propositional signature Ξ with respect to the bunch X of variables", are generated by a calculus which relates the structure

IDENTITIES

$$\frac{\mathrm{p}(X) \in \Xi}{X \vdash_{\Sigma,\Xi} \mathrm{p}(X) : \mathbf{Prop}} \text{ Atoms}$$

$$\frac{X(x \in A) \vdash_{\Sigma,\Xi} \phi(X) : \mathbf{Prop} \quad Y \vdash_{\Sigma} t : A}{X(Y) \vdash_{\Sigma,\Xi} \phi(X[t/x])} \text{ Substitution}$$

STRUCTURALS

$$\frac{X(Y) \vdash_{\Sigma,\Xi} \phi : \mathbf{Prop}}{X(Y\,;\,Y') \vdash_{\Sigma,\Xi} \phi : \mathbf{Prop}} \; W$$

$$\frac{X(Y\,;\,Y') \vdash_{\Sigma,\Xi} \phi : \mathbf{Prop}}{X(Y) \vdash_{\Sigma,\Xi} \phi[i(Y)/i(Y')] : \mathbf{Prop}} \; (Y' \cong Y) \quad C$$

MULTIPLICATIVES

$$\frac{X \vdash_{\Sigma,\Xi} \phi(X) : \mathbf{Prop} \quad Y \vdash \psi(Y) : \mathbf{Prop}}{X\,,\,Y \vdash_{\Sigma,\Xi} \phi(X) * \psi(Y) : \mathbf{Prop}} \; *$$

$$\frac{X \vdash_{\Sigma,\Xi} \phi(X) : \mathbf{Prop} \quad Y \vdash_{\Sigma,\Xi} \psi(Y) : \mathbf{Prop}}{X\,,\,Y \vdash \phi(X) \mathrel{-\!\!*} \psi(Y) : \mathbf{Prop}} \; -\!\!*$$

ADDITIVES

$$\frac{X \vdash_{\Sigma,\Xi} \phi(X) : \mathbf{Prop} \quad Y \vdash_{\Sigma,\Xi} \psi(Y) : \mathbf{Prop}}{X\,;\,Y \vdash_{\Sigma,\Xi} \phi(X) \wedge \psi(Y) : \mathbf{Prop}} \; \wedge$$

$$\frac{X \vdash_{\Sigma,\Xi} \phi(X) : \mathbf{Prop} \quad Y \vdash_{\Sigma,\Xi} \psi(Y) : \mathbf{Prop}}{X\,;\,Y \vdash_{\Sigma,\Xi} \phi(X) \rightarrow \psi(Y) : \mathbf{Prop}} \; \rightarrow$$

$$\frac{X \vdash_{\Sigma,\Xi} \phi(X) : \mathbf{Prop} \quad Y \vdash_{\Sigma,\Xi} \psi(Y) : \mathbf{Prop}}{X\,;\,Y \vdash_{\Sigma,\Xi} \phi(X) \vee \psi(Y) : \mathbf{Prop}} \; \vee$$

Table 11.2. Rules for Well-formed Propositions

of bunches of variables and the structure of propositions, given in Table 11.2. The multiplicative and additive units are well-formed in their respective unit contexts.

LEMMA 11.1 (ADDITIVE RULES) *The following additive versions of the* $\wedge$, $\vee$ *and* $\rightarrow$ *rules are admissible:*

$$\frac{X \vdash_{\Sigma,\Xi} \phi(X) : \mathbf{Prop} \quad X \vdash_{\Sigma,\Xi} \psi(X) : \mathbf{Prop}}{X \vdash_{\Sigma,\Xi} \phi(X) \wedge \psi(X) : \mathbf{Prop}} \; \wedge$$

$$\frac{X \vdash_{\Sigma,\Xi} \phi(X) : \mathbf{Prop} \quad X \vdash_{\Sigma,\Xi} \psi(X) : \mathbf{Prop}}{X \vdash_{\Sigma,\Xi} \phi(X) \to \psi(X) : \mathbf{Prop}} \to$$

$$\frac{X \vdash_{\Sigma,\Xi} \phi(X) : \mathbf{Prop} \quad X \vdash_{\Sigma,\Xi} \psi(X) : \mathbf{Prop}}{X \vdash_{\Sigma,\Xi} \phi(X) \vee \psi(X) : \mathbf{Prop}} \vee.$$

PROOF Straightforward. □

2. Variations on Predication

Given our formulation of predication, we can see that for each propositional connective we must choose whether to adopt additive or multiplicative maintenance of variables.

For example, as we have seen, the binary rule for introducing $*$ may be formulated either as

$$\frac{(X)\Gamma \vdash \phi \quad (Y)\Delta \vdash \psi}{(X,Y)\Gamma,\Delta \vdash \phi * \psi} \; *I(m),$$

in which the two premisses use distinct bunches of variables, combined multiplicatively in the conclusion, or as

$$\frac{(X)\Gamma \vdash \phi \quad (Y)\Delta \vdash \psi}{(X;Y)\Gamma,\Delta \vdash \phi * \psi} \; *I(a),$$

using ";" to combine bunches of variables. It follows, via Contraction,

$$\frac{Y(X;X')\Gamma \vdash \phi}{Y(X)\Gamma[X/X'] \vdash \phi[X/X']} \; (X \cong X') \quad \mathit{Contraction},$$

that the following, familiar, form of $*I(a)$ is sound:

$$\frac{(X)\Gamma \vdash \phi \quad (X)\Gamma \vdash \psi}{(X)\Gamma,\Delta \vdash \phi * \psi} \; *I(a).$$

This latter form is the one taken in linear logic [Girard, 1987].

In the sequel, we confine our development to the case in which (additive) multiplicative variable maintenance is paired with the (additive) multiplicative propositional connectives.

Whilst the development of the case in which all variable maintenance is additive, as in linear logic, seems straightforward, it remains unclear how best to formulate a system in which the additive propositional connectives are paired with multiplicative variable maintenance.

Chapter 12

NATURAL DEDUCTION AND SEQUENT CALCULUS

1. Propositional Rules

The propositional rules of predicate **BI** are not merely copies of their counterparts in propositional **BI**. Each proposition, ϕ, occurring in a sequent in an inference must be well-formed, *i.e.*, the sequent must include the variables X such that $X \vdash_{\Sigma,\Xi} \phi : \mathbf{Prop}$, as determined by the calculus of well-formed propositions. We extend well-formedness to bunches of propositions, Γ, as follows:

$$X \vdash_{\Sigma,\Xi} \Gamma : \mathbf{Prop} \quad \text{iff} \quad X \vdash_{\Sigma,\Xi} \phi_\Gamma : \mathbf{Prop},$$

where ϕ_Γ is the proposition obtained from Γ by replacing each semicolon with $\wedge$ and each comma with $*$, with association respecting the tree structure of Γ. We then have:

LEMMA 12.1 (WELL-FORMED BUNCHES) *If*

$$X \vdash_{\Sigma,\Xi} \Gamma : \mathbf{Prop} \textit{ and } Y \vdash_{\Sigma,\Xi} \Delta : \mathbf{Prop},$$

then

1 $X\, , Y \vdash_{\Sigma,\Xi} \Gamma\, , \Delta : \mathbf{Prop}$*;*

2 $X\, ; Y \vdash_{\Sigma,\Xi} \Gamma\, ; \Delta : \mathbf{Prop}$.

PROOF A straightforward induction on the structure of bunches. □

Where no confusion may arise, we shall refer to $(X)\Gamma$, provided it is such that $X \vdash_{\Sigma,\Xi} \Gamma : \mathbf{Prop}$, as a bunch and will refer to it simply as the bunch Γ.

We define

$$(X)\Gamma\,,\,(X')\Gamma' = (X, X')\Gamma, \Gamma' \quad \text{and} \quad (X)\Gamma\,;\,(X')\Gamma' = (X; X')\Gamma; \Gamma'$$

and extend coherent equivalence to bunches $(X)\Gamma$ as follows:

$$(X)\Gamma \equiv (X')\Gamma' \quad \text{iff} \quad (X) \equiv (X') \quad \text{and} \quad \Gamma \equiv \Gamma' \tag{12.1}$$

It follows that predicate **BI**'s sequents should be of the form

$$(X)\Gamma \vdash_{\Sigma,\Xi} \phi,$$

where $X \vdash_{\Sigma,\Xi} \Gamma : \mathbf{Prop}$ and $X \vdash_{\Sigma,\Xi} \phi : \mathbf{Prop}$. Before giving predicate **BI**'s natural deduction system, **NBI**, we explain how such sequents work, using a comparison between the $-\!\!*$ L rule of predicate **BI**'s sequent calculus and the $-\!\!*$ E rule of predicate **NBI**.

The form of the left rule for $-\!\!*$ in predicate **BI** is, essentially, simple:

$$\frac{(X)\Gamma, \psi \vdash_{\Sigma,\Xi} \chi \qquad (Y)\Delta \vdash_{\Sigma,\Xi} \phi}{(X, Y)\Gamma, \Delta, \phi -\!\!* \psi \vdash_{\Sigma,\Xi} \chi} \; -\!\!* L.$$

Here the bunches of variables X and Y from the two premisses are combined using "," to form the bunch required for the conclusion. This simple form is possible because of the "pure multiplicativity" of propositional **BI**'s $-\!\!*$ L rule: each proposition occuring in the conclusion occurs in exactly one of the premisses. The $-\!\!*$ E rule in propositional **BI**,

$$\frac{\Gamma \vdash \phi -\!\!* \psi \qquad \Delta \vdash \phi}{\Gamma, \Delta \vdash \psi} \; -\!\!* E,$$

does not have this property: the proposition ϕ occurs in both premisses. Accordingly, the form of the $-\!\!*$ E rule in predicate **BI** must reflect the duplication of ϕ by duplicating the variables required to establish that ϕ be a proposition. The rule goes as follows:

$$\frac{(X, Z)\Gamma \vdash_{\Sigma,\Xi} \phi -\!\!* \psi \qquad (Z, Y)\Delta \vdash_{\Sigma,\Xi} \phi}{(X, Y)\Gamma, \Delta \vdash_{\Sigma,\Xi} \psi} \quad Z \vdash_{\Sigma,\Xi} \phi : \mathbf{Prop} \quad -\!\!* E.$$

The form of this rule reflects the fact that the translation of **BI**'s natural deduction system into its sequent calculus maps the implication elimination rules to derivations which compose the corresponding implication left rules with a Cut on the principal formula of the elimination. To see this, consider the following simplified form of the Cut rule in predicate **BI**:

$$\frac{(X, Z)\Gamma, \phi \vdash_{\Sigma,\Xi} \psi \qquad (Z, Y)\Delta \vdash_{\Sigma,\Xi} \phi}{(X, Y)\Gamma, \Delta \vdash_{\Sigma,\Xi} \psi} \quad Z \vdash_{\Sigma,\Xi} \phi : \mathbf{Prop} \quad Cut.$$

The propositional Cut on ϕ is accompanied by the Cut on the bunch Z, which must be included in $-\!\!*$ E.

A similar analysis obtains for $*$E; it is formulated as follows:

$$\frac{(X,Z)\Delta \vdash_{\Sigma,\Xi} \phi * \psi \qquad (Z,Y)\Gamma(\phi,\psi) \vdash_{\Sigma,\Xi} \chi}{(X,Y)\Gamma(\Delta) \vdash_{\Sigma,\Xi} \chi} \quad Z \vdash_{\Sigma,\Xi} \phi*\psi : \mathbf{Prop} \quad *E.$$

The form of the axiom rule is, at first sight, quite odd. However, it is forced by our need to maintain multiplicative bunches of variables. To see this, consider, for example, a proof of

$$(\emptyset_m)\forall_{\mathbf{new}}x.\mathrm{p}(x) \vdash_{\Sigma,\Xi} \forall_{\mathbf{new}}x.\mathrm{p}(x)\,.$$

A proof begins with an inference,

$$\frac{(x,y)\mathrm{p}(y) \vdash_{\Sigma,\Xi} \mathrm{p}(x)}{(x)\forall_{\mathbf{new}}x.\mathrm{p}(x) \vdash_{\Sigma,\Xi} \mathrm{p}(x)},$$

in which, unlike the corresponding situation in intuitionistic logic, x and y must be distinct.

The price that must be paid is that we must record the fact that X_1 and X_2 have been treated as being the same in the axiom sequent. (It follows that $X_1 \cong X_2$.) This equality must be maintained in the proofs starting from such axioms and this is recorded by the symmetric and transitive relation, the *axiom pairing* $Axiom(X_1, X_2)$, on bunches, which is subsequently maintained up to α-conversion, *i.e.*, if some variable in X_1 or X_2 subsequently becomes bound and then α-converted, then $Axiom(X_1, X_2)$ must be similarly updated or α-converted. Similar concerns apply in the Substitution, Cut and quantifier rules, *i.e.*, the "identity group". Similarly, we must permit Contraction for multiplicatively combined *but axiom-related* bunches. It follows that we can recover the more familiar form of axiom; for example:

$$\frac{(x,y)\mathrm{p}(x) \vdash_{\Sigma,\Xi} \mathrm{p}(y)}{(x)\mathrm{p}(x) \vdash_{\Sigma,\Xi} \mathrm{p}(x)} \quad Axiom(x,y) \quad C.$$

An alternative solution to this problem would be to permit multiple occurrences of variables in bunches. Such a solution is adopted in the formulation of a dependent type theory, the $\lambda\Lambda$-calculus, in Chapter 15.

The remaining rules are now quite straightforward. Just as for propositional **BI**, we call the natural deduction system for predicate **BI**, given in Table 12.1, **NBI**.

IDENTITY AND STRUCTURE

$$\frac{X_1 \vdash_{\Sigma,\Xi} \phi(X_1) : \mathbf{Prop} \quad X_2 \vdash_{\Sigma,\Xi} \phi(X_2) : \mathbf{Prop}}{(X)\phi(X_1) \vdash_{\Sigma,\Xi} \phi(X_2)} \; (X = X_1, X_2 \text{ or } X_1; X_2,\ \mathit{Axiom}(X_1, X_2))$$

$$\frac{(X)\Gamma \vdash_{\Sigma,\Xi} \phi}{(Y)\Delta \vdash_{\Sigma,\Xi} \phi} \; ((Y)\Delta \equiv (X)\Gamma)\ E$$

$$\frac{(Y(X))\Gamma(\Delta) \vdash_{\Sigma,\Xi} \phi}{(Y(X; X'))\Gamma(\Delta; \Delta') \vdash_{\Sigma,\Xi} \phi} \; (X' \vdash_{\Sigma,\Xi} \Delta' : \mathbf{Prop}) \;\; W \qquad \frac{(Y(X))\Gamma \vdash_{\Sigma,\Xi} \phi}{(Y(X; X'))\Gamma \vdash_{\Sigma,\Xi} \phi} \; W$$

$$\frac{(Y(X; X'))\Gamma(\Delta; \Delta') \vdash_{\Sigma,\Xi} \phi}{(Y(X))\Gamma(\Delta)[X/X'] \vdash_{\Sigma,\Xi} \phi[X/X']} \; (\Delta' \cong \Delta, X' \cong X, X' \vdash_{\Sigma,\Xi} \Delta' : \mathbf{Prop}) \;\; C$$

$$\frac{(Y(X; X'))\Gamma \vdash_{\Sigma,\Xi} \phi}{(Y(X))\Gamma[X/X'] \vdash_{\Sigma,\Xi} \phi[X/X']} \; (X' \cong X)$$

$$\frac{(Y(X, X'))\Gamma \vdash_{\Sigma,\Xi} \phi}{(Y(X))\Gamma[X/X'] \vdash_{\Sigma,\Xi} \phi[X/X']} \; (\mathit{Axiom}(X, X'), X' \cong X) \;\; C$$

$$\frac{(X(x \in A))\Gamma \vdash_{\Sigma,\Xi} \phi \quad Y \vdash_{\Sigma} t : A}{(X(Y))\Gamma[t/x] \vdash_{\Sigma,\Xi} \phi[t/x]} \; \mathit{Axiom}(Y, x') \quad \mathit{Substitution},$$

for all x' such that $\mathit{Axiom}(x, x')$ (here we abuse notation and write just $\mathit{Axiom}(Y, x')$ rather than pick out ys from Y).

MULTIPLICATIVES

$$\frac{}{(\emptyset_m)\emptyset_m \vdash_{\Sigma,\Xi} I} \; I\,I \qquad \frac{(X)\Gamma(\emptyset_m) \vdash_{\Sigma,\Xi} \chi \quad (Y)\Delta \vdash_{\Sigma,\Xi} I}{(X, Y)\Gamma(\Delta) \vdash_{\Sigma,\Xi} \chi} \; I\,E$$

$$\frac{(X)\Gamma, \phi \vdash_{\Sigma,\Xi} \psi}{(X)\Gamma \vdash_{\Sigma,\Xi} \phi \mathrel{-\!\!*} \psi} \; \mathrel{-\!\!*} I \qquad \frac{(X, Z)\Gamma \vdash_{\Sigma,\Xi} \phi \mathrel{-\!\!*} \psi \quad (Z, Y)\Delta \vdash_{\Sigma,\Xi} \phi}{(X, Y)\Gamma, \Delta \vdash_{\Sigma,\Xi} \psi} \; \mathrel{-\!\!*} E$$

where $Z \vdash_{\Sigma,\Xi} \phi : \mathbf{Prop}$

$$\frac{(X)\Gamma \vdash \phi \quad (Y)\Delta \vdash_{\Sigma,\Xi} \psi}{(X, Y)\Gamma, \Delta \vdash_{\Sigma,\Xi} \phi * \psi} \; *I \qquad \frac{(X, Z)\Gamma(\phi, \psi) \vdash_{\Sigma,\Xi} \chi \quad (Z, Y)\Delta \vdash_{\Sigma,\Xi} \phi * \psi}{(X, Y)\Gamma(\Delta) \vdash_{\Sigma,\Xi} \chi} \; *E$$

where $Z \vdash_{\Sigma,\Xi} \phi * \psi : \mathbf{Prop}$

ADDITIVES

$$\frac{}{(\emptyset_a)\emptyset_a \vdash_{\Sigma,\Xi} 1} \; 1\,I \qquad \frac{(X)\Gamma(\emptyset_a) \vdash_{\Sigma,\Xi} \chi \quad (Y)\Delta \vdash_{\Sigma,\Xi} 1}{(X; Y)\Gamma(\Delta) \vdash_{\Sigma,\Xi} \chi} \; 1\,E$$

$$\frac{(X)\Gamma; \phi \vdash_{\Sigma,\Xi} \psi}{(X)\Gamma \vdash_{\Sigma,\Xi} \phi \to \psi} \; \to I \qquad \frac{(X; Z)\Gamma \vdash_{\Sigma,\Xi} \phi \to \psi \quad (Z; Y)\Delta \vdash_{\Sigma,\Xi} \phi}{(X; Y)\Gamma; \Delta \vdash_{\Sigma,\Xi} \psi} \; \to E$$

where $Z \vdash_{\Sigma,\Xi} \phi : \mathbf{Prop}$

$$\frac{(X)\Gamma \vdash_{\Sigma,\Xi} \phi \quad (Y)\Delta \vdash_{\Sigma,\Xi} \psi}{(X; Y)\Gamma; \Delta \vdash_{\Sigma,\Xi} \phi \wedge \psi} \; \wedge I \qquad \frac{(X; Z)\Gamma(\phi; \psi) \vdash_{\Sigma,\Xi} \chi \quad (Z; Y)\Delta \vdash_{\Sigma,\Xi} \phi \wedge \psi}{(X; Y)\Gamma(\Delta) \vdash_{\Sigma,\Xi} \chi} \; \wedge E$$

where $Z \vdash_{\Sigma,\Xi} \phi \wedge \psi : \mathbf{Prop}$

$$\frac{(X)\Gamma \vdash_{\Sigma,\Xi} \phi_i}{(X)\Gamma \vdash_{\Sigma,\Xi} \phi_1 \vee \phi_2} \; (i = 1, 2) \quad \vee I$$

where $X \vdash_{\Sigma,\Xi} \phi_1 \vee \phi_2 : \mathbf{Prop}$

$$\frac{(X)\Gamma \vdash_{\Sigma,\Xi} \bot}{(X)\Gamma \vdash_{\Sigma,\Xi} \phi} \; \bot E$$

$$\frac{(X; Z_1; Z_2)\Gamma \vdash_{\Sigma,\Xi} \phi_1 \vee \phi_2 \quad (Y; Z_1)\Delta(\phi_1) \vdash_{\Sigma,\Xi} \chi \quad (Y; Z_2)\Delta(\phi_2) \vdash_{\Sigma,\Xi} \chi}{(Y; X)\Delta(\Gamma) \vdash_{\Sigma,\Xi} \chi} \; \vee E$$

where $Z_i \vdash_{\Sigma,\Xi} \phi_i$ $(i = 1, 2)$.

Table 12.1. Predicate **NBI**

Notice that the introduction and elimination rules for additive and multiplicative implications, conjunctions and units are identical in form, following Prawitz's prescription [Prawitz, 1971]. The difference between them is the antecedent-combining operations they use. We can replace the $\wedge$E rule with the simpler, and perhaps more familiar, form

$$\frac{(X)\Gamma \vdash_{\Sigma,\Xi} \phi_1 \wedge \phi_2}{(X)\Gamma \vdash_{\Sigma,\Xi} \phi_i} \quad i = 1, 2.$$

LEMMA 12.2 *If* $(X)\Gamma \vdash_{\Sigma,\Xi} \phi$ *in* **NBI**, *then* $X \vdash_{\Sigma,\Xi} \phi :$ **Prop**.

PROOF A straightforward induction on the structure of proofs in **NBI**. □

LEMMA 12.3 *1* $(X)\Gamma(\phi_1, \phi_2) \vdash \psi$ *iff* $(X)\Gamma(\phi_1 * \phi_2) \vdash \psi$*;*

2 $(X)\Gamma(\phi_1; \phi_2) \vdash \psi$ *iff* $(X)\Gamma(\phi_1 \wedge \phi_2) \vdash \psi$.

PROOF Again, straightforward induction on the structure of proofs in **NBI**. □

Our need to maintain the pairings given by the relation *Axiom* extends to the Cut rule. The reason is that predicate **BI**'s Cut must build in some substitution.

LEMMA 12.4 (ADMISSIBILITY OF CUT) *The following Cut rule is admissible in* **NBI***:*

$$\frac{(X(X_1))\Gamma(\psi(X_1)) \vdash_{\Sigma,\Xi} \psi \quad (Y(X_2))\Delta \vdash_{\Sigma,\Xi} \phi(X_2)}{(X(Y[X_1/X_2])[U/x'])\Gamma(\Delta)[X_1/X_2] \vdash_{\Sigma,\Xi} \psi} \quad \begin{array}{l} Axiom(X_1, X_2) \\ X_i \vdash_{\Sigma,\Xi} \phi_i(X_i) : \mathbf{Prop} \end{array} \quad Cut,$$

for all x' *in* $Y(-)$ *such that* $Axiom(x', X)$*, where* U *denotes* $\emptyset_m$ *or* $\emptyset_a$ *according to the occurrences of* x'.

PROOF By induction on the structure of proofs in **NBI**. □

EXAMPLE 12.5 *An example of Cut, which illustrates simply the use of the relation Axiom, is as follows (omitting types, for brevity):*

$$\frac{(v, w)\mathrm{p}(v) \vdash_{\Sigma,\Xi} \mathrm{p}(w) \quad (x, y)\mathrm{p}(x) \vdash_{\Sigma,\Xi} \mathrm{p}(y)}{(v, w)\mathrm{p}(v) \vdash_{\Sigma,\Xi} \mathrm{p}(w)},$$

in which we have $Axiom(v, w)$ *from the left-hand premiss and* $Axiom(x, y)$ *from the right-hand premiss* (i.e., *the Cut sequent*)*. The rule introduces*

Axiom(y, v), *with the consequence, as expected, that both* x *and* y *are absent in the conclusion the conclusion. In this instance,* X_1 *is* v, X_2 *is* y, *so that* $\Gamma(\Delta)[X_1/X_2]$ *is just* $p(v)$ *and* $(X(Y[X_1/X_2])[U/x'])$ *is just* $(v, x, w)[\emptyset_m/x]$, i.e., *just* (v, w), *since Axiom*(x, y) *and Axiom*(y, v) *imply Axiom*(x, v).

Alternatively, suppose we have premisses in which the axiom-related variables have been contracted, e.g.,

$$\frac{(y,z)\mathrm{q}(y), \mathrm{q}(y) \mathrel{-\!\!*} \mathrm{p}(z) \vdash_{\Sigma,\Xi} \mathrm{p}(z) \quad (x)\mathrm{q}(x) \vdash_{\Sigma,\Xi} \mathrm{q}(x)}{((y,z)[U/x'])(\mathrm{q}(x), \mathrm{q}(y) \mathrel{-\!\!*} \mathrm{p}(z))[y/x] \vdash_{\Sigma,\Xi} \mathrm{p}(z)},$$

i.e.,

$$\frac{(y,z)\mathrm{q}(y), \mathrm{q}(y) \mathrel{-\!\!*} \mathrm{p}(z) \vdash_{\Sigma,\Xi} \mathrm{p}(z) \quad (x)\mathrm{q}(x) \vdash_{\Sigma,\Xi} \mathrm{q}(x)}{(y,z)\mathrm{q}(y), \mathrm{q}(y) \mathrel{-\!\!*} \mathrm{p}(z) \vdash_{\Sigma,\Xi} \mathrm{p}(z)}.$$

In fact, we shall refer to the systems we have given for each of the judgements (i) $(X)\Gamma \vdash_{\Sigma,\Xi} \phi : \mathbf{Prop}$, (ii) $(X)\Gamma \vdash_{\Sigma,\Xi} \phi$ and (iii) $X \vdash_{\Sigma} t : A$ collectively as **NBI**. The interaction between (i) and (ii) is of little further importance. The interaction, via Substitution, between (ii) and (iii) is more important. Both are subject to reduction rules, with normalization properties as discussed Chapter 2 and, for the quantifiers in particular, which may be treated independently without affecting the instances of Substitution.

2. Quantifier Rules

We have seen that, as well as the familiar intuitionistic quantifiers, $\forall$ and $\exists$, predicate **BI** also has multiplicative, or intensional, quantifiers, obtained by observing structural restrictions on the level of terms as well as propositions.

We begin by extending the well-formedness judgement, as follows:

$$\frac{X; x!A \vdash_{\Sigma,\Xi} \phi : \mathbf{Prop}}{X \vdash_{\Sigma,\Xi} \forall x!A.\phi : \mathbf{Prop}}\ \forall \qquad \frac{X; x!A \vdash_{\Sigma,\Xi} \phi : \mathbf{Prop}}{X \vdash_{\Sigma,\Xi} \exists x!A.\phi : \mathbf{Prop}}\ \exists$$

and

$$\frac{X, x : A \vdash_{\Sigma,\Xi} \phi : \mathbf{Prop}}{X \vdash_{\Sigma,\Xi} \forall_{\mathbf{new}} x : A.\phi : \mathbf{Prop}}\ \forall_{\mathbf{new}} \qquad \frac{X, x : A \vdash_{\Sigma,\Xi} \phi : \mathbf{Prop}}{X \vdash_{\Sigma,\Xi} \exists_{\mathbf{new}} x : A.\phi : \mathbf{Prop}}\ \exists_{\mathbf{new}}.$$

Note that the multiplicatives are distinguished from the additives by the use of "," in place of ";" in the extension of X by the eigenvariable $x : A$ or $x!A$, respectively. Otherwise, these rules all simply treat quantifiers as binders.

MULTIPLICATIVES

$$\frac{(X, x : A)\Gamma \vdash_{\Sigma,\Xi} \phi}{(X)\Gamma \vdash_{\Sigma,\Xi} \forall_{\mathbf{new}} x : A \,.\, \phi}\ \forall_{\mathbf{new}} I$$

$$\frac{(X)\Gamma \vdash_{\Sigma,\Xi} \forall_{\mathbf{new}} x : A \,.\, \phi \quad Y \vdash_{\Sigma} t : A}{(X, Y)\Gamma \vdash_{\Sigma,\Xi} \phi[t/x]}\ \forall_{\mathbf{new}} E$$

$$\frac{(X, Y)\Gamma \vdash_{\Sigma,\Xi} \phi[t/x] \quad Y \vdash_{\Sigma} t : A \quad X \vdash_{\Sigma,\Xi} \Gamma : \mathbf{Prop}}{(X)\Gamma \vdash_{\Sigma,\Xi} \exists_{\mathbf{new}} x : A.\phi}\ \exists_{\mathbf{new}} I$$

$$\frac{(X)\Gamma \vdash_{\Sigma,\Xi} \exists_{\mathbf{new}} x : A.\phi \quad (Y, x : A)\Delta, \phi \vdash_{\Sigma,\Xi} \psi}{(X, Y)\Gamma, \Delta \vdash_{\Sigma,\Xi} \psi}\ \exists_{\mathbf{new}} E$$

ADDITIVES

$$\frac{(X; x!A)\Gamma \vdash_{\Sigma,\Xi} \phi}{(X)\Gamma \vdash_{\Sigma,\Xi} \forall x!A \,.\, \phi}\ \forall I \qquad \frac{(X)\Gamma \vdash_{\Sigma,\Xi} \forall x!A \,.\, \phi \quad Y \vdash_{\Sigma} t : A}{(X; Y)\Gamma \vdash_{\Sigma,\Xi} \phi[t/x]}\ \forall E$$

$$\frac{(X; Y)\Gamma \vdash_{\Sigma,\Xi} \phi[t/x] \quad Y \vdash_{\Sigma} t : A}{(X)\Gamma \vdash_{\Sigma,\Xi} \exists x!A.\phi}\ \exists I$$

$$\frac{(X)\Gamma \vdash_{\Sigma,\Xi} \exists x!A.\phi \quad (X; x!A)\Delta; \phi \vdash_{\Sigma,\Xi} \psi}{(X)\Gamma; \Delta \vdash_{\Sigma,\Xi} \psi}\ \exists E$$

Table 12.2. Quantifier Rules

Turning to **NBI**, we extend it with the following introduction and elimination rules for the quantifiers. There is a small technical point concerning their interaction with the form of axiom sequents which requires some care. We must prevent the variables which are used to form axioms, *i.e.*, in the conditions $Axiom(X_1, X_2)$, from being bound by both existential elimination (or left) *and* universal introduction (or right) rules. This is enforced in side-conditions on these rules, similar to the usual side-conditions on the quantifier rules in intuitionistic logic.

The quantifier rules, given in Table 12.2, do not compromise the admissibility of Cut, Lemma 12.4.

The existential introduction rules have the condition that the variables in Y do not occur in Γ. The universal introduction rules have the usual restriction that x must not occur free in Γ and that $x \notin \{ y \mid \mathit{Axiom}(y,z) \text{ and } z \in \mathrm{FV}(\Gamma) \}$. Similarly, the existential elimination rules have the restriction that x must not occur free in Δ or ψ and that $x \notin \{ y \mid \mathit{Axiom}(y,z) \text{ and } z \in \mathrm{FV}(\psi) \text{ or } \mathrm{FV}(\Delta), \mathrm{FV}(\phi) \}$ (here also we abuse notation and write just $\mathit{Axiom}(y,z)$ rather than pick out y and z from arbitrary bunches). We call these conditions, collectively, the *eigenvariable conditions.*

The idea of the introduction rule for $\forall_{\mathbf{new}}$ is that we may infer $\forall_{\mathbf{new}} x : A\,.\,\phi$ in the usual way for universal quantification, except that the variable x must sit in multiplicative combination with all of the other variables. Moreover, it must not be possible to change the relative status of a variable via sequences of unit operations on bunches. This requirement is enforced by the declaration of a variable as being multiplicative ($x : A$) or additive ($x!A$) in a bunch.[1] In the elimination rule we must be careful not to substitute an arbitrary term for x but only one that is respective of the multiplicative relationship between x and other variables in X. In particular, t cannot contain any of these other variables appearing in X; this requirement is implemented by the linearity restriction. At this point, we emphasize two points from the introduction to Part II:

- $\forall_{\mathbf{new}}$ in not read literally as "for all". Rather, the multiplicative relationship between x and other variables must be observed; and

- $\forall_{\mathbf{new}}$ is not an infinitary multiplicative conjunction. Rather, $\forall_{\mathbf{new}}$ is closely related to the multiplicative implication $-\!\!*$, just as $\forall$ is closely related to $\rightarrow$. These relationships may be understood in the context of dependent types (*q.v.* Chapter 15).

We conclude this section by remarking that it is easy to see that predicate **BI** is a conservative extension of intuitionistic predicate logic. As an example, consider that the (slightly abbreviated, with types omitted) proof in Figure 12.1, in which for brevity we anticipate the left rule for the universal quantifier to be introduced in the sequel, represents the usual intuitionistic proof of the sequent $\mathrm{p}(x); \forall y.\mathrm{p}(y) \rightarrow \mathrm{q}(y) \vdash_{\Sigma,\Xi} \mathrm{q}(x)$, except that, in order to illustrate their rôles, we distinguish the Substitution and Contraction steps (on variables).

[1]In the propositional rules, the ***binary*** connective itself provides the syntactic enforcement of this property. For the ***unary*** quantifers, we must enforce it explicitly.

$$\dfrac{\dfrac{\dfrac{\dfrac{\dfrac{\dfrac{\dfrac{}{(x;y)\mathrm{p}(x) \vdash_{\Sigma,\Xi} \mathrm{p}(y)}\;Axiom(x,y) \qquad \dfrac{}{(w;z)\mathrm{p}(w) \vdash_{\Sigma,\Xi} \mathrm{p}(z)}\;Axiom(w,z)}{(x;y;w;z)\mathrm{p}(x);\mathrm{p}(y) \to \mathrm{q}(w) \vdash_{\Sigma,\Xi} \mathrm{q}(z)}\;\to L(\text{see § 4})}{(x;x';y;w)\mathrm{p}(x);\mathrm{p}(y) \to \mathrm{q}(w) \vdash_{\Sigma,\Xi} \mathrm{q}(x')}\;Substitution}{(x;y;w)\mathrm{p}(x);\mathrm{p}(y) \to \mathrm{q}(w) \vdash_{\Sigma,\Xi} \mathrm{q}(x)}\;Contraction}{(x;y;y')\mathrm{p}(x);\mathrm{p}(y) \to \mathrm{q}(y') \vdash_{\Sigma,\Xi} \mathrm{q}(x)}\;Substitution}{(x;y)\mathrm{p}(x);\mathrm{p}(y) \to \mathrm{q}(y) \vdash_{\Sigma,\Xi} \mathrm{q}(x)}\;Contraction}{(x;\emptyset_a)\mathrm{p}(x);\forall y.\mathrm{p}(y) \to \mathrm{q}(y) \vdash_{\Sigma,\Xi} \mathrm{q}(x)}\;\forall L(\text{ see § 4})$$

Figure 12.1. Substitution and Contraction

In contrast, much weaker arguments obtain for, say, $\forall_{\mathbf{new}}$. For example, omitting types, we have the following derivation:

$$\dfrac{\dfrac{\dfrac{}{(x,y)\mathrm{p}(x) \vdash_{\Sigma,\Xi} \mathrm{p}(y)}\;Axiom(x,y)}{(\emptyset_m,y)\forall_{\mathbf{new}}x.\mathrm{p}(x) \vdash_{\Sigma,\Xi} \mathrm{p}(y)}\;\forall_{\mathbf{new}}L \qquad \dfrac{\vdots}{Y \vdash_{\Sigma} t}}{(\emptyset_m,Y)\forall_{\mathbf{new}}x.\mathrm{p}(x) \vdash_{\Sigma,\Xi} \mathrm{p}(t)}\;Substitution.$$

We can now clearly see that predicate **BI** is an extension of predicate multiplicative intuitionistic linear logic: Linear logic combines additive variable maintenance only (and so additive quantifiers only) with multiplicative connectives. So to obtain a suitable version of **BI** we must, for example, modify the $-\!\!*$ L rule (*q.v.* § 4) to be

$$\dfrac{(X)\Gamma \vdash_{\Sigma,\Xi} \phi \quad (Y)\Delta(\psi) \vdash_{\Sigma,\Xi} \chi}{(X;Y)\Delta(\Gamma, \phi -\!\!* \psi) \vdash_{\Sigma,\Xi} \chi}\;-\!\!* L.$$

The other multiplicative rules must be modified similarly. Such a system conservatively extends predicate multiplicative intuitionistic linear logic.

We say that $(X)\Gamma$ is *consistent* (with respect to Σ and Ξ) if

$$(X)\Gamma \not\vdash_{\Sigma,\Xi} \bot\,.$$

Finally, note that **BI**'s two implications each give rise to a notion of theorem: a proposition τ is a *theorem* if either

- $(U)\emptyset_a \vdash \tau$ or

- $(U)\emptyset_m \vdash \tau$,

where U denotes $\emptyset_a$ or $\emptyset_m$, is provable. However, if either $(\emptyset_a)\emptyset_a \vdash \tau$, $(\emptyset_m)\emptyset_a \vdash \tau$ or $(\emptyset_a)\emptyset_m \vdash \tau$ is provable, then $(\emptyset_m)\emptyset_m \vdash \tau$, so that we need just $(\emptyset_m)\emptyset_m \vdash \tau$.

3. Strong Normalization and Subject Reduction

In Chapter 2, we have shown how to establish the strong normalization (SN) and subject reduction (SR) properties for propositional **BI**'s natural deduction system (propositional **NBI**), which may be viewed as the $\alpha\lambda$-calculus. Our technique [Troelstra and Schwichtenberg, 1996], a standard one in the theory of term rewriting, was to translate propositional **BI** into propositional **IL**.

In this section, we extend our result to predicate **NBI**. Firstly, we give a sketch of the necessary extension, including reduction rules, for the quantifiers. In the spirit of Chapter 2, we explain the reductions using simplified graphical forms of **NBI**'s natural deduction rules, suppressing, for simplicity, the manipulation of bunches of variables.

REDUCTIONS

β-reductions

$\forall$: Here the bound variable $x!A$ corresponds to an extension $X; x!A$ of the bunch of variables X.

$$\dfrac{\dfrac{\begin{matrix}\Phi \\ \vdots \\ \phi(x)\end{matrix}}{\forall x!A.\phi(x)}\ \forall I}{\phi[t/x]}\ \forall E \quad \succcurlyeq \quad \begin{matrix}\Phi[t/x] \\ \vdots \\ \phi[t/x]\end{matrix}$$

$\exists$: Again, here the bound variable $x!A$ corresponds to an extension $X; x!A$ of the bunch of variables X.

$$\dfrac{\dfrac{\begin{matrix}\Phi \\ \vdots \\ \phi[t/x]\end{matrix}}{\exists x!A.\phi(x)}\ \exists I \qquad \begin{matrix}[\phi] \\ \Phi' \\ \vdots \\ \chi\end{matrix}}{\chi}\ \exists E \quad \succcurlyeq \quad \begin{matrix}\Phi \\ \vdots \\ [\phi[t/x]] \\ \Phi'[t/x] \\ \vdots \\ \chi\end{matrix}$$

Here we use the "discharge notation", $[-]$, to denote graphically the removal of a propostion from the antecedent of a consequence.

$\forall_{\mathbf{new}}$: Here the bound variable $x : A$ corresponds to an extension $X, x : A$ of the bunch of variables X:

$$\dfrac{\dfrac{\begin{matrix}\Phi\\ \vdots\\ \phi(x)\end{matrix}}{\forall_{\mathbf{new}} x : A.\phi(x)}\,\forall_{\mathbf{new}} I}{\phi[t/x]}\,\forall_{\mathbf{new}} E \quad \succcurlyeq \quad \begin{matrix}\Phi[t/x]\\ \vdots\\ \phi[t/x]\end{matrix}$$

$\exists_{\mathbf{new}}$: Again, here the bound variable $x : A$ corresponds to an extension $X, x : A$ of the bunch of variables X:

$$\dfrac{\dfrac{\begin{matrix}\Phi\\ \vdots\\ \phi[t/x]\end{matrix}}{\exists_{\mathbf{new}} x : A.\phi(x)}\,\exists_{\mathbf{new}} I \qquad \begin{matrix}[\phi]\\ \Phi'\\ \vdots\\ \chi\end{matrix}}{\chi}\,\exists_{\mathbf{new}} E \quad \succcurlyeq \quad \begin{matrix}\Phi\\ \vdots\\ [\phi[t/x]]\\ \Phi'[t/x]\\ \vdots\\ \chi\end{matrix}$$

Again, here we use the "discharge notation", $[-]$, to denote graphically the removal of a propostion from the antecedent of a consequence.

ζ-reductions

$\exists$: As in the β reductions, the bound variable for $\exists$ corresponds to a bunch of variables of the form $X; x!A$:

$$\dfrac{\dfrac{\begin{matrix}\Phi\\ \vdots\\ \exists x!A.\phi\end{matrix} \quad \begin{matrix}\Phi'\\ \vdots\\ \chi\end{matrix}}{\chi}\,\exists E \qquad \begin{matrix}\Phi''\\ \vdots\end{matrix}}{\psi}\,Elim \quad \succcurlyeq \quad \dfrac{\begin{matrix}\Phi\\ \vdots\\ \exists x!A.\phi\end{matrix} \qquad \dfrac{\begin{matrix}\Phi'\\ \vdots\\ \chi\end{matrix} \quad \begin{matrix}\Phi''\\ \vdots\end{matrix}}{\psi}\,Elim}{\psi}\,\exists E$$

where *Elim* stands for any additive elimination rule.

$\exists_{\mathbf{new}}$: As in the β reductions, the bound variable for $\exists_{\mathbf{new}}$ corresponds to a bunch of variables of the form $X, x : A$:

$$\dfrac{\dfrac{\begin{matrix}\Phi\\ \vdots\\ \exists_{\mathbf{new}} x : A.\phi\end{matrix} \quad \begin{matrix}\Phi'\\ \vdots\\ \chi\end{matrix}}{\chi}\,\exists_{\mathbf{new}} E \qquad \begin{matrix}\Phi''\\ \vdots\end{matrix}}{\psi}\,Elim \quad \succcurlyeq \quad \dfrac{\begin{matrix}\Phi\\ \vdots\\ \exists_{\mathbf{new}} x : A.\phi\end{matrix} \qquad \dfrac{\begin{matrix}\Phi'\\ \vdots\\ \chi\end{matrix} \quad \begin{matrix}\Phi''\\ \vdots\end{matrix}}{\psi}\,Elim}{\psi}\,\exists_{\mathbf{new}} E$$

where *Elim* stands for any multiplicative elimination rule.

We note that multiplicative additive cross cases do not reduce.[2] We suppress the degenerate "simplification" reductions.

We extend the mapping $\mapsto$, defined for propositional **BI** in Chapter 2, to predicate **BI**'s four quantifiers as follows:

$$\begin{array}{ccccccc} \forall & \mapsto & \forall & \quad & \forall_{\mathbf{new}} & \mapsto & \forall \\ \exists & \mapsto & \exists & \quad & \exists_{\mathbf{new}} & \mapsto & \exists. \end{array}$$

It should be clear that all of **NBI**'s reductions have images in those of the natural deduction system for intuitionistic predicate logic, so that the proof by translation, as described in Chapter 2, is sound.

THEOREM 12.6 (STRONG NORMALIZATION) *Proofs in predicate* **NBI** *are strongly normalizing.*

PROOF SKETCH By induction on the structure of proofs in **NBI**, using the translation to natural deduction for intuitionistic logic via the mapping $\mapsto$. The proof relies on the techniques used in [Prawitz, 1965, Ghani, 1995, Pym and Ritter, 2001]. □

THEOREM 12.7 (SUBJECT REDUCTION) *If* Φ *is a proof in* **NBI** *of* $(X)\Gamma \vdash_{\Sigma,\Xi} \phi$ *and if* $\Phi \succeq \Phi'$, *then* Φ' *is a proof in* **NBI** *of* $(X)\Gamma \vdash_{\Sigma,\Xi} \phi$.

PROOF By induction on the structure of the derivation of the reduction $\Phi \succeq \Phi'$. □

4. Predicate BI as a Sequent Calculus

In Chapter 6, we have shown how to reformulate propositional **NBI** as a sequent calculus [Gentzen, 1934], in which the elimination rule for each connective, #, is replaced by a corresponding left rule which introduces # to the left-hand side of a sequent. We called the calculus **LBI**. We showed that Cut-elimination holds for **LBI** and that **LBI** and **NBI** are equivalent for logical consequence.

Here, we sketch the extension of **LBI** to predicate **BI**. For example, the rule for $-\!\!*$ on the left goes as follows:

$$\frac{(X)\Gamma \vdash_{\Sigma,\Xi} \phi \quad (Y(Z))\Delta(\psi) \vdash_{\Sigma,\Xi} \chi}{(Y(X,Z))\Delta(\Gamma, \phi -\!\!* \psi) \vdash_{\Sigma,\Xi} \chi} \quad Z \vdash_{\Sigma,\Xi} \psi : \mathbf{Prop} \quad -\!\!*\, L;$$

[2]Because "," and ";" do not distribute over one another.

similarly, the rule for $\to$ on the left goes as follows:

$$\frac{(X)\Gamma \vdash_{\Sigma,\Xi} \phi \quad (Y(Z))\Delta(\psi) \vdash_{\Sigma,\Xi} \chi}{(Y(X;Z))\Delta(\Gamma; \phi \to \psi) \vdash_{\Sigma,\Xi} \chi} \quad Z \vdash_{\Sigma,\Xi} \psi : \mathbf{Prop} \quad \to L.$$

Note that we take care to ensure the correct substitution pattern for the first-order variables: The bunch X of variables, required for Γ and ϕ is substituted into the bunch required for the major premiss .

The formulation of the left rules for the four quantifiers, in which we take term, Σ, and predicate, Ξ, signatures, goes as follows:

$\forall$:

$$\frac{(Y(X))\Gamma(\phi(t)) \vdash_{\Sigma,\Xi} \psi}{(Y(\emptyset_a))\Gamma(\forall x!A.\phi(x)) \vdash_{\Sigma,\Xi} \psi} \quad (X \vdash_{\Sigma} t : A) \quad \forall L,$$

where we assume that the variables in X do not occur in Γ, ψ. The right rule is

$$\frac{(X; x!A)\Gamma \vdash_{\Sigma,\Xi} \phi}{(X)\Gamma \vdash_{\Sigma,\Xi} \forall x!A.\phi} \quad \forall R,$$

subject to the same conditions as $\forall I$;

$\exists$:

$$\frac{(Y(x!A))\Gamma(\phi(x)) \vdash_{\Sigma,\Xi} \psi}{(Y(\emptyset_a))\Gamma(\exists x!A.\phi(x)) \vdash_{\Sigma,\Xi} \psi} \quad \exists L,$$

where $x \notin \mathrm{FV}(\Gamma, \psi)$. The right rule translates directly from **NBI**;

$\forall_{\mathbf{new}}$:

$$\frac{(Y(X))\Gamma(\phi(t)) \vdash_{\Sigma,\Xi} \psi}{(Y(\emptyset_m))\Gamma(\forall_{\mathbf{new}} x : A.\phi(x)) \vdash_{\Sigma,\Xi} \psi} \quad (X \vdash_{\Sigma} t : A) \quad \forall_{\mathbf{new}} L,$$

where we assume that the variables in X do not occur in Γ, ψ. The right rule is

$$\frac{(X; x!A)\Gamma \vdash_{\Sigma,\Xi} \phi}{(X)\Gamma \vdash_{\Sigma,\Xi} \forall x!A.\phi} \quad \forall_{\mathbf{new}} R,$$

subject to the same conditions as $\forall_{\mathbf{new}} I$;

$\exists_{\mathbf{new}}$:

$$\frac{(Y(x : A))\Gamma(\phi(x)) \vdash_{\Sigma,\Xi} \psi}{(Y(\emptyset_m))\Gamma(\exists_{\mathbf{new}} x : A.\phi(x)) \vdash_{\Sigma,\Xi} \psi} \quad \exists_{\mathbf{new}} L,$$

where $x \notin \mathrm{FV}(\Gamma, \psi)$. The right rule translates directly from **NBI**.

The need for the distinction between : and ! may be seen particularly clearly in the sequent calculus. To see this, compare the left rules for the quantifiers with the left rules (propositional will do) for the implications,

$$\frac{\Gamma \vdash \phi \quad \Delta(\psi) \vdash \chi}{\Delta(\Gamma, \phi \mathrel{-\!\!*} \psi) \vdash \chi} \; \mathrel{-\!\!*} L \quad \text{and} \quad \frac{\Gamma \vdash \phi \quad \Delta(\psi) \vdash \chi}{\Delta(\Gamma; \phi \to \psi) \vdash \chi} \; \to L,$$

in which the multiplicative/additive status of the rule is built into the form of operator (connective/quantifier). The quantifier rules have this property only if we impose it via a distinction such as the one provided by :/!.

As we have seen, the left rules for the quantifiers give access to variables occurring in arbitrary positions inside a bunch, on the left-hand side of a sequent. Consequently, the form of the axiom must be modified to permit the extraction of a (sub-)bunch of variables from an arbitrary position within a bunch:

$$\frac{X_1 \vdash_{\Sigma,\Xi} \phi(X_1) : \mathbf{Prop} \quad X[U/X_1] \vdash_{\Sigma,\Xi} \phi(X_2) : \mathbf{Prop}}{(X)\phi(X_1) \vdash_{\Sigma,\Xi} \phi(X_2)} \ (U = \emptyset_m \text{ or } \emptyset_a,\ \mathit{Axiom}(X_1, X_2)).$$

The Cut-elimination theorem and equivalence of **LBI** and **NBI**, proved for propositional **BI** in Chapter 6, extend to predicate **BI** straightforwardly: We omit the details, contenting ourselves with a brief example of the argument for each of them.

Starting with Cut-elimination, consider the following example of (a simple version of) the inductive case for a Cut on $\phi \mathrel{-\!\!*} \psi$: the proof figure

$$\frac{\dfrac{(X, X')\Gamma \vdash_{\Sigma,\Xi} \phi \quad (Y, Y')\Delta(\psi) \vdash_{\Sigma,\Xi} \chi}{(X, X', Y, Y')\Delta(\Gamma, \phi \mathrel{-\!\!*} \psi) \vdash_{\Sigma,\Xi} \chi} \mathrel{-\!\!*} L \quad \dfrac{(X', Y', Z)\Theta, \phi \vdash_{\Sigma,\Xi} \psi}{(Z, X', Y')\Theta \vdash_{\Sigma,\Xi} \phi \mathrel{-\!\!*} \psi} \mathrel{-\!\!*} R}{(X, Y, Z)\Delta(\Gamma, \Theta) \vdash_{\Sigma,\Xi} \chi} \mathit{Cut},$$

where $X' \vdash_{\Sigma,\Xi} \phi : \mathbf{Prop}$ and $Y' \vdash_{\Sigma,\Xi} \psi : \mathbf{Prop}$, reduces to the proof figure

$$\frac{\dfrac{(X', Y', Z)\Theta, \phi \vdash_{\Sigma,\Xi} \psi \quad (Y, Y')\Delta(\psi) \vdash_{\Sigma,\Xi} \chi}{(X', Y, Z)\Delta(\Theta, \phi) \vdash_{\Sigma,\Xi} \chi} \mathit{Cut} \quad (X, X')\Gamma \vdash_{\Sigma,\Xi} \phi}{(X, Y, Z)\Delta(\Gamma, \Theta) \vdash_{\Sigma,\Xi} \chi} \mathit{Cut}.$$

The other cases follow similar patterns.

Turning to the equivalence of **NBI** and **LBI**, the argument is illustrated by the following simplified example of the derivation of $\forall_{\mathbf{new}}\mathrm{L}$ in **NBI**:

$$
\dfrac{\dfrac{\dfrac{\begin{array}{c}\mathit{Axiom}\\ \vdots\end{array}}{(\emptyset_m)\forall_{\mathbf{new}}x : A.\phi \vdash_{\Sigma,\Xi} \forall_{\mathbf{new}}y : A.\phi} \quad Y \vdash_{\Sigma} t : A}{(Y, \emptyset_m)\forall_{\mathbf{new}}x : A.\phi \vdash_{\Sigma,\Xi} \phi(t)}\ \forall_{\mathbf{new}}E \qquad \dfrac{\vdots}{(X, Y)\Gamma, \phi(t) \vdash_{\Sigma,\Xi} \chi}}{(X, \emptyset_m)\Gamma, \forall_{\mathbf{new}}x : A.\phi \vdash_{\Sigma,\Xi} \chi}\ \mathit{Cut}.
$$

We refrain from presenting the details of the translations between **NBI** and **LBI**.

Chapter 13

KRIPKE SEMANTICS FOR PREDICATE BI

1. Predicate Kripke Models

We extend the Kripke semantics of propositional **BI** to predicates and quantifiers. Let $\mathcal{M} = (M, e, \cdot, \sqsubseteq)$ be a preordered commutative monoid, viewed as a preordered monoidal category. Recall, from Chapter 4, that a Kripke model of propositional **BI**, and hence of $\alpha\lambda$, is a triple

$$\langle [\mathcal{M}^{op}, \mathbf{Set}], \models, [\![-]\!] \rangle,$$

in which $[\![-]\!]$ is a partial function from $\mathcal{P}(L)$, the collection of **BI** propositions, or $\alpha\lambda$-types, over a language L of propositional letters (or atomic types) to $obj([\mathcal{M}^{op}, \mathbf{Set}])$ and in which $\models \subseteq M \times \bigcup_{\phi \in \mathcal{P}(L)}$ is a satisfaction relation. Recall also that the definition of a Kripke model of propositional **BI**, given in Chapter 4, makes essential use of Day's tensor product construction and its right adjoint [Day, 1970].

In order to define models of predicate **BI**, we must have a mechanism for interpreting predicates, with predicate letters in a predicate signature Ξ. In order to interpret predicates, we must first have an interpretation of the term language, *i.e.*, $\alpha\lambda$ extended with (constants and) function symbols declared in a term signature Σ. For this, we must first define a notion of *environment*, which specifies a binding of variables to individuals. We suppose that we are given a functor D in $[\mathcal{M}^{op}, \mathbf{Set}]$; this functor is the *domain of individuals*. As in the possible worlds semantics of intuitionistic logic, the use of a functor instead of a constant set allows different collections of individuals to exist at different worlds.

EXAMPLE 13.1 *Consider the monoid $\mathcal{N} = (\mathbb{N}, 0, +, \leq)$ of lifted natural numbers, under addition. Then D may be taken to be the functor sending n to the set $\{0, ..., n-1\}$, with the morphism part being inclusion.*

Our term language, $\alpha\lambda$-calculus, is typed. Accordingly, we require an interpretation of types. Recall that types A are either atoms, units or constructed from $*$, $-\!*$, $\wedge$, $\rightarrow$ and $\vee$.

DEFINITION 13.2 (INTERPRETATION OF TYPES) *Atomic types α are interpreted by chosen functors $[\![\alpha]\!]$ elements of obj$([\mathcal{M}^{op}, \mathbf{Set}])$, where $[\![-]\!]$ is a partial function from bunches to obj$([\mathcal{M}^{op}, \mathbf{Set}])$. The units and types constructed from $*$, $-\!*$, $\wedge$, $\rightarrow$ and $\vee$ are given by the corresponding constructions in $[\mathcal{M}^{op}, \mathbf{Set}]$.* □

The interpretation of bunches, and so the definition of environments, should recall a similar construction in Chapter 8.

DEFINITION 13.3 (INTERPRETATION OF BUNCHES OF VARIABLES) *Let $D \in$ obj$([\mathcal{M}^{op}, \mathbf{Set}])$ be a domain of individuals. We define the* interpretation functor with respect to D, $[\![X]\!]^D$, *for the bunch X, as follows:*

$$\begin{array}{lcllcl} [\![x \in A]\!]^D & = & D & & & \\ [\![\emptyset_m]\!]^D & = & I & [\![X, Y]\!]^D & = & [\![X]\!]^D \otimes [\![Y]\!]^D \\ [\![\emptyset_a]\!]^D & = & 1 & [\![X; Y]\!]^D & = & [\![X]\!]^D \times [\![Y]\!]^D, \end{array}$$

where $[\![-]\!]^D$ is a partial function from bunches to obj$([\mathcal{M}^{op}, \mathbf{Set}])$. If Axiom$(X, Y)$, then $[\![X]\!]^D = [\![Y]\!]^D$ and

$$[\![X, Y]\!]^D \;=\; [\![X]\!]^D \times [\![Y]\!]^D.$$

We require that, where defined, $[\![x : A]\!]^D(m) \in [\![A]\!](m)$, for each world m. □

The interpretation of a bunch X, Y such that Axiom(X, Y) as a cartesian product reflects the fact that Contraction is permitted for such bunches, just as it is for bunches of the form $X; X'$, where $X \cong X'$.

Definition 13.3 defines the interpretation of a variable $x \in A$ at world m to be a set $D(m)$. Semantically, the distinction between $x : A$ and $x!A$ is not significant, as may be seen from the forms of the forcing clauses for the quantifiers in Definition 13.4. An *environment for $x : A$ with respect to D at m* is given by a choice of element $u \in D(m)$.[1]

In Definition 13.3, $\otimes$ is Day's tensor product and $\times$ is cartesian product of functors. As a result, each $[\![X]\!]^D$ is a functor, so that when $n \sqsubseteq m$

[1] Alternatively, we could require that $D(m)$ be a singleton, $\{u\}$. The choice between these two formulations has little effect on our development.

and $u \in [\![X]\!]^D(m)$ we obtain an element $[\![X]\!]^D(n \sqsubseteq m)(u) \in [\![X]\!]^D(n)$. We say that $u \in [\![X]\!]^D(m)$ is an *environment for X with respect to D at m* and (abusing notation) write u_x to denote the binding for the variable x in X determined by u.

Given a choice of environment $u \in [\![X]\!]^D(m)$, we can define the interpretation of the terms of predicate **BI**, the terms of $\alpha\lambda$ over a signature Σ as defined by the inference rules for $\alpha\lambda$, provided we have given interpretations of the constant symbols in Σ.

The constants in Σ are interpreted by chosen global elements. At each world m and for each environment u, there are elements, $\hat{c}$, as follows:

$$[\![c!A]\!]^D(m) = \hat{c}(m) \in [\![A]\!]^D(m) \tag{13.1}$$

The combinators of $\alpha\lambda$, corresponding to the type-constructors $-\!\!*$, $\rightarrow$, $*$, $\wedge$, $\vee$ and the units, are interpreted in the normal way (see Chapter 4). Given these definitions, we have the following familiar property, for appropriately typed function symbols, f:

$$[\![f(\bar{t})]\!]_u^D(m) \simeq \hat{f}([\![\bar{t}]\!]_u^D(m)) \tag{13.2}$$

Here, $\simeq$ denotes Kleene equality and we write $[\![t]\!]_u^D$ to denote the interpretation of a term t with respect to domain of individuals D and choice of environment u. For simplicity, we write just the term t, *etc.*, rather that the $\alpha\lambda$-sequent within which it is typed.

We must also require that the interpretation of a predicate letter $\mathrm{p} : A \in \Xi$ be well-typed:

$$[\![\mathrm{p}]\!](m) \subseteq [\![A]\!](m) \tag{13.3}$$

We are now ready to define Kripke models of predicate **BI**. The definition is the natural generalization of Kripke models of propositional **BI**, Definition 4.1 in Chapter 4, to predicates and quantifiers, using the notion of environment established in Definition 13.3. The forcing relation must be extended not only to account for quantifiers but also to account for predicates. The essential set-up is as follows:

$$(X)u \mid m \models \phi,$$

where m is a world and $u \in [\![X]\!](m)$ is an environment for X at m.

If we were to adopt a "cross case", such as the pairing of additive predicate formation with multiplicative propositional connectives found in predicate linear logic [Girard, 1987], then the propositional clauses would have to be altered accordingly. For example, predicate linear logic's multiplicative conjunction, $\otimes$, which uses additive predication, would require the following clause:

$$(X)u \mid m \models_{\Sigma,\Xi} \phi \otimes \psi \quad \text{iff} \quad \begin{array}{l} \text{for some} \quad n, n' \in M \quad (m \sqsubseteq n \cdot n', \\ (Y)\, u_Y \mid n \models_{\Sigma,\Xi} \phi \text{ and } (Z)\, u_Z \mid n' \models_{\Sigma,\Xi} \psi), \\ \text{where } X = Y; Z \text{ and } u = (u_Y, u_Z).^2 \end{array}$$

Notice that, corresponding to the combination of bunches of variables using ";", we have used the cartesian pairing operation, $(-,-)$, instead of Day's pairing operation $[-,-]$. Consequently, we must require that (u_Y, u_Z) be defined at $n \cdot n'$. One way to enforce this condition is to require that the domain functor D be constant. Such a situation amounts, essentially, to having *Beth models* [van Dalen, 1986].

In contrast, **BI**'s standard choice pairs multiplicative predication with the multiplicative propositional connectives, so that the multiplicative conjunction $*$ requires the following clause:

$$(X)\, u \mid m \models_{\Sigma,\Xi} \phi * \psi \quad \text{iff} \quad \begin{array}{l} \text{for some} \quad n, n' \in M, \quad m \sqsubseteq n \cdot n', \\ (Y)\, u_Y \mid n \models_{\Sigma,\Xi} \phi \text{ and } (Z)\, u_Z \mid n' \models_{\Sigma,\Xi} \psi, \\ \text{where } X = Y, Z \text{ and } u = [u_Y, u_Z]. \end{array}$$

Here we must use Day's tensor product to combine environments appropriate to each component of the conjunction to form an environment appropriate to the conjunction.

With these ideas, the definition of Kripke models of predicate **BI** now goes as follows:

DEFINITION 13.4 (KRIPKE MODELS) *Let $\mathcal{M} = (M, e, \cdot, \sqsubseteq)$ be a preordered commutative monoid, viewed as a preordered monoidal category, and let $\mathcal{P}(L(\Sigma, \Xi))$ denote the collection of* **BI** *propositions over the language $L(\Sigma,\Xi)$ given by a term signature Σ and predicate signature Ξ. Let $D \in obj([\mathcal{M}^{op}, \mathbf{Set}])$ be a domain of individuals, with interpretation functor $[\![-]\!]^D$, and let $[\![-]\!] : \mathcal{P}(L(\Sigma,\Xi), \mathcal{X}) \rightharpoonup obj([\mathcal{M}^{op}, \mathbf{Set}])$ be a partial function from the* **BI** *propositions over $L(\Sigma, \Xi)$ to the objects of $[\mathcal{M}^{op}, \mathbf{Set}]$. A* Kripke model *is a quadruple*

$$\langle [\mathcal{M}^{op}, \mathbf{Set}], \models_{\Sigma,\Xi}, D, [\![-]\!] \rangle,$$

where $[\mathcal{M}^{op}, \mathbf{Set}]$ is the category of presheaves over the preorder category $\mathcal{M}$ to **Set**, *$\models^{\mathcal{M}}_{\Sigma,\Xi} \subseteq M \times \mathcal{P}(L(\Sigma,\Xi))$ is a satisfaction relation such that, for each $\mathrm{p} \in \Xi$ and each world m, $[\![\mathrm{p}]\!](m) \subseteq [\![A]\!](m)$ and satisfying, where defined, the constraints in Table 13.1.*

$(X)u \mid m \models^{\mathcal{M}}_{\Sigma,\Xi} \mathrm{p}(\bar{t}(X))$	iff	$[\![\bar{t}(X)]\!]^D_u(m) \in [\![\mathrm{p}]\!](m)$, $\mathrm{p} \in \Xi$ $X \vdash_{\Sigma,\Xi} \mathrm{p}(\bar{t}(X)) : \mathbf{Prop}$
$(X)u \mid m \models^{\mathcal{M}}_{\Sigma,\Xi} \forall_{\mathbf{new}} x : A.\phi$	iff	for all n and all $u_x \in D(n)$ $(X, x : A)\ [u, u_x] \mid m \cdot n \models^{\mathcal{M}}_{\Sigma,\Xi} \phi$
$(X)u \mid m \models^{\mathcal{M}}_{\Sigma,\Xi} \exists_{\mathbf{new}} x : A.\phi$	iff	for some n and some $u_x \in D(n)$ $(X, x : A)\ [u, u_x] \mid m \cdot n \models^{\mathcal{M}}_{\Sigma,\Xi} \phi$
$(X)u \mid m \models^{\mathcal{M}}_{\Sigma,\Xi} \forall x!A.\phi$	iff	for all $n \sqsubseteq m$ and all $u_x \in D(n)$ $(X; x!A)\ ([\![X]\!]^D(n \sqsubseteq m)u, u_x) \mid$ $n \models^{\mathcal{M}}_{\Sigma,\Xi} \phi$
$(X)u \mid m \models^{\mathcal{M}}_{\Sigma,\Xi} \exists x!A.\phi$	iff	for some $u_x \in D(m)$ $(X; x!A)\ (u, u_x) \mid m \models^{\mathcal{M}}_{\Sigma,\Xi} \phi$
$(X)u \mid m \models^{\mathcal{M}}_{\Sigma,\Xi} \phi * \psi$	iff	for some $n, n' \in M$ such that $m \sqsubseteq n \cdot n'$, $(Y)u_Y \mid n \models^{\mathcal{M}}_{\Sigma,\Xi} \phi$ and $(Z)u_Z \mid n' \models^{\mathcal{M}}_{\Sigma,\Xi} \psi$, where $X = Y, Z$ and $u = [u_Y, u_Z]$
$(X)u \mid m \models^{\mathcal{M}}_{\Sigma,\Xi} \phi \twoheadrightarrow \psi$	iff	for all $n \in M$, all Z and all $v \in [\![Z]\!]^D(n)$, $(Z, Y')[v, u_{Y'}] \mid n \cdot m_{Y'} \models^{\mathcal{M}}_{\Sigma,\Xi} \phi$ implies $(Y, Z)[u_Y, v] \mid m_Y \cdot n \models^{\mathcal{M}}_{\Sigma,\Xi} \psi$, where $X = Y, Y'$, $u = [u_Y, u_{Y'}]$ and $m = m_Y \cdot m_{Y'}$
$(X)u \mid m \models^{\mathcal{M}}_{\Sigma,\Xi} \phi \wedge \psi$	iff	$(X)u \mid m \models^{\mathcal{M}}_{\Sigma,\Xi} \phi$ and $(X)u \mid m \models^{\mathcal{M}}_{\Sigma,\Xi} \psi$
$(X)u \mid m \models^{\mathcal{M}}_{\Sigma,\Xi} \phi \vee \psi$	iff	$(X)u \mid m \models^{\mathcal{M}}_{\Sigma,\Xi} \phi$ or $(X)u \mid m \models^{\mathcal{M}}_{\Sigma,\Xi} \psi$
$(X)u \mid m \models^{\mathcal{M}}_{\Sigma,\Xi} \phi \rightarrow \psi$	iff	for all $n \sqsubseteq m$, $(X)u \mid n \models^{\mathcal{M}}_{\Sigma,\Xi} \phi$ implies $(X)[\![X]\!]^D(n \sqsubseteq m)u \mid n \models^{\mathcal{M}}_{\Sigma,\Xi} \psi$
$(X)u \mid m \models^{\mathcal{M}}_{\Sigma,\Xi} \top$	for all	$m \in M$
$(X)u \mid m \models^{\mathcal{M}}_{\Sigma,\Xi} I$	iff	$m \sqsubseteq e$
$(X)u \mid m \not\models^{\mathcal{M}}_{\Sigma,\Xi} \bot$	for any	m.

Table 13.1. Predicate Kripke Semantics

We require the following three conditions:

1 *Enough points for Σ: for every $c : A \in \Sigma$, there is a $\hat{c} \in [\mathcal{M}^{op}, \mathbf{Set}]$ such that, for every world m, $\hat{c}(m) \in [\![A]\!](m)$ and, for all worlds m and n, $[\![c]\!]^D(m) = \hat{c}(m) = \hat{c}(n) = [\![c]\!]^D(n)$,* i.e., *constants are interpreted in the same way at every world;*

2 Kripke monotonicity:

$$(X)u \mid m \models^{\mathcal{M}}_{\Sigma,\Xi} \phi \text{ and } n \sqsubseteq m \text{ implies } (X)([\![X]\!](m \sqsubseteq n))u \mid n \models^{\mathcal{M}}_{\Sigma,\Xi} \phi$$

3 Axioms: a pairing of the form $Axiom(X_1, X_2)$ for each pair of bunches of variables occurring in the $Axiom(X_1, X_2)$ side-conditions of propositional axioms interpreted in $\langle[\mathcal{M}^{op}, \mathbf{Set}], \models, [\![-]\!]\rangle$. (By Definition 13.3, we have $[\![X_1]\!]^D = [\![X_2]\!]^D$ for such pairings.)

Wherever no confusion will arise, we shall refer to a model

$$\langle[\mathcal{M}^{op}, \mathbf{Set}], \models, [\![-]\!]\rangle$$

simply as $\mathcal{M}$. □

We remark that it is a straightforward matter, using Day's construction and the techniques laid out in [Lambek and Scott, 1986], to generalize Kripke models to presheaf categories of the form $\mathbf{Set}^{\mathcal{C}^{op}}$, where $\mathcal{C}$ is any small symmetric monoidal category.

Let $\langle[\mathcal{M}^{op}, \mathbf{Set}], \models^{\mathcal{M}}_{\Sigma,\Xi}, D, [\![-]\!]\rangle$ be a Kripke model. $\phi(X)$ is true at m, with respect to environment u, if $(X)u \mid m \models^{\mathcal{M}}_{\Sigma,\Xi} \phi$. For relative truth, we write

$$(X)u \mid m \models^{\mathcal{M}}_{\Sigma,\Xi} \Gamma \quad \text{iff} \quad (X)u \mid m \models^{\mathcal{M}}_{\Sigma,\Xi} \phi_\Gamma,$$

where, as usual, ϕ_Γ is the formula obtained from Γ by replacing each semi-colon with $\wedge$ and each comma with $*$, with association respecting the tree structure of Γ.

Similarly, we write

$$(X)\Gamma \models^{\mathcal{M}}_{\Sigma,\Xi} \phi,$$

if and only if, for all $m \in M$ and all $u \in [\![X]\!]^D(m)$,

$$(X_1)u \mid m \models^{\mathcal{M}}_{\Sigma,\Xi} \Gamma \quad \text{implies} \quad (X_2)u \mid m \models^{\mathcal{M}}_{\Sigma,\Xi} \phi.$$

where $X = X_1 \circ X_2$ with, as usual, $\circ$ denoting either ";" or ",", and $Axiom(X_1, X_2)$.

ϕ is valid if, for all m and all u in all $\mathcal{M}$,

$$(X)u \mid m \models^{\mathcal{M}}_{\Sigma,\Xi} \phi.$$

For relative validity, we write

$$(X)\Gamma \models_{\Sigma,\Xi} \phi$$

if and only if, for all such models, $\mathcal{M}$,

$$(X)\Gamma \models^{\mathcal{M}}_{\Sigma,\Xi} \phi.$$

In the context of a given model $\mathcal{M}$, and where no confusion is likely, we sometimes write just $\models_{\Sigma,\Xi}$ for $\models^{\mathcal{M}}_{\Sigma,\Xi}$.

To see that this definition is consistent with the sub-object classifier semantics of intuitionistic logic [Lambek and Scott, 1986] consider the pullback diagram in $\mathbf{Sh}(\mathcal{X})$),

$$\begin{array}{ccc} h^m & \xrightarrow{\mu} & [\![\mathrm{p}]\!] \\ {\scriptstyle \bigcirc_{h^m}}\downarrow & & \downarrow{\scriptstyle \chi_m} \\ 1 & \xrightarrow[\top]{} & \Omega \end{array}$$

and note that an arrow $h^m \xrightarrow{\mu} [\![\mathrm{p}]\!]$ is, by the Yoneda lemma, determined uniquely by an element $\check{\mu} \in [\![\mathrm{p}]\!](U)$.

LEMMA 13.5 (SUBSTITUTION) *Let $\langle [\mathcal{M}^{op}, \mathbf{Set}], \models, D, [\![-]\!] \rangle$ be a Kripke model. If $Y \vdash_\Sigma t : A$ and if $[\![t]\!]^D_u \in [\![Y]\!]^D(m)$ is defined, then, provided all the required interpretations are defined,*

$$(X(x : A))u(u_x) \mid m \models^{\mathcal{M}}_{\Sigma,\Xi} \phi(x)$$

if and only if

$$(X(Y))u(u_Y) \mid m \models^{\mathcal{M}}_{\Sigma,\Xi} \phi[t/x].$$

PROOF By induction on the structure of terms and propositions. □

LEMMA 13.6 (QUANTIFIERS AS INSTANCES) *Let*

$$\langle [\mathcal{M}^{op}, \mathbf{Set}], \models, D, [\![-]\!] \rangle$$

be a Kripke model. Then, provided all the required interpretations are defined:

1 $(X)u \mid m \models^{\mathcal{M}}_{\Sigma,\Xi} \forall_{\mathbf{new}} x : A.\phi$ *if and only if, for all n and all t such that $Y \vdash_\Sigma t : A$ and $[\![t]\!]^D_v(n)$, $(X, Y)[u, v] \mid m \cdot n \models^{\mathcal{M}}_{\Sigma,\Xi} \phi[t/x]$;*

2 $(X)u \mid m \models^{\mathcal{M}}_{\Sigma,\Xi} \exists_{\mathbf{new}} x : A.\phi$ *if and only if, for some n, there is a t such that $(X, Y)[u, v] \mid m \cdot n \models^{\mathcal{M}}_{\Sigma,\Xi} \phi[t/x]$, where $Y \vdash_\Sigma t : A$ and $[\![t]\!]^D_v(n)$;*

3 $(X)u \mid m \models^{\mathcal{M}}_{\Sigma,\Xi} \forall x!A.\phi$ *if and only if, for all* $n \sqsubseteq m$ *and all* t *such that* $X \vdash_{\Sigma} t : A$ *and* $[\![t]\!]^{D}_{v}(n)$, $(X)(u,v) \mid n \models^{\mathcal{M}}_{\Sigma,\Xi} \phi[t/x]$*;*

4 $(X)u \mid m \models^{\mathcal{M}}_{\Sigma,\Xi} \exists x!A.\phi$ *if and only if, for some* n*, there is a* t *such that* $(X)(u,v) \mid m \models^{\mathcal{M}}_{\Sigma,\Xi} \phi[t/x]$*, where* $X \vdash_{\Sigma} t : A$ *and* $[\![t]\!]^{D}_{v}(m)$*.*

PROOF By induction on the structure of propositions. □

2. Elementary Soundness and Completeness for Predicate BI

The soundness theorem for predicate **BI** without $\perp$ is an extension of the soundness theorem for propositional **BI** without $\perp$, given in Chapter 4. Whereas the soundness theorem in Chapter 4 is stated for all models, the soundness theorem for predicate **BI** is stated for all Σ, Ξ-Kripke models.

LEMMA 13.7 (CUT) *The Cut rule is sound in Kripke models.*

PROOF We sketch a simple version,[3] Let $\mathcal{M}$ be any (Σ, Ξ)-Kripke model of **BI** without $\perp$. Suppose that

$$(X, x)[u, u_x] \mid m \models_{\Sigma,\Xi} \Delta, \phi$$

implies

$$(X, x)[u, u_x] \mid m \models_{\Sigma,\Xi} \psi,$$

where $x \notin \mathrm{FV}(\Delta, \psi)$, and

$$(Y)v \mid n \models_{\Sigma,\Xi} \phi,$$

then

$$(X, Y)[u, v] \mid m \cdot n \models_{\Sigma,\Xi} \psi.$$

To see this, observe that

$$(X, x)[u, u_x] \mid m \cdot n \models_{\Sigma,\Xi} \Delta, \phi$$

[3]Note that although the Cut rule must handle axiom-related variables, the semantics does not distinguish axiom-related variables, so the simple version we sketch represents just the slight restriction of the general case to Cuts on single variables at top-level in the bunch, *i.e.*,

$$\frac{(X, x)\Delta, \phi(x) \vdash_{\Sigma,\Xi} \psi \quad (Y)\Gamma \vdash_{\Sigma,\Xi} \phi}{(X, Y)\Delta, \Gamma \vdash_{\Sigma,\Xi} \psi}.$$

if and only if, by the definition of forcing for $*$, and hence for ",",

$$(X)u \mid m \models_{\Sigma,\Xi} \Delta$$

and

$$(x)u_x \mid n \models_{\Sigma,\Xi} \phi\,.$$

Now use Lemma 13.5. □

THEOREM 13.8 (SOUNDNESS OF PREDICATE **BI** FOR KRIPKE MODELS) *Let $\mathcal{M}$ be any (Σ, Ξ)-Kripke model of* **BI**. *If $(X)\Gamma \vdash_{\Sigma,\Xi} \phi$ is provable in* **NBI** *without $\bot$, then, provided all the required interpretations are defined, $(X)\Gamma \models^{\mathcal{M}}_{\Sigma,\Xi} \phi$.*

PROOF We assume an arbitrary model $\mathcal{M}$ and suppress the definedness provisos. The cases for axioms and the propositional connectives proceed just as for propositional Kripke models, extended to account for the formation of predicates. The cases for the additive quantifiers proceed just as in the corresponding proof for **IL**; see, for example, [van Dalen, 1986]. Accordingly, we consider just the cases for the multiplicative connectives, $*$ and $-\!\!*$, multiplicative quantifiers, $\forall_{\mathbf{new}}$ and $\exists_{\mathbf{new}}$.

Axiom: If the last inference in Φ is an axiom inference, then, by induction on the structure of the principal formula of the axiom, we must show that, for any world m, in any model $\mathcal{M}$, and any environment u appropriate to X at m,

$$(X)u \mid m \models^{\mathcal{M}}_{\Sigma,\Xi} \mathrm{p}(X_1) \quad \text{implies} \quad (X)u \mid m \models^{\mathcal{M}}_{\Sigma,\Xi} \mathrm{p}(X_2),$$

where p is a predicate letter, X splits into X_1 and X_2, where we have that $\mathrm{Axiom}(X_1, X_2)$. This is an immediate consequence of our Axioms condition.

$*$I: Suppose that $(X)\Gamma \vdash_{\Sigma,\Xi} \phi * \psi$. By the induction hypothesis, we have that, for all $n \in M$, Y and u_Y, and all $n' \in M$, Z and u_Z, such that $X = Y, Z$,

$$(Y)u_Y \mid n \models^{\mathcal{M}}_{\Sigma,\Xi} \phi \text{ and } (Z)u_Z \mid n' \models^{\mathcal{M}}_{\Sigma,\Xi} \psi.$$

We must establish that $(X)\Gamma \models^{\mathcal{M}}_{\Sigma,\Xi} \phi * \psi$, *i.e.*, that, for all $m \in M$, there exist $n, n' \in M$ such that $m \sqsubseteq n \cdot n'$, $(Y)u_Y \mid n \models^{\mathcal{M}}_{\Sigma,\Xi} \phi$ and $(Z)u_Z \mid n' \models^{\mathcal{M}}_{\Sigma,\Xi} \psi$, where $X = Y, Z$ and $u = [u_Y, u_Z]$ is an arbitrary environment appropriate to X at m. The result follows straightforwardly by partitioning m and appealing to the induction hypothesis.

$*$E: Straightforward.

$-\!\!*$I: By the induction hypothesis, we have that, for all m and all u appropriate to X at m,

$$(X)u \mid m \models^{\mathcal{M}}_{\Sigma,\Xi} \Gamma, \phi$$

implies

$$(X)u \mid m \models^{\mathcal{M}}_{\Sigma,\Xi} \psi.$$

We must show that if

$$(X)u \mid m \models^{\mathcal{M}}_{\Sigma,\Xi} \Gamma,$$

then

$$(X)u \mid m \models^{\mathcal{M}}_{\Sigma,\Xi} \phi -\!\!* \psi.$$

We begin by noting that, since $(X)\Gamma \vdash_{\Sigma,\Xi} \phi -\!\!* \psi$, it must, according to the formation rule for $*$, and hence for "," in Γ, ϕ, be that $X = X', X''$, $u = [u_{X'}, u_{X''}]$ and $m = m_{X'} \cdot m_{X''}$, with $X' \vdash_{\Sigma,\Xi} \phi_\Gamma$: **Prop** and $X'' \vdash_{\Sigma,\Xi} \phi$: **Prop**. Therefore we must show that if

$$(X')u_{X'} \mid m_{X'} \models^{\mathcal{M}}_{\Sigma,\Xi} \Gamma,$$

then

$$(X)u_X \mid m_X \models^{\mathcal{M}}_{\Sigma,\Xi} \phi -\!\!* \psi.$$

By the definition of satisfaction, Definition 13.4, we have

$$(X', X'')[u_{X'}, u_{X''}] \mid m_{X'} \cdot m_{X''} \models^{\mathcal{M}}_{\Sigma,\Xi} \phi -\!\!* \psi$$

iff, for all n, all Z and all $v \in [\![Z]\!]^D(n)$,

$$(Z, X'')[v, u_{X''}] \mid n \cdot m_{X''} \models^{\mathcal{M}}_{\Sigma,\Xi} \phi$$

implies

$$(X', Z)[u_{X'}, v] \mid m_{X'} \cdot n \models^{\mathcal{M}}_{\Sigma,\Xi} \psi.$$

If $(Z, X'')[v, u_{X''}] \mid n \cdot m_{X''} \models^{\mathcal{M}}_{\Sigma,\Xi} \phi$, then

$$(X', X'', Z)[u_{X'}, u_{X''}, v] \mid m_{X'} \cdot m_{X''} \cdot n \models^{\mathcal{M}}_{\Sigma,\Xi} \Gamma, \phi.$$

By the induction hypothesis,

$$(X', X'', Z)[u_{X'}, u_{X''}, v] \mid m_{X'} \cdot m_{X''} \cdot n \models^{\mathcal{M}}_{\Sigma,\Xi} \psi,$$

i.e., $(X, Y)[u_X, v] \mid m_X \cdot n \models^{\mathcal{M}}_{\Sigma,\Xi} \psi$. Therefore $(X)u \mid m \models^{\mathcal{M}}_{\Sigma,\Xi} \phi -\!\!* \psi$.

$-\!\!*$E: Straightforward.

$\forall_{\mathbf{new}}$I: By the induction hypothesis and by the structure of Day's tensor product in $[\mathcal{M}^{op}, \mathbf{Set}]$, we have that

$$(X, x : A)[v, v(x)] \mid n \cdot n' \models_{\Sigma,\Xi} \Gamma$$

implies

$$(X, x : A)[v, v(x)] \mid n \cdot n' \models_{\Sigma,\Xi} \phi,$$

where $x : A \notin \mathrm{FV}(\Gamma)$, $x : A$ satisfies the condition on its use in axioms and where $v(x) \in D(n')$. But both n and n' are arbitrary, so we have immediately that

$$(X)v \mid n \models_{\Sigma,\Xi} \forall_{\mathbf{new}} x : A.\phi.$$

$\forall_{\mathbf{new}}$E: Straightforward, using substitution.

$\exists_{\mathbf{new}}$I: Straightforward.

$\exists_{\mathbf{new}}$E: The $\exists_{\mathbf{new}} E$ rule is

$$\frac{(X)\Gamma \vdash_{\Sigma,\Xi} \exists_{\mathbf{new}} x : A.\phi \quad (Y, x : A)\Delta, \phi \vdash_{\Sigma,\Xi} \psi}{(X, Y)\Gamma, \Delta \vdash_{\Sigma,\Xi} \psi},$$

where $x : A \notin \mathrm{FV}(\Delta, \psi)$ and $x : A$ satisfies the condition on its use in axioms. So by the induction hypothesis, we have, in the evident notation, that

$$(X)u \mid m \models_{\Sigma,\Xi} \Gamma$$

implies

$$(X)u \mid m \models_{\Sigma,\Xi} \exists_{\mathbf{new}} x : A.\phi$$

and that

$$(Y, x : A)[v, v(x)] \mid n \cdot n' \models_{\Sigma,\Xi} \Delta, \phi$$

implies

$$(Y, x : A)[v, v(x)] \mid n \cdot n' \models_{\Sigma,\Xi} \psi.$$

We must establish that

$$(X, Y)[u, v] \mid m \cdot n \models_{\Sigma,\Xi} \Gamma, \Delta$$

implies

$$(X, Y)[u, v] \mid m \cdot n \models_{\Sigma,\Xi} \psi.$$

By the definition of $\models_{\Sigma,\Xi}$,

$$(X)u \mid m \models_{\Sigma,\Xi} \exists_{\mathbf{new}} x : A.\phi$$

if and only if, for some m' and some $u_x \in D(m')$,

$$(X, x : A)[u, u_x] \mid m \cdot m' \models_{\Sigma,\Xi} \phi .$$

By Lemma 13.7, and using $x \notin \mathrm{FV}(\Delta, \psi)$, we get that if

$$(X, Y)[u, v] \mid m \cdot n \models_{\Sigma,\Xi} \Gamma, \Delta,$$

then

$$(X, Y)[u, v] \mid m \cdot n \models_{\Sigma,\Xi} \psi.$$

The remaining cases are similar. □

As in Chapter 4, we can argue that soundness holds even in the presence of $\bot$.

Turning to completeness, we extend the notion of a *prime bunch*, introduced in Chapter 4, to handle predicates and quantifiers. Just as in the proof systems **NBI** and **LBI**, the cases for the quantifiers and Substitution are subject to the constraints determined by the relation Axiom (we refrain from stating them explicitly a second time). Just as for propositional **BI**, we must exclude $\bot$ from **BI**.

DEFINITION 13.9 (PRIME BUNCH) *A well-formed bunch $(X)\Gamma$ is* prime *if*

1 $(X)\Gamma$ is closed under consequences $\vdash$ generated by $-\!\!*\, E$, $\to E$, $\forall_{\mathbf{new}} E$, $\forall E$ and Substitution, i.e.,

(a) $(X)\Gamma$ is of the form $(X)\Gamma(\phi,\, \phi -\!\!* \psi)$, where $(X)\Gamma \vdash_{\Sigma,\Xi} \psi$, implies $(X)\Gamma$ is also of the form $(X)\Gamma(\psi)$;

(b) $(X)\Gamma$ is of the form $(X)\Gamma(\phi;\, \phi \to \psi)$ where $(X)\Gamma \vdash_{\Sigma,\Xi} \psi$, implies $(X)\Gamma$ is also of the form $(X)\Gamma(\psi)$;

(c) $(X)\Gamma$ is of the form $(X)\Gamma(\tau\text{-?}\phi)$, where τ is a theorem, -? denotes either $\to$ or $-\!\!*$ and $(X)\Gamma \vdash_{\Sigma,\Xi} \phi$, implies $(X)\Gamma$ is also of the form $(X)\Gamma(\phi)$;

(d) $(Z(X))\Gamma$ is of the form $(Z(X(x \in A)))\Gamma(\phi(x))$, where $Y \vdash_{\Sigma} t : A$, implies $(Z(X))\Gamma$ is also of the form $(Z(X(x \in A), Y))\Gamma(\phi[t/x])$;

2 $(X)\Gamma$ is of the form $(X)\Gamma(\phi * \psi)$, where $\phi * \psi$ is a principal subformula of some sub-bunch of $(X)\Gamma$, implies $(X)\Gamma$ is also of the form $(X)\Gamma(\phi, \psi)$;

3 $(X)\Gamma$ is of the form $(X)\Gamma(\phi \wedge \psi)$ implies is of the form $(X)\Gamma(\phi; \psi)$;

4 $(X)\Gamma$ *of the form* $(X)\Gamma(\phi \vee \psi)$ *implies* $(X)\Gamma$ *is also either of the form* $(X)\Gamma(\phi)$ *or of the form* $(X)\Gamma(\psi)$*;*

5 $(Z(X))\Gamma$ *is of the form* $(Z(X))\Gamma(\forall_{\mathbf{new}}x : A.\phi)$*, where* $Y \vdash_\Sigma t : A$ *and* $(Z(X,Y))\Gamma(\forall_{\mathbf{new}}x : A.\phi) \vdash_{\Sigma,\Xi} \phi([t/x])$*, implies* $(Z(X))\Gamma$ *is also of the form* $(Z(X,Y))\Gamma(\phi([t/x])$*;*

6 $(Z(X))\Gamma$ *is of the form* $(Z(X))\Gamma(\forall x!A.\phi)$*, where* $Y \vdash_\Sigma t : A$ *and* $(Z(X;Y))\Gamma(\forall x!A.\phi) \vdash_{\Sigma,\Xi} \phi([t/x])$*, implies* $(Z(X))\Gamma$ *is also of the form* $(Z(X;Y))\Gamma(\phi([t/x])$*;*

7 $(Y(X))\Gamma$ *is of the form* $(Y(X))\Gamma(\exists_{\mathbf{new}}x : A.\phi)$ *implies* $(Y(X))\Gamma$ *is also of the form* $(Y(X, x : A))\Gamma(\phi(x))$*;*

8 $(Y(X))\Gamma$ *is of the form* $(Y(X))\Gamma(\exists x!A.\phi)$ *implies* $(Y(X))\Gamma$ *is also of the form* $(Y(X; x!A))\Gamma(\phi(x))$.

□

LEMMA 13.10 (FINITENESS) *Let* X *be a bunch of variables and* Σ *be a term signature. There are finitely many terms* $t : A$ *such that* $X \vdash_\Sigma t : A$*.*
□

Next we extend the notion of evaluation, also introduced in Chapter 4, to well-formed predicate bunches.

DEFINITION 13.11 (EVALUATION) *Let* Γ *be a bunch. Each of the following, of the form* redex $\rightsquigarrow$ reduct, *is called an* evaluation *in* Γ*:*

1(a) $(X)\Gamma(\phi, \phi \multimap \psi) \rightsquigarrow (X')\Gamma(\psi)$*, where* X' *is* X *with any variables required only for the well-formedness of* ϕ *replaced by units;*

1(b) $(X)\Gamma(\phi; \phi \to \psi) \rightsquigarrow (X)\Gamma(\phi; \phi \to \psi; \psi)$*;*

1(c) $(X)\Gamma(\tau \multimap \phi) \rightsquigarrow (X')\Gamma(\psi)$*, where* X' *is* X *with any variables required only for the well-formedness of* τ *replaced by units, and* $(X)\Gamma(\tau \to \phi) \rightsquigarrow (X)\Gamma(\tau \to \phi; \phi)$*;*

1(d) $((X(x \in A)\Gamma(\phi(x)), Y \vdash_\Sigma t : A) \rightsquigarrow (X(Y))\Gamma(\phi(t))$*, subject to the side-condition on the Substitution rule. Simultaneous substitutions are combined using ";", so that if we also have* $Y' \vdash_\Sigma t' : A$*, then we evaluate to* $(X(Y;Y'))\Gamma(\phi(t);\phi(t'))$*;*

2 $(X)\Gamma(\phi * \psi) \rightsquigarrow (X)\Gamma(\phi, \psi)$*;*

3 $(X)\Gamma(\phi \wedge \psi) \rightsquigarrow (X)\Gamma(\phi; \psi)$*;*

4 $(X)\Gamma(\phi \vee \psi) \rightsquigarrow (X')\Gamma(\phi)$ *or* $(X)\Gamma(\phi \vee \psi) \rightsquigarrow (X'')\Gamma(\psi)$, *where* X' *and* X'' *are* X *with any variables required only for the well-formedness of* ϕ *and* ψ, *respectively, replaced by units;*

5 $((X)\Gamma(\forall_{\mathbf{new}} x : A.\phi(x)), Y \vdash_\Sigma t : A) \rightsquigarrow (X, Y)\Gamma(\phi(t))$;

6 $((X)\Gamma(\forall x!A.\phi(x)), Y \vdash_\Sigma t : A) \rightsquigarrow (X; Y)\Gamma(\forall x!A.\phi(x); \phi(t))$;

7 $(X)\Gamma(\exists_{\mathbf{new}} x : A.\phi) \rightsquigarrow (X, x : A)\Gamma(\phi(x))$;

8 $(X)\Gamma(\exists x!A.\phi) \rightsquigarrow (X; x!A)\Gamma(\exists x!A.\phi; \phi(x))$.

We say that $(X')\Gamma'$ *is an* evaluation *of* $(X)\Gamma$ *and write* $(X)\Gamma \rightsquigarrow (X')\Gamma'$, *if* $(X')\Gamma'$ *is obtained by evaluating any of the redexes in* $(X)\Gamma$. *If* $(X')\Gamma'$ *is also prime, we say that* $(X')\Gamma'$ *is a* prime *evaluation of* $(X)\Gamma$.

We say that $(X')\Gamma'(\Delta') \preceq (X)\Gamma(\Delta)$ *if* $(X')\Gamma' = (X)\Gamma[\Delta'/\Delta]$ *for some* Δ, Δ' *such that* Δ' *is a reduct of* Δ. □

Just as in the propositional case, a prime evaluation of a given $(X)\Gamma$ may be constructed inductively. For example, suppose we have

$$\begin{aligned}(Y)\Delta = (x : A, y : B) \quad & (\phi(x)\,,\ \phi(x) \mathrel{-\!\!*} \psi(y)\,, \\ & \forall_{\mathbf{new}} v : A.\mathrm{p}(v)\,,\ \exists_{\mathbf{new}} w : B.\mathrm{q}(w))\end{aligned}$$

and suppose that $\Sigma = a : A\,,\, b : B$, with A and B atomic. Evaluating $\mathrel{-\!\!*} E$ and Substitution, we get

$$\begin{aligned}(\emptyset_m) \quad & (\psi(a)\,;\ \psi(b))\,, \\ & \forall_{\mathbf{new}} v : A.\mathrm{p}(v)\,,\ \exists_{\mathbf{new}} w : B.\mathrm{q}(w).\end{aligned}$$

Now, reducing the quantifiers, we get to

$$\begin{aligned}(w : B) \quad & (\psi(a)\,;\ \psi(b))\,, \\ & \mathrm{p}(a)\,,\ (\mathrm{q}(w)\,;\ \mathrm{q}(b)).\end{aligned}$$

and making a subsequent substitution for w, we get to

$$\begin{aligned}(\emptyset_m) \quad & (\psi(a)\,;\ \psi(b))\,, \\ & \mathrm{p}(a)\,,\ \mathrm{q}(b).\end{aligned}$$

Note that, as in Chapter 4, we have respected the original structure of the bunch.

We now form the prime evaluation of $(Y)\Delta$ as

$$(Y)(\Delta; ((\psi(a)\,;\ \psi(b))\,,\ (\mathrm{p}(a)\,,\ \mathrm{q}(b)).$$

This may readily be verified to satisfy, for example, condition (5) of Definition 13.9.

In general, in predicate **BI**, permitting countably many variables available at each type, the evaluation of bunches need not terminate. Therefore, for completeness, we must introduce the following notion of generalized bunch:

DEFINITION 13.12 (GENERALIZED BUNCHES) *A* generalized bunch *is a bunch with possibly (countably) infinitely many occurrences of "," or ";".* □

Let $(X)\Gamma$ be a generalized bunch. We say that $(X)\Gamma \vdash_{\Sigma,\Xi} \phi$ is provable in **NBI** without $\bot$ if $(X)\Gamma_f \vdash_{\Sigma,\Xi} \phi$ is provable in **NBI** without $\bot$, where $(X)\Gamma_f$ is some finite sub-bunch of $(X)\Gamma$. Similarly, if X is a generalized bunch of variables, we write $(X)u \mid m \models \phi$ if $(X_f)u \mid m \models \phi$, where (X_f) is some finite sub-bunch of (X).

LEMMA 13.13 (PRIME EVALUATION) *If* $(X)\Gamma \not\vdash_{\Sigma,\Xi} \phi$ *in* **NBI** *without* $\bot$*, then there is a prime evaluation,* $\lceil (X)\Gamma \rceil$*, of* $(X)\Gamma$ *such that*

$$\lceil (X)\Gamma \rceil \not\vdash_{\Sigma,\Xi} \phi$$

in **NBI** *without* $\bot$*, where* $\lceil X \rceil \vdash_{\Sigma,\Xi} \lceil \Gamma \rceil$.

PROOF The method presented in Chapter 4 for propositional **BI**, an adaptation of van Dalen's method [van Dalen, 1986] for intuitionistic logic, similar to the construction in [Dummett, 1977], in which $\lceil \Gamma \rceil$ is constructed as the co-limit of a sequence,

$$\Gamma = \Gamma_0 \preceq \Gamma_1 \preceq \Gamma_2 \preceq \ldots \preceq \lceil \Gamma \rceil,$$

of extensions of Γ may be further adapted to predicate **BI** by considering (generalized) bunches $(X)\Gamma$. We assume we have a denumerable collection of generalized bunches of variables.

As for propositional **BI**, closing under $\vdash$ and the clause for $\vee$ are essentially identical to the intuitionistic case, *q.v.* [van Dalen, 1983, Dummett, 1977], although note that we have restricted the closure to that generated by evaluating $-\!\!* E$, $\rightarrow E$, $\forall_{\mathbf{new}} E$, $\forall E$ and Substitution, and have added the necessary reductions for the remaining connectives. Note also that we must take a little more care over variables.

We successively reduce the remaining unreduced redexes: reduce $\wedge$, close, reduce $\vee$, close, reduce $*$, close, and so on, respecting the original structure of the bunch.[4] We discuss some of the cases, numbered as in Definition 13.11, in detail.

[4]Note that we intend no fixed order of reductions.

- Evaluating a redex of the form (1(a)) at stage k, we look for the first redex $(X)\Gamma_k(\phi_1, \phi_2 \mathrel{-\!\!*} \phi_2)$ such that $(X)\Gamma \vdash \phi_2$ which has yet to be reduced, where ϕ_1 and $\phi_1 \mathrel{-\!\!*} \phi_2$ are extensions of multiplicatively combined redices. It cannot be that $(X)\Gamma_k(\phi_2) \vdash \phi$, for then we should have $(X)\Gamma_k(\phi_1, \phi_2 \mathrel{-\!\!*} \phi_2) \vdash \phi$, so we can define

$$((X)\Gamma)_{k+1} = (X')\Gamma_k(\phi_2),$$

  where X' is X with any variables required only for the well-formedness of ϕ replaced by units.

- Case (1(b)) is similar, with variables unchanged. In this case, as for propositional **BI**, the redex is marked as *reduced.*

- Evaluating a redex of the form (1(c)) at stage k, we look, for example, for the first redex $(X)\Gamma_k(\tau \mathrel{-\!\!*} \psi)$, where τ is a theorem, and put $(X)\Gamma_{k+1} = (X')\Gamma_k(\psi)$, where X' is X with any variables required only for the well-formedness of τ replaced by units. In the case of $\rightarrow$, the redex is marked as reduced.

- Evaluating a redex of the form (2) at stage k, we look for the first redex $((X)\Gamma)_k(\phi_1 * \phi_2)$ such that $((X)\Gamma)_k(\phi_1 * \phi_2) \vdash \phi$ which has yet to be evaluated. It cannot be that $((X)\Gamma)_k(\phi_1, \phi_2) \vdash \phi$, for then we should have $((X)\Gamma)_k(\phi_1 * \phi_2) \vdash \phi$, so we can define

$$((X)\Gamma)_{k+1} = (X)\Gamma_k(\phi_1 \, , \, \phi_2).$$

  Note that we have *not* evaluated $(X)\Gamma_k$ non-trivially in this case: we have simply added to a dependency on $\phi * \psi$ a dependency on the bunch ϕ, ψ, *i.e.*, we have, essentially, performed a $*L$-reduction (*q.v.* Chapter 12) and a contraction, thereby preserving consequences.

- Evaluating a redex of the form (5) at stage k, we define

$$((X)\Gamma)_{k+1} = (X, Y)\Gamma_k(\phi[t/x]).$$

- Upon evaluating a redex of the form (6), the redex is marked as reduced.

- Evaluating a redex of the form (7) at stage k, we define

$$((X)\Gamma)_{k+1} = (X, x : A)\Gamma_k(\phi(x)).$$

The last two cases illustrate why we must work with generalized bunches. The case for the substitution rule is similar. Each of the remaining cases is similar to one of those discussed above.

We take $\overset{\rightsquigarrow}{(X)}\Gamma$ to be the limit of the $((X)\Gamma)_k$s over $k \geq 0$. Formally, we must consider an inductive definition over trees ordered by $\preceq$ and show that the co-limit, which is a generalized bunch, exists; the argument is evident, though note that looping is prevented by marking, upon reduction, the additive redexes as reduced. Then we set $\lceil (X)\Gamma \rceil = (X)\Gamma; \overset{\rightsquigarrow}{(X)}\Gamma$.

Given this construction, we must then check that (i) $\lceil (X)\Gamma \rceil \not\vdash_{\Sigma,\Xi} \phi$, and (ii) $\lceil (X)\Gamma \rceil$ is a prime evaluation of $(X)\Gamma$.

For (i), we show, by induction on i, that $((X)\Gamma)_i \not\vdash_{\Sigma,\Xi} \phi$, starting with our assumption that $(X)\Gamma \not\vdash_{\Sigma,\Xi} \phi$. For example, for the predicate $*$ case, suppose that $((X)\Gamma)_{i+1} \vdash_{\Sigma,\Xi} \phi$, *i.e.*, that $((X)\Gamma)_i(\phi_1 * \phi_2) \vdash_{\Sigma,\Xi} \phi$. Then it must be, by Lemma 12.3, that $((X)\Gamma)_i(\phi_1, \phi_2) \vdash_{\Sigma,\Xi} \phi$, a contradiction.

For (ii), that $\lceil (X)\Gamma \rceil$ is prime, we proceed just as in Chapter 4, [van Dalen, 1983], with additional cases for the variables and quantifiers.

For existence, note that we permit the availability of an infinite set of variables of each type from which to form bunches of variables and hence terms, so that, in general we construct generalized bunches.[5] □

As in the propositional case, in the presence of disjunction, prime evaluations are not unique: evaluating a redex of the form $\Gamma(\phi \vee \psi)$ yields the choice between $\Gamma(\phi \vee \psi; \phi)$ and $\Gamma(\phi \vee \psi; \psi)$. Formally, this choice is handled by working with finite sets of evaluations.

LEMMA 13.14 (MODEL EXISTENCE) *There exists a Kripke model*

$$\langle [\mathcal{T}^{op}, \mathbf{Set}], \models^{\mathcal{T}}_{\Sigma,\Xi}, D, [\![-]\!] \rangle,$$

where $\mathcal{T} = (T, e, \cdot, \sqsubseteq)$, *and a world* $t \in T$ *such that if* $(X)\Gamma \vdash_{\Sigma,\Xi} \phi$ *is not derivable in* **NBI** *without* $\bot$, *then, for all* u_X *at* t, $(X)u_X \mid t \models^{\mathcal{T}}_{\Sigma,\Xi} \Gamma$ *and* $(X)u_X \mid t \not\models^{\mathcal{T}}_{\Sigma,\Xi} \phi$.

PROOF We construct, using the normal proofs defined by **NBI** without $\bot$, a term model with the desired property. We begin by noting that, by Lemma 4.5, with the requisite assumptions about variables, any bunch Γ over X may be evaluated to a prime bunch $\lceil \Gamma \rceil$ over $\lceil X \rceil$, such that if $(X)\Gamma \not\vdash \phi$, then $\lceil (X)\Gamma \rceil \not\vdash \phi$.

Let $\mathcal{T} = (T, e, \cdot, \sqsubseteq)$ be the monoid of finite sets of prime evaluations of bunches (*i.e.*, generalized bunches) in the language $\mathcal{P}(L(\Sigma, \Xi))$ defined as follows:

[5] For cardinality, note that similar analyses, which may readily be adapted to our setting, have been treated formally in, for example, [Dummett, 1977, van Dalen, 1986, Lambek and Scott, 1986].

- T is $B/\approx$, where B is the set of finite sets of prime evaluations $(X)\Gamma$ of bunches and $\approx \; = \; \dashv\vdash \; \cup \; \simeq$, where $\dashv\vdash$ is the evident equality generated by derivability and $\simeq$ is the evident equality generated by the tree isomorphism of bunches of variables;

- $\cdot$ is given by the prime evaluation of the combination of bunches using the comma, ",": $(X)\Gamma \cdot (Y)\Delta \cong \lceil (X,Y)\Gamma \, , \, \Delta \rceil$, so that

$$\{ (X_1)\Gamma_1, \ldots , (X_m)\Gamma_m \} \cdot \{ (Y_1)\Delta_1, \ldots , (Y_n)\Delta_n \} =$$

$$\begin{array}{llllll} \{ & (X_1)\Gamma_1 \cdot (Y_1)\Delta_1 & , & (X_1)\Gamma_1 \cdot (Y_2)\Delta_2 & , & \ldots \\ & (X_2)\Gamma_2 \cdot (Y_1)\Delta_1 & , & (X_2)\Gamma_2 \cdot (Y_2)\Delta_2 & , & \ldots \\ & \vdots & , & \vdots & , & \ldots \; \}; \end{array}$$

- e is $\{ (\emptyset_m)\emptyset_m \}$;

- $\sqsubseteq$ is given by extension of bunches by semi-colon, ";", so that

$$\{ (X_1)\Gamma_1, \ldots , (X_m)\Gamma_m \} \sqsubseteq \{ (Y_1)\Delta_1, \ldots , (Y_n)\Delta_n \}$$

  iff, for all $1 \leq i \leq m$ and all $1 \leq j \leq n$, $(X_i)\Gamma_i \sqsubseteq (Y_j)\Delta_j$.

We write $\lceil (X)\Gamma \rceil \vdash \phi$ to denote that $(Y)\Delta \vdash \phi$, for some finite $\lceil \Gamma \rceil \sqsubseteq (Y)\Delta$.

The interpretation of types: atomic types α are interpreted as

$$[\![\alpha]\!](\{(X)\Gamma s\}) = \hat{\alpha}(\{(X)\Gamma s\}) = \{\alpha\}$$

with the interpretation of other types defined via the DCC structure of $[\mathcal{T}^{op}, \mathbf{Set}]$ in the usual way. The interpretation of variables, via a domain of individuals, $D \in obj([\mathcal{T}^{op}, \mathbf{Set}])$, is given by

$$D(\{(X_i)\Gamma_i s\}) = \{ x \in A \mid X \vdash_\Sigma x : A \quad \text{for some } X_i \}$$

so that

$$[\![x \in A]\!]^D \;=\; D((x \in A)\Gamma) \;=\; \{ x \}$$

for every Γ in $\{x\}$; so environments u for variables X are given by $[\![x \in A]\!]^D_u((x \in A)\Gamma) = x$, with environments for bunches being built up using $(-,-)$ and $[-,-]$. The interpretation of constants, $c : A$, is given by

$$[\![c!A]\!]((X)\Gamma) \;=\; \hat{c}((X)\Gamma) \;=\; \{c\}$$

Now define $[\![-]\!] : \mathcal{P}(L) \rightharpoonup obj([\mathcal{T}^{op}, \mathbf{Set}])$ as follows, in which proofs in **NBI** without $\perp$ are represented as $\alpha\lambda$-terms (with types at the level of predicate **BI**'s propositions, without $\perp$, analogously to the treatment of propositional **BI**'s proofs, without $\perp$, in Chapter 4):

1 $[\![I]\!]((X)\Gamma) = \mathcal{T}((X)\Gamma, (\emptyset_m)\emptyset_m)$;

2 $[\![\top]\!]((X)\Gamma) = \{\, * \,\}$;

3 For non-unit propositions,

$$[\![\phi]\!](\{(X_1)\Gamma_1, \ldots, (X_m)\Gamma_m\}) \;=\; \{\, \Phi \mid \Phi \text{ is a normal proof of } (X_i)\Gamma_i \vdash_{\Sigma,\Xi} \phi \text{ for some } 1 \leq i \leq m \,\}.$$

(Here, again, we presume SN for **NBI** with $\vee$ and here also with $\exists$ and $\exists_{\mathbf{new}}$, obtained using the techniques discussed in Chapters 2 and 12.) Henceforth, we will abuse this notation and write just $[\![\phi]\!]((X)\Gamma)$, reflecting our use of sets of bunches only to formalize the choices generated by disjunctive evaluation redexes.

It is important to note that the proofs Φ referred to here are not restricted to the judgement of logical consequence but must also include proofs, as determined by the rules for well-typed terms, that a given term has a given type relative to a bunch of variables, with the two judgements interacting via the Substitution rule.

Since

$$\{\, \Psi \mid \Psi \text{ is a normal proof in } \mathbf{NBI} \text{ without } \bot \text{ of } X \vdash_{\Sigma} t : A \,\}$$

is in bijective "propositions-as-types" correspondence with

$$\{\, t \mid \text{there is a normal proof in } \mathbf{NBI} \text{ without } \bot \text{ of } X \vdash_{\Sigma} t : A \,\},$$

it follows that we can define

$$[\![\mathrm{p}]\!]((X)\Gamma) \;=\; \{\, \Psi \mid \Psi \text{ is a normal proof of } X \vdash_{\Sigma} t : A \text{ and } (X)\Gamma \vdash_{\Sigma,\Xi} \mathrm{p}(t) \,\}.$$

It is then easy to check that

$$(X)\Gamma \vdash_{\Sigma,\Xi} \mathrm{p}(t) \quad \text{iff} \quad [\![t]\!]((X)\Gamma) \in [\![\mathrm{p}]\!]((X)\Gamma).$$

Note that if $\mathrm{p} : A$ is a predicate symbol of type $\emptyset_m$, *i.e.*, a proposition symbol, this definition collapses appropriately.

Now we define $\models^{\mathcal{M}}_{\Sigma,\Xi}$ by

$$(X)u \mid (Y)\Gamma \models^{\mathcal{T}}_{\Sigma,\Xi} \phi \quad \text{iff} \quad [\![\phi]\!]((X)\Gamma) \quad \text{is non-empty.}$$

It remains for us to check that this forcing relation satisfies its constraints, defined by induction on the structure of propositions. We exploit our use of normal proofs and prime bunches in the definition of $\mathcal{T}$.

Just as in the propositional model existence construction, Lemma 4.6 in Chapter 4, we sketch just a few cases, omitting some routine details:

Atoms: $(X)u \mid (Y)\Gamma \models^{\mathcal{T}}_{\Sigma,\Xi} \mathrm{p}(t)$ if and only if $[\![\mathrm{p}(t)]\!]((X)\Gamma)) \neq \emptyset$ if and only if $[\![t]\!]((X)\Gamma) \in [\![\mathrm{p}]\!]((X)\Gamma)$, which follows immediately from the definition of $\mathcal{T}$;

$*$: $(X)u \mid (Y)\Gamma \models^{\mathcal{T}}_{\Sigma,\Xi} \phi * \psi$ if and only if $[\![\phi * \psi]\!]((X)\Gamma)) \neq \emptyset$ if and only if, for some $(Y)\Delta$ and $(Y')\Delta'$ such that $\Gamma = \Delta, \Delta'$, $Z, Z' = X$ and $u = [u_Z, u_{Z'}]$, $[\![\phi]\!]((Z)\Delta)) \neq \emptyset$ and $[\![\psi]\!]((Z')\Delta')) \neq \emptyset$ if and only if $(Z)u_Z \mid (Y)\Delta \models^{\mathcal{T}}_{\Sigma,\Xi} \phi$ and $(Z')u_{Z'} \mid (Z')\Delta' \models^{\mathcal{T}}_{\Sigma,\Xi} \psi$;

$\forall$: $(X)u \mid (Y)\Gamma \models^{\mathcal{T}}_{\Sigma,\Xi} \forall x!A.\phi$ if and only if $[\![\forall x!A.\phi]\!]((X)\Gamma) \neq \emptyset$ if and only if

$$\{\Phi \mid \Phi \quad \text{is a normal proof of} \quad (X)\Gamma \vdash_{\Sigma,\Xi} \forall x!A.\phi\} \neq \emptyset$$

if and only if

$$\{\Phi \mid \Phi \quad \text{is a normal proof of} \quad (X; x!A)\Gamma \vdash_{\Sigma,\Xi} \phi\} \neq \emptyset.$$

if and only if, for all $(X)\Gamma \sqsubseteq (Z)\Delta$ and all $u_x \in D((Z)\Delta)$,

$$(X; x!A)([\![X]\!]^D((X)\Gamma \sqsubseteq (Z)\Delta)u, u_x) \mid (Z)\Delta \models^{\mathcal{T}}_{\Sigma,\Xi} \phi;$$

$\exists$: Similar;

$\forall_{\mathbf{new}}$: Similar;

$\exists_{\mathbf{new}}$: $(X)u \mid (Y)\Gamma \models^{\mathcal{T}}_{\Sigma,\Xi} \exists_{\mathbf{new}} x : A.\phi$ if and only if $[\![\exists_{\mathbf{new}} x : A.\phi]\!]((X)\Gamma) \neq \emptyset$ if and only if there is some $Y \vdash_\Sigma t : A$ and $[\![\phi[t/x]]\!]((X, Y)\Gamma) \neq \emptyset$ if and only if, since $(X, Y)\Gamma \equiv (X)\Gamma \cdot (Y)\emptyset_m$ and where $u_x \in D((X, Y)\Gamma)$,

$$(X, x : A)[u, u_x] \mid (X, Y)\Gamma \models^{\mathcal{T}}_{\Sigma,\Xi} \phi.$$

Finally, we must check that satisfaction corresponds to derivability as required. We claim that $(X)\Gamma \models^{\mathcal{T}}_{\Sigma,\Xi} \phi$ holds — note, as above, that $(X)\Gamma$ is read here as an element of the commutative preordered monoid $\mathcal{T}$ — if and only if $(X)\Gamma \vdash_{\Sigma,\Xi} \phi$ is derivable in **NBI** without $\perp$. Now, $(X)\Gamma \models^{\mathcal{T}}_{\Sigma,\Xi} \phi$ if and only if

$$\{\Phi \mid \Phi \quad \text{is a normal proof in } \mathbf{NBI} \text{ without } \perp \text{ of} \quad (X)\Gamma \vdash_{\Sigma,\Xi} \phi\}$$

is non-empty, which holds if and only if there is a proof of $(X)\Gamma \vdash_{\Sigma,\Xi} \phi$.

The monoidal isomorphisms are given by equality and the preorder is given by inclusion, just as in Chapter 4.

So, given the data $(X)\Gamma \not\vdash_{\Sigma,\Xi} \phi$, we set $t = \lceil (X)\Gamma \rceil$ and observe that clearly $\lceil (X)\Gamma \rceil \models^{\mathcal{T}}_{\Sigma,\Xi} \Gamma$, since the construction of $\lceil (X)\Gamma \rceil$ performs — essentially — left reductions, at both propositional and term levels, for $\wedge$, $\vee$ and $*$ and closes using extension via ";", whilst $\lceil (X)\Gamma \rceil \not\models^{\mathcal{T}}_{\Sigma,\Xi} \phi$.
□

THEOREM 13.15 (COMPLETENESS OF **BI** FOR KRIPKE MODELS) **NBI** *without* $\bot$ *is complete for Kripke models: if* $(X)\Gamma \models^{\mathcal{M}}_{\Sigma,\Xi} \phi$ *in* **BI** *without* $\bot$*, then* **NBI** *without* $\bot$ *proves* $(X)\Gamma \vdash_{\Sigma,\Xi} \phi$.

PROOF Suppose $(X)\Gamma \not\vdash_{\Sigma,\Xi} \phi$. Then Lemma 13.14 yields a contradiction. □

Note that we have not considered an (extensional) equality, corresponding to the usual set-theoretic equality in models. Had we done so, we should have had to take equivalence classes of provably equal bunches as the basis for $\mathcal{T}$.

Chapter 14

TOPOLOGICAL KRIPKE SEMANTICS FOR PREDICATE BI

1. Topological Kripke Models of Predicate BI with $\perp$

Just as for propositional **BI**, soundness and completeness for **BI** with $\perp$ may be obtained for a semantics based not on presheaves (or **Set**-valued functor categories) on a preordered monoid but rather based on sheaves on a commutative topological monoid. We define topological Kripke models of predicate **BI** and sketch the soundness and completeness arguments.

Just as in the Kripke semantics propositional **BI**, the definition of a topological Kripke model must be generalized to account for predication and quantification. The modifications are similar, *mutatis mutandis*, to those required for Kripke models. The key cases are for predicate letters, disjunction and its unit, $\perp$ (or inconsistency), and the existential quantifiers, leading to the following definition of a model:

DEFINITION 14.1 (TOPOLOGICAL KRIPKE MODEL) *Let* $(\mathcal{X}, *, e)$ *be an open commutative topological monoid and let* $\mathcal{P}(L(\Sigma, \Xi))$ *denote the collection of* **BI** *propositions over the language* $L(\Sigma, \Xi)$ *given by a term signature* Σ *and predicate signature* Ξ. *Let* $D \in obj(\mathbf{Sh}(\mathcal{X}))$ *be a domain of individuals, with interpretation functor* $[\![-]\!]^D$, *and let* $[\![-]\!] : \mathcal{P}(L(\Sigma, \Xi), \mathcal{X}) \rightharpoonup obj(\mathbf{Sh}(\mathcal{X}))$ *be a partial function from the* **BI** *propositions over* $L(\Sigma, \Xi)$ *to the objects of* $\mathbf{Sh}(\mathcal{X})$. *A* topological Kripke model *is a quadruple*

$$\langle \mathbf{Sh}(\mathcal{X}), \models_{\Sigma,\Xi}, D, [\![-]\!] \rangle,$$

where $\mathbf{Sh}(\mathcal{X})$ *is the category of sheaves over the preorder category* $\mathcal{M}$ *to* **Set**, $\models^{\mathcal{M}}_{\Sigma,\Xi} \subseteq M \times \mathcal{P}(L(\Sigma, \Xi), \mathcal{X})$ *is a forcing relation such that, for* *each* $\mathrm{p} \in \Xi$ *and each world* m, $[\![\mathrm{p}]\!](m) \subseteq [\![A]\!](m)$ *and satisfying, where defined, the constraints in Table 14.1.*

We require the following three conditions:

1. *Enough points for* Σ: *for every* $c : A \in \Sigma$, *there is a* $\hat{c} \in \mathbf{Sh}(S)$ *such that, for every open set* U, $\hat{c}(U) \in [\![A]\!](U)$ *and, for all worlds* U *and* V, $[\![c]\!]^D(U) = \hat{c}(U) = \hat{c}(V) = [\![c]\!]^D(V)$;
2. *Kripke monotonicity:* $(X)u \mid U \models^{\mathcal{M}}_{\Sigma,\Xi} \phi$ *and* $V \subseteq U$ *implies*
$$(X)[\![X]\!]^D(V \subseteq U)u \mid V \models^{\mathcal{M}}_{\Sigma,\Xi} \phi;$$
3. *Axioms: a pairing of the form* $Axiom(X_1, X_2)$ *for pair of bunches of variables occurring in the* $Axiom(X_1, X_2)$ *side-conditions of propositional axioms interpreted in* $\langle \mathbf{Sh}(\mathcal{X}), \models^{\mathcal{M}}_{\Sigma,\Xi}, \rangle$.

Wherever no confusion may arise, we shall refer to a model

$$\langle \mathbf{Sh}(\mathcal{X}), \models^{\mathcal{M}}_{\Sigma,\Xi}, D, [\![-]\!] \rangle$$

simply as $\mathcal{X}$. □

To see that this definition is consistent with the sub-object classifier semantics of intuitionistic logic [Lambek and Scott, 1986] consider the pullback diagram in $\mathbf{Sh}(\mathcal{X})$),

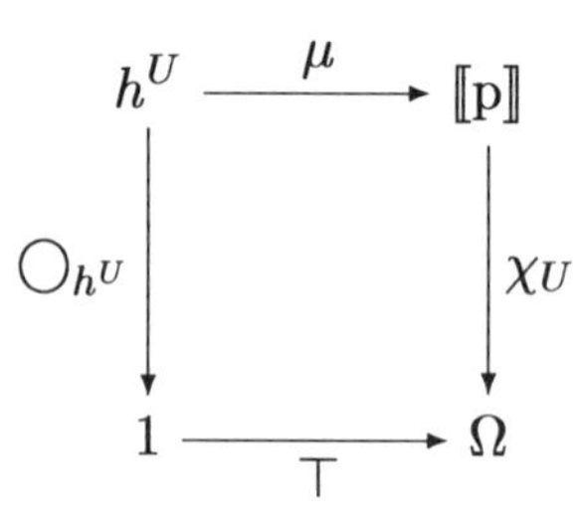

and note that an arrow $h^U \xrightarrow{\mu} [\![\mathrm{p}]\!]$ is, by the Yoneda lemma, determined uniquely by an element $\check{\mu} \in [\![\mathrm{p}]\!](U)$.

We define truth and validity for topological Kripke models of predicate **BI** just as in Chapter 13.

2. Soundness and Completeness for predicate BI with $\perp$

THEOREM 14.2 (SOUNDNESS) *Predicate* **BI** *with* $\perp$ *is sound for topological Kripke models: if* $\Gamma \vdash_{\Sigma,\Xi} \phi$ *is provable in* **NBI**, *then, for all topological Kripke models,* $\mathcal{M}$, $\Gamma \models^{\mathcal{M}}_{\Sigma,\Xi} \phi$.

$(X)u_X \mid U \models^{\mathcal{M}}_{\Sigma,\Xi} \mathrm{p}(\bar{t}(X))$	iff	$[\![\bar{t}(X)]\!]^D_{u_X}(U) \in [\![\mathrm{p}]\!](U), \quad \mathrm{p} \in \Xi$ $X \vdash_{\Sigma,\Xi} \mathrm{p}(\bar{t}(X)) : \mathbf{Prop}$
$(X)u_X \mid U \models^{\mathcal{M}}_{\Sigma,\Xi} \phi(X) \wedge \psi(X)$	iff	$(X)u_X \mid U \models^{\mathcal{M}}_{\Sigma,\Xi} \phi(X)$ and $(X)u_X \mid U \models^{\mathcal{M}}_{\Sigma,\Xi} \psi(X)$
$(X)u_X \mid U \models^{\mathcal{M}}_{\Sigma,\Xi} \phi(X) \rightarrow \psi(X)$	iff	for all $V \subseteq U$ $(X)u_X \mid V \models^{\mathcal{M}}_{\Sigma,\Xi} \phi(X)$ $(X)[\![X]\!]^D(V \subseteq U)u_X \mid V \models^{\mathcal{M}}_{\Sigma,\Xi} \psi(X)$
$(X)u_X \mid U \models^{\mathcal{M}}_{\Sigma,\Xi} \perp$	iff	$U = \emptyset$
$(X)u_X \mid U \models^{\mathcal{M}}_{\Sigma,\Xi} \phi(Y) \vee \psi(Y')$	iff	$U = V \cup V'$ for some $V, V' \in \Omega(S)$ such that $(Y)u_Y \mid V \models^{\mathcal{M}}_{\Sigma,\Xi} \phi(Y)$ and $(Y')u_{Y'} \mid V' \models^{\mathcal{M}}_{\Sigma,\Xi} \psi(Y')$, where $X = Y; Y'$ and $u_X = (u_Y, u'_Y)$
$(X)u_X \mid U \models^{\mathcal{M}}_{\Sigma,\Xi} \phi(Y) * \psi(Y')$	iff	$U = V \cdot V'$ for some $V, V' \in \Omega(S)$ such that $(Y)u_Y \mid V \models^{\mathcal{M}}_{\Sigma,\Xi} \phi(Y)$ and $(Y')u_{Y'} \mid V' \models^{\mathcal{M}}_{\Sigma,\Xi} \psi(Y')$, where $X = Y, Y'$ and $u = [u_Y, u'_Y]$
$(X)u_X \mid U \models^{\mathcal{M}}_{\Sigma,\Xi} \phi(Y') \mathrel{-\!\!*} \psi(Y)$	iff	for all V, all Z and all $v \in [\![Z]\!]^D(V)$ $(Z, Y')[v, u_{Y'}] \mid V \cdot U_{Y'} \models^{\mathcal{M}}_{\Sigma,\Xi} \phi(Y')$ implies $(Y, Z)[u_Y, v] \mid U_Y \cdot V \models^{\mathcal{M}}_{\Sigma,\Xi} \psi(Y)$ where $X = Y, Y'$ and $u = [u_Y, u_{Y'}]$
$(X)u_X \mid U \models^{\mathcal{M}}_{\Sigma,\Xi} \forall_{\mathbf{new}} x : A.\phi$	iff	for all V and all $u_x \in D(V)$ $(X, x : A)[u_X, u_x] \mid U \cdot V \models^{\mathcal{M}}_{\Sigma,\Xi} \phi$
$(X)u_X \mid U \models^{\mathcal{M}}_{\Sigma,\Xi} \forall x!A.\phi$	iff	for all $V \subseteq U$ and all $u_x \in D(V)$ $(X; x!A)[\![X]\!]^D(V \subseteq U)(u_X, u_x) \mid V \models^{\mathcal{M}}_{\Sigma,\Xi} \phi$
$(X)u_X \mid U \models^{\mathcal{M}}_{\Sigma,\Xi} \exists_{\mathbf{new}} x : A.\phi$	iff	there is an open cover $U = \bigcup_i U_i$ such that for all i there is some $u_x \in D(U)$ such that $(X, x : A)[\![X, x : A]\!]^D(U_i \subset U)[u_X, u_x] \mid U$ $\models^{\mathcal{M}}_{\Sigma,\Xi} \phi$
$(X)u_X \mid U \models^{\mathcal{M}}_{\Sigma,\Xi} \exists x!A.\phi$	iff	there is an open cover $U = \bigcup_i U_i$ such that for all i there is some $u_x \in D(U_i)$ such that $(X; x!A)[\![X; x!A]\!]^D(U_i \subseteq U)(u_X, u_x) \mid U \models^{\mathcal{M}}_{\Sigma,\Xi} \phi$

Table 14.1. Predicate Semantics in Sheaves

PROOF SKETCH The cases for the propositional connectives follows the pattern of the topological Kripke semantics of propositional **BI** given in Chapter 5. The cases for the quantifiers are similar to those for the Kripke semantics of given in Chapter 13 (see [Lambek and Scott, 1986] for a discussion of similar issues). □

Turning to completeness, we follow the same line of argument as in Chapter 13 but constructing from the syntax and proofs of predicate **BI** not a Kripke model but rather a topological Kripke model. Just as in Chapter 13, we need to construct the prime evaluation of a predicate bunch but the same construction, Definition 13.11, works in this setting. Moreover, Lemma 13.13 is unchanged. So it remains only to construct a commutative topological monoid from (prime) predicate bunches.

LEMMA 14.3 (MODEL EXISTENCE) *There is a topological Kripke model*

$$\langle [\mathbf{Sh}(\mathcal{T}), \models^{\mathcal{T}}_{\Sigma,\Xi}, D, [\![-]\!] \rangle,$$

and a world $t \in T$ such that if $(X)\Gamma \vdash_{\Sigma,\Xi} \phi$ is not derivable in **NBI**, *then, for all u_X at t, $(X) \quad u_X \mid t \models^{\mathcal{T}}_{\Sigma,\Xi} \Gamma$ and $(X) \quad u_X \mid t \not\models^{\mathcal{T}}_{\Sigma,\Xi} \phi$.*

PROOF SKETCH We can define a term topological Kripke **BI**-model, $\mathcal{T}$, in which we suppress the routine definition of $[\![-]\!]$, as follows:

- $\mid \mathcal{T} \mid$ is $B/ \dashv\vdash$, where B is the set of sets of consistent bunches, $(X)\Gamma$ and where $\dashv\vdash$ is the evident equality generated by derivability, *i.e.*, if S and S' are sets of consistent bunches, then $S \vdash S'$ iff, for any $\Gamma \in S$, there exists $\Gamma' \in S'$ such that $\phi_\Gamma \vdash \phi_{\Gamma'}$;

- Open sets are elements of $\mathcal{T}$ closed under prime evaluation of bunches.

- The monoid operation, $\cdot$, is given by the consistent prime evaluation of the combination of bunches using the comma, ",": $(X)\Gamma * (Y)\Delta \cong \lceil (X)\Gamma, (Y)\Delta \rceil$, where $\cong$ denotes isomorphism of labelled trees, so that

$$\{ (X_1)\Gamma_1, \ldots, (X_m)\Gamma_m \} \cdot \{ (Y_1)\Delta_1, \ldots, (Y_1)\Delta_n \} =$$

$$\begin{array}{lllll} \{ & (X_1)\Gamma_1 * (Y_1)\Delta_1 & , & (X_1)\Gamma_1 * (Y_1)\Delta_2 & , \ \ldots \\ & (X_2)\Gamma_2 * (Y_1)\Delta_1 & , & (X_2)\Gamma_2 * (Y_1)\Delta_2 & , \ \ldots \\ & \vdots & , & \vdots & , \ \ldots \ \} \end{array}$$

$$\setminus \perp((X)\Gamma, (Y)\Delta)(= \{ (X_i)\Gamma_i * (Y_j)\Delta_j \mid (X_i)\Gamma_i * (Y_j)\Delta_j \vdash \perp \});$$

- The unit, e, is given by $\{ (\emptyset_m)\emptyset_m \}$, where $\emptyset_m$ is the unit of ",";

- $[\![\phi]\!]((X)\Gamma) = \{\, \Phi \mid \Phi$ is a normal proof of $(X)\Gamma \vdash_{\Sigma,\Xi} \phi \,\}$. (Here, again, we presume SN for **NBI** with all of $\vee$, $\exists$ and $\exists_{\mathbf{new}}$, and here also $\bot$. As usual, the proof requires the techniques mentioned in Chapters 2 and 12.);
- Define, for non-empty Γ, the forcing relation, $\models^{\mathcal{T}}_{\Sigma,\Xi}$, by

$$(X)u \mid (Y)\Gamma \models^{\mathcal{T}}_{\Sigma,\Xi} \phi \quad \text{iff} \quad [\![\phi]\!]((X)\Gamma) \quad \text{is non-empty.}$$

The completeness property for $\mathcal{T}$, *i.e.*, if $(X)\Gamma \vdash_{\Sigma,\Xi} \phi$ is not provable in **NBI**, then $(X)\Gamma \not\models^{\mathcal{T}}_{\Sigma,\Xi} \phi$, follows by induction on the structure of formulæ. □

Completeness now follows by the usual argument:[1]

THEOREM 14.4 (COMPLETENESS) *Predicate* **NBI** *with* $\bot$ *is complete for topological Kripke models: if* $\Gamma \models_{\Sigma,\Xi} \phi$ *then* $\Gamma \vdash_{\Sigma,\Xi} \phi$ *is provable in* **NBI**. □

We conclude by remarking that the semantics in Grothendieck sheaves on a preordered commutative monoid, described for propositional **BI** in Chapter 4, can be extended to predicate **BI**, following the pattern for intuitionistic logic [Mac Lane and Moerdijk, 1992]. Indeed, just as for propositional **BI**, an alternative construction of a term model would follow the pattern of Proposition 5.15 and use open sets of formulæ, closed under $\vdash$. In the more general setting of Grothendieck sheaves, developed in Chapter 4 for propositional **BI** and which may be extended to predicate **BI**, disjunction is handled via the Grothendieck topology. However, as in the propositional case, the construction of prime bunches (*cf. prime theories* [van Dalen, 1983]) is of independent interest.

[1]Note that here, as in Chapter 13, we have not considered an extensional equality.

Chapter 15

RESOURCE SEMANTICS, TYPE THEORY AND FIBRED CATEGORIES

1. Predicate BI

So far, we have given a Tarski/Kripke-style semantics for predicate **BI**, in presheaf categories on a preordered commutative monoid and in sheaves on an open topological monoid. However, we should also like to have a BHK-style semantics for predicate **BI**'s proofs, extending the BHK-style semantics of propositional **BI**, based on doubly closed categories. In this section, we give a sketch of such a semantics.

Although presheaf/sheaf DCCs are adequate for the Tarski-style semantics of predicate **BI**, they do not yield a good interpretation of its proofs. A good interpretation requires, essentially, a fibred structure, as represented in Figure 15.1: Here we provide a sketch of how it might be done. The basic idea, which follows the pattern established in intuitionistic logic and associated type theories [Lawvere, 1969, Seely, 1983, Seely, 1984, Bénabou, 1985, Jacobs, 1998, Pitts, 1992, Pym, 2000a, Pym, 2000b, Pym, 2000c], is that a sequent $(X)\Gamma \vdash \phi$ is interpreted as follows:

- A base category $\mathcal{B}$ which is a DCC, in which we interpret bunches of variables and substitutions in the term language (which we can take to be the full $\alpha\lambda$-calculus) of predicate **BI** (*cf.* [Seely, 1983, Ambler, 1992]):

Objects: interpretations $[\![X]\!]$ of bunches X;

Arrows: interpretations $[\![\sigma]\!] : [\![X]\!] \longrightarrow [\![X']\!]$ of substitutions σ, in which the structure of substitutions corresponds to the structure of bunches:

$$\sigma ::= I \mid 1 \mid t \mid \sigma , \sigma \mid \sigma ; \sigma$$

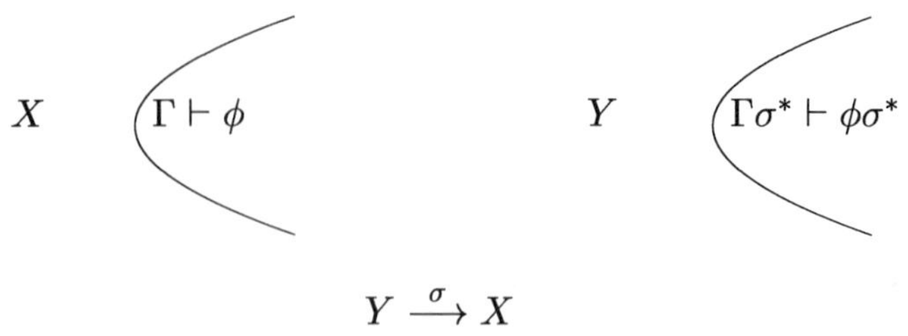

Figure 15.1. Fibred Models

where t is a term and I and 1 denote the monoidal isomorphisms; see [Jay, 1989a, Jay, 1989b, Jay, 1990] for similar ideas;

- Over each object $B = [\![X]\!]$ of $\mathcal{B}$, a DCC $\mathcal{E}(B)$, in which we interpret sequents $\Gamma \vdash \phi$ over a bunch of variables X;
- Functors corresponding to arrows in the base are handled in the usual doctrinal way, *e.g.*, [Seely, 1983, Pitts, 1992, Jacobs, 1998], with the images of the monoidal isomorphisms in the base being isomorphisms between the corresponding fibres;
- The *additive* quantifiers are interpreted, as in [Seely, 1983, Pitts, 1992, Jacobs, 1998], by left and right adjoints to the "Weakening" functors:

$$w^*_{[\![x!A]\!]} = \mathcal{E}(w) \qquad w : [\![X]\!] \longrightarrow [\![X; x!A]\!];$$

- The *multiplicative* quantifiers cannot be interpreted in terms of Weakening because the required projection maps do not exist. Instead, they are interpreted by *requiring* the existence of certain natural isomorphisms. For example, we require a natural isomorphism

$$[\![\forall_{\mathbf{new}} x : A]\!] : \mathcal{E}([\![X \, , \, x : A]\!]) \longrightarrow \mathcal{E}([\![X]\!]).$$

Rather than define this semantics in any detail, we develop a closely related semantics for the $\lambda\Lambda$-calculus, a dependently-typed λ-calculus which stands in a propositions-as-types correspondence, presented in [Ishtiaq, 1999, Ishtiaq and Pym, 2000][1] with a structural variant, *i.e.*,

[1]See § 9 for a discussion of this correspondence.

with Dereliction, of a fragment of **BI**. The $\lambda\Lambda$-calculus is motivated by the design of a logical framework which is able to represent some substructural logics adequately.

Most of the remainder of this chapter is joint work with Samin Ishtiaq. A detailed account of the $\lambda\Lambda$-calculus (and the RLF logical framework) is given in [Ishtiaq and Pym, 1998, Ishtiaq and Pym, 1999, Ishtiaq, 1999, Ishtiaq and Pym, 2000]. The work there develops ideas originally presented in [Pym, 1992]. Here, we present a summary of the type theory, motivated by the logical framework RLF.

2. Logical Frameworks

Logical frameworks are formal meta-logics which, *inter alia*, provide languages for describing logics in a manner that is suitable for mechanical implementation. The LF logical framework [Avron et al., 1992, Harper et al., 1993, Pym, 1990] provides such a metatheory and is suitable for logics which have at least the structural strength of minimal propositional logic. We wish to study a logical framework for describing relevant logics. Now, in order to describe a logical framework one must:

1 Characterize the class of object-logics to be represented;

2 Give a meta-logic or language, together with its meta-logical status *vis-à-vis* the class of object-logics; and

3 Characterize the representation mechanism for object-logics.

The above prescription may conveniently be summarized by the slogan

$$Framework = Language + Representation.$$

We remark that these components are not entirely independent of each other [Pym, 1996]. We will point out some interdependencies later in this section.

One representation mechanism is that of judgements-as-types, which originates from Martin-Löf's [Martin-Löf, 1996] development of Kant's [Kant, 1800] notion of judgement. The two higher-order judgements, the hypothetical $J \vdash J'$ and the general $\bigwedge_{x \in C} . J(x)$, correspond to ordinary and dependent function spaces, respectively. The methodology of judgements-as-types is that judgements are represented as the type of their proofs. A logical system is represented by a signature which assigns kinds and types to a finite set of constants that represent its syntax, its judgements and its rule schemes. An object-logic's rules and proofs are seen as proofs of hypothetico-general judgements $\bigwedge_{x_1 \in C_1} \cdots \bigwedge_{x_m \in C_m} . J \vdash J'$. Representation theorems relate con-

$(X, J_1(\phi_1), \ldots, J_m(\phi_m)) \vdash_L \delta : J(\phi)$	*object – consequence*
$\Downarrow$	*encoding*
$\Gamma_X, x_1{:}J_1(\phi_1), \ldots, x_m{:}J_n(\phi_n) \vdash_{\Sigma_L} M_\delta : J(\phi)$	*meta – consequence,*

Figure 15.2. Representing Object-logics in a Meta-logic

sequence in an object-logic $\vdash_L$ to consequence in a represented, or encoded, logic $\vdash_{\Sigma_L}$, as pictured in Figure 15.2, where X is the set of variables that occur in ϕ_i, ϕ; J_i, J are judgements; δ is a proof-object (*e.g.*, a λ-term); Γ_X corresponds to X; each x_i corresponds to a place-holder for the encoding of J_i; and M_δ is a meta-logic term corresponding to the encoding of δ.

In the sequel, we do not consider the complete apparatus of judged object-logics. The example encodings in [Ishtiaq and Pym, 1998] are pathological in the sense that they require only one judgement. For example, the encoding of a fragment of intuitionistic linear logic requires the judgement of $(J_i = J =)$ $proof$. This is in contrast to the general multi-judgement representation techniques [Avron et al., 1998]. We conjecture that our studies may be applied to the general case, although we defer this development to another occasion.

A certain class of uniform representations is identified by considering surjective encodings between consequences of the object-logic $\vdash_L$ and consequences of the meta-logic $\vdash_{\Sigma_L}$ [Harper et al., 1994].[2] So, all judgements in the meta-logic have corresponding judgements in the object-logic. The judgement-as-types methodology has the property that encoded systems inherit the structural properties of the meta-logic. It is for this reason that LF — whose language, the $\lambda\Pi$-calculus, admits Weakening and Contraction — cannot uniformly encode linear and other relevant logics. To illustrate this point, suppose Σ_{ILL} is a uniform encoding of intuitionistic linear logic in LF, and that $\Gamma_X, \Gamma_\Delta \vdash_{\Sigma_{ILL}} M_\delta{:}J(\phi)$ is the image of the object-consequence $(X, \Delta) \vdash_{ILL} \delta{:}J(\phi)$. If $\Gamma_X, \Gamma_\Delta \vdash_{\Sigma_{ILL}} M_\delta{:}J(\phi)$ is provable, then so is $\Gamma_X, \Gamma_\Delta, \Gamma_\Theta \vdash_{\Sigma_{ILL}} M_\delta{:}J(\phi)$. By uniformity, the latter is the image of an object-logic consequence $(X, \Delta, \Theta) \vdash_{ILL} \delta'{:}J(\phi)$, which implies Weakening in linear logic, a contradiction.

[2]The specification in [Harper et al., 1994] is a stronger one, requiring uniformity over all "presentations" of a given logic. Such concerns are beyond our present scope.

Thus we seek a language in which Weakening and Contraction are not forced. We motivate the connectives of the language by considering the natural deduction form of rules for weak logics. We do this in a general way, by considering Prawitz's general form of schematic introductions from a more relevant point of view. Prawitz [Prawitz, 1978] gives these for intuitionistic logic. A schematic introduction rule for an n-ary sentential operator $\#$ is represented by an introduction rule of the form below. In the figure, only the bound assumptions for G_j are shown; we elide those for G_k, where $k \neq j$, for the sake of readability:

$$\frac{\begin{array}{ccccc} & & [H_{j,1}]\cdots[H_{j,h_j}] & & \\ \vdots & & \vdots & & \vdots \\ G_1 & \cdots & G_j & \cdots & G_p \end{array}}{\#(F_1,\ldots,F_n)}$$

In the above rule, $1 \leq j \leq p$. The Fs, Gs and Hs are formulæ constructed in the usual way. An inference infers a formula $\#(F_1, \ldots, F_n)$ from p premisses $G_1, \ldots, G_p$ and may bind assumptions of the form $H_{j,1}, \ldots, H_{j,h_j}$ that occur above the premiss G_j. We let the assumptions be multi-sets, thus keeping the structural rule of Exchange. We require that discharge be compulsory. In the case of the natural deduction presentation of intuitionistic linear logic, for instance, we require that $\{F_1, \ldots, F_n\} = \{G_j, H_{j,1} \ldots, H_{j,h_j}\}$. For example, in the rule for $-\!\!\circ$-introduction, whose conclusion is $\phi -\!\!\circ\, \psi$, we have $\{F_1, F_2\} = \{\phi, \psi\}$, $G_1 = \psi$ and $H_{1,1} = \phi$.

We annotate the introduction schema below to indicate our method of encoding. The Λ is a linear universal quantifier, o is the type of propositions and $\in$ ranges over both linear (F:o) and exponential (F!o) declarations. Each inference — *i.e.*, the binding of assumptions $H_{j,1}, \ldots, H_{j,h_j}$ above premiss G_j and the inference of formula $\#(F_1, \ldots, F_n)$ from premisses $G_1, \ldots, G_p$ — is represented by a $-\!\!\!-\!\!\circ$:

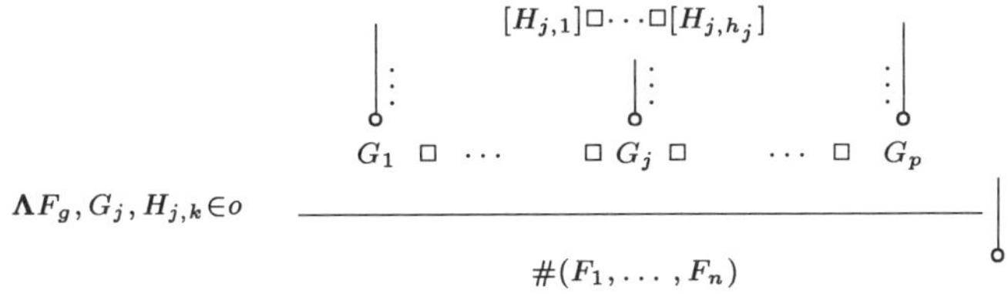

The premisses $G_1, \ldots, G_p$ are combined either multiplicatively or additively, depending on whether their contexts are disjoint or not. We distinguish between these combinations by the use of two conjunctions; the multiplicative $\otimes$ ("tensor") and the additive & ("with") and so force the structural rules. We use $\Box$ as meta-syntax for both $\otimes$ and &, though mindful of the relationship between the two operators. Full expressivity is recovered by introducing the modality ! ("bang") into our language. The premiss $!G$ allows us to depart from relevant inference, and to choose the number of times we use G in the conclusion.

In the meta-logic, then, the schematic introduction rule would be represented by a constant of the following type:

$$\Lambda F_g, G_j, H_{j,k} \in o. \ldots \Box(\Box_{l \leq h_j}(H_{j,l}) \multimap G_j)\Box \ldots \multimap \#(F_1, \ldots, F_n),$$

where $1 \leq l \leq h_j$ and $\Box_{l \leq h_j}$ represents an iterated $\Box$. From the general encoding formula above, it may be seen that the connectives $\Box$ (*i.e.*, $\otimes$ and &) and ! occur only negatively. In the tensor's case, this allows us to curry away the $\otimes$, modulo a permutation of the premisses. For example, in the following type, we are able to replace the occurrence of $\otimes$,

$$(\Box_{l \leq h_j}(H_{j,l}) \multimap G_j) \otimes (\Box_{l' \leq h_{j'}}(H_{j',l'}) \multimap G_{j'}) \multimap \#(F_1, \ldots, F_n),$$

by a $\multimap$,

$$(\Box_{l \leq h_j}(H_{j,l}) \multimap G_j) \multimap (\Box_{l' \leq h_{j'}}(H_{j',l'}) \multimap G_{j'}) \multimap \#(F_1, \ldots, F_n).$$

We can also consider a currying away of the & by a non-dependent version of the additive function space.

We recapitulate exactly how we have used the three logical constants in the framework: the & is used to undertake additive conjunction; the Λ is used to quantify and (in its non-dependent form $\multimap$) to represent implication; and the ! is used to represent dereliction from relevant inference. We should then be able to formulate a precise idea regarding the completeness of the set $\{\&, \Lambda, !\}$ with respect to all sentential operators that have explicit schematic introduction rules [Prawitz, 1978, Schroeder-Heister, 1983].

A similar analysis may be undertaken for the corresponding elimination rule, to yield a general schema together with its representation in RLF.

Our analysis allows us two degrees of freedom. The first is at the structural level of types. In this section, our main intention has been to motivate a language in which the structural rules of Weakening and

Contraction are not forced, and so to be able to uniformly encode linear and other substructural logics. Choosing a different language, with its particular structural and distributivity properties, would allow us to uniformly encode another class of logics. The family of relevant logics determined by these choices is very interesting from a representational perspective, though we pursue it no further here.

The second, orthogonal, degree of freedom, and one that we do concentrate on in the sequel, concerns the corresponding range of structural choices at the level of terms (as opposed to types). Considering this aspect from the logical point of view, we consider multiple occurrences of the same proof. The degree to which a proof may be shared by propositions is a structural property which determines, via the propositions-as-types, or Curry-Howard-de Bruijn, correspondence, a type theory whose functions and arguments share variables to a corresponding degree.

3. The $\lambda\Lambda$-calculus

The $\lambda\Lambda$-calculus is a first-order dependent type theory with both linear and intuitionistic function types. The calculus is used for deriving typing judgements. There are three entities in the $\lambda\Lambda$-calculus: *objects*, *types* and *families of types*, and *kinds*. Objects (denoted by M, N) are classified by types. Families of types (denoted by A, B) may be thought of as functions which map objects to types. Kinds (denoted by K) classify families. In particular, there is a kind Type which classifies the types. We will use U, V to denote any of the entities. The abstract syntax of the $\lambda\Lambda$-calculus is given by the following grammar:

$$
\begin{array}{lcl}
K & ::= & \mathsf{Type} \mid \Lambda x{:}A\,.K \mid \Lambda x!A\,.K \\
A & ::= & a \mid \Lambda x{:}A\,.B \mid \Lambda x!A\,.B \mid \lambda x{:}A\,.B \mid \\
  &   & \quad \lambda x!A\,.B \mid AM \mid A\&B \\
M & ::= & c \mid x \mid \lambda x{:}A\,.M \mid \lambda x!A\,.M \mid \langle M,N\rangle \mid \pi_i(M).
\end{array}
$$

We write $x{\in}A$ to range over both linear ($x{:}A$) and intuitionistic ($x!A$) variable declarations. The λ and Λ bind the variable x. The object $\lambda x{:}A\,.M$ is an inhabitant of the linear dependent function type $\Lambda x{:}A\,.B$. The object $\lambda x!A\,.M$ is an inhabitant of the type $\Lambda x!A\,.B$. This form of abstraction may also be written as $\Pi x{:}A\,.B$), where Π denotes the usual intuitionistic dependent function space.[3] The notion of linear free and

[3]Indeed, Π could be taken as a primitive constructor, obtained from Λ via a Dereliction rule in the sense of **BI**:

$$\frac{\Gamma, x{:}A \vdash_\Sigma M{:}B}{\Gamma; x!A \vdash_\Sigma M{:}B}.$$

bound variables (LFV, LBV) and substitution may be defined accordingly. When x is not free in B we write $A \multimap B$ and $A \rightarrow B$ for $\Lambda\, x{:}A\,.B$ and $\Lambda\, x!A\,.B$, respectively. We do not distinguish between α-convertible terms.

We can define the notion of *linear occurrence* by extending the general idea of occurrence for the λ-calculus [Barendregt, 1984], although we note that other, potentially superior, definitions may be possible.

DEFINITION 15.1 *1 x linearly occurs in x;*

2 If x linearly occurs in U or V (or both), then x linearly occurs in $\lambda\, y{\in}U\,.V$, in $\Lambda\, y{\in}U\,.V$, and in UV, where $x \neq y$;

3 If x linearly occurs in both U and V, then x linearly occurs in $\langle U, V\rangle$, $U\&V$ and $\pi_i(U)$.

□

The definition of occurrence is extended to an inhabited type and kind by stating that x occurs in $U{:}V$ if it occurs in U, in V, or in both. These notions are useful in the proof of the subject reduction property of the type theory.

We remark, though, that the above definitions are not "linear" in the Girard sense [Benton, 1994, Barber, 1996]. However, they seem quite natural in the bunches setting. O'Hearn and Pym give examples of **BI** terms — the $\lambda\Lambda$-calculus is in propositions-as-types correspondence with a non-trivial fragment of a variant of **BI**— where linear variables appear more than once or not at all [O'Hearn and Pym, 1999].

EXAMPLE 15.2 *The linear variable x occurs in the terms $cx{:}Bx$, $fx{:}d$ and $\lambda\, y{:}cx\,.y \,:\, Cx \multimap Cx$.*

We refer informally to the concept of a linearity constraint. Essentially this means that all linear variables declared in the context are used — a notion of *relevance*. Given this, the judgement $x{:}A, y{:}cx \vdash_\Sigma y{:}cx$ in which the linear x is consumed by the (type of) y declared after it and the y itself is consumed in the succedent, is a valid one.

In the $\lambda\Lambda$-calculus, signatures are used to keep track of the types and kinds assigned to constants. Contexts are used to keep track of the types, both linear and intuitionistic, assigned to variables. The abstract syntax for signatures and contexts is given by the following grammar:

$$\begin{array}{lrcl} \text{Signatures} & \Sigma & ::= & \langle\rangle \mid \Sigma, a!K \mid \Sigma, c!A \\ \text{Contexts} & \Gamma & ::= & \langle\rangle \mid \Gamma, x{:}A \mid \Gamma, x!A. \end{array}$$

The $\lambda\Lambda$-calculus is a formal system for deriving the following judgements:

$\vdash \Sigma$ sig	Σ is a valid signature
$\vdash_\Sigma \Gamma$ context	Γ is a valid context in Σ
$\Gamma \vdash_\Sigma K$ kind	K is a valid kind in Σ and Γ
$\Gamma \vdash_\Sigma A{:}K$	A has a kind K in Σ and Γ
$\Gamma \vdash_\Sigma M{:}A$	M has a type A in Σ and Γ.

The definition of the type theory depends crucially on several notions to do with the joining and maintenance of contexts; these are the notions of *context joining*, *variable sharing* and *multiple occurrences*. These notions are crucial in allowing the formation of sufficiently complex linear dependent types. The rules for deriving judgements in the type theory are given in Tables 15.1 and 15.2 below. These are conveniently separated into a linear and an intuitionistic set, the latter being directly related to the intuitionistic $\lambda\Pi$-calculus [Pym, 1995b].

We now discuss some of the rules of the type theory which exhibit the various notions we have introduced. The rules for extending contexts are as follows:

$$\frac{\vdash_\Sigma \Gamma \text{ context} \quad \Gamma \vdash_\Sigma A{:}\mathsf{Type}}{\vdash_\Sigma \Gamma, x{:}A \text{ context}} \qquad \frac{\vdash_\Sigma \Gamma \text{ context} \quad \Gamma \vdash_\Sigma A{:}\mathsf{Type}}{\vdash_\Sigma \Gamma, x!A \text{ context}}$$

The main point about these rules is that a context may be extended with either linear or intuitionistic variables. (There is no zone or "stoup" separating the linear from the intuitionistic parts of the context.)

Some of the more interesting type formation rules are discussed next (here, we simplify the systematic naming of the rules used in Tables 15.1 and 15.2). The C rule exports type constants from the signature. This may only be allowed in an entirely intuitionistic context.

$$C\ \frac{\vdash_\Sigma !\Gamma \text{ sig} \quad a!K \in \Sigma}{!\Gamma \vdash_\Sigma a{:}K}$$

$$\Lambda\mathcal{I}2\ \frac{\Gamma \vdash_\Sigma A{:}\mathsf{Type} \quad \Delta \vdash_\Sigma B{:}\mathsf{Type}}{\Xi \vdash_\Sigma A \multimap B{:}\mathsf{Type}} \qquad \begin{array}{c}[\Xi';\Gamma;\Delta] \\ \Xi = \Xi' \backslash ((lin(\Gamma) \cap lin(\Delta))\end{array}$$

$$\Lambda\mathcal{I}1\ \frac{\Gamma, x{:}A \vdash_\Sigma B{:}\mathsf{Type}}{\Gamma \vdash_\Sigma \Lambda\, x{:}A\,.B \ : \ \mathsf{Type}} \qquad \Lambda!\mathcal{I}\ \frac{\Gamma, x!A \vdash_\Sigma B{:}\mathsf{Type}}{\Gamma \vdash_\Sigma \Lambda\, x!A\,.B \ : \ \mathsf{Type}}$$

Valid Signatures

$$\frac{}{\vdash \langle\rangle \ \mathsf{sig}}\ (\Sigma)$$

$$\frac{\vdash \Sigma\ \mathsf{sig} \quad \vdash_\Sigma K\ \mathsf{kind} \quad a \notin \Sigma}{\vdash \Sigma, a!K\ \mathsf{sig}}\ (\Sigma K!) \qquad \frac{\vdash \Sigma\ \mathsf{sig} \quad \vdash_\Sigma A{:}\mathsf{Type} \quad c \notin \Sigma}{\vdash \Sigma, c!A\ \mathsf{sig}}\ (\Sigma A!)$$

Valid Contexts

$$\frac{\vdash \Sigma\ \mathsf{sig}}{\vdash_\Sigma \langle\rangle\ \mathsf{context}}\ (\Gamma)$$

$$\frac{\vdash_\Sigma \Gamma\ \mathsf{context} \quad \Delta \vdash_\Sigma A{:}\mathsf{Type} \quad [\Xi;\Gamma;\Delta] \quad (x \notin dom(\Xi) \text{ or } x{:}A \in \Xi)}{\vdash_\Sigma \Xi, x{:}A\ \mathsf{context}}\ (\Gamma A)$$

$$\frac{\vdash_\Sigma \Gamma\ \mathsf{context} \quad \Delta \vdash_\Sigma A{:}\mathsf{Type} \quad [\Xi;\Gamma;\Delta] \quad (x \notin dom(\Xi) \text{ or } x{:}A \in \Xi)}{\vdash_\Sigma \Xi, x!A\ \mathsf{context}}\ (\Gamma A!)$$

Valid Kinds

$$\frac{\vdash_\Sigma \Gamma\ \mathsf{context}}{\Gamma \vdash_\Sigma\ \mathsf{Type}\ \mathsf{kind}}\ (KAx) \qquad \frac{\Gamma, x{:}A \vdash_\Sigma K\ \mathsf{kind}}{\Gamma \vdash_\Sigma \Lambda\, x{:}A\,.K\ \mathsf{kind}}\ (K\Lambda\mathcal{I}1)$$

$$\frac{\Gamma \vdash_\Sigma A{:}\mathsf{Type} \quad \Delta \vdash_\Sigma K{:}\mathsf{Kind} \quad [\Xi';\Gamma;\Delta] \quad \Xi = \Xi' \backslash (lin(\Gamma) \cap lin(\Delta))}{\Xi \vdash_\Sigma A \multimap K{:}\mathsf{Kind}}\ (K\Lambda\mathcal{I}2)$$

$$\frac{\Gamma, x!A \vdash_\Sigma K\ \mathsf{kind}}{\Gamma \vdash_\Sigma \Lambda\, x!A\,.K\ \mathsf{kind}}\ (K\Lambda!\mathcal{I})$$

Table 15.1. $\lambda\Lambda$-calculus

The $\Lambda\mathcal{I}1$ and $\Lambda\mathcal{I}2$ rules form linear types. The second of these introduces the notion of context joining for binary multiplicative rules. The join must respect the ordering of the premiss contexts and the type of linear–intuitionistic variables. A method to join Γ and Δ to form Ξ, denoted by $[\Xi;\Gamma;\Delta]$, is defined in § 4 below. (The second side-condition will be explained using Example 15.3 below.)

Some of the more interesting object-level rules are discussed next (here, again, we simplify the systematic naming of the rules used in Tables 15.1 and 15.2). The two variable declaration rules, Var and $Var!$, declare linear and intuitionistic variables, respectively. These rules should not be seen as Weakening in the context Γ as, by induction, the variables declared in Γ are "used" in the construction of the type A. In

Valid Families of Types

$$\frac{\vdash_\Sigma !\Gamma \text{ context} \quad a!K \in \Sigma}{!\Gamma \vdash_\Sigma a{:}K}\ (Ac)$$

$$\frac{\Gamma, x{:}A \vdash_\Sigma B{:}\mathsf{Type}}{\Gamma \vdash_\Sigma \Lambda\, x{:}A\,.B : \mathsf{Type}}\ (A\Lambda\mathcal{I}1) \qquad \frac{\Gamma \vdash_\Sigma A{:}\mathsf{Type} \quad \Delta \vdash_\Sigma B{:}\mathsf{Type} \quad [\Xi';\Gamma;\Delta] \quad \Xi = \Xi' \backslash (lin(\Gamma) \cap lin(\Delta))}{\Xi \vdash_\Sigma A \multimap B{:}\mathsf{Type}}\ (A\Lambda\mathcal{I}2)$$

$$\frac{\Gamma, x!A \vdash_\Sigma B{:}\mathsf{Type}}{\Gamma \vdash_\Sigma \Lambda\, x!A\,.B : \mathsf{Type}}\ (A\Lambda!\mathcal{I})$$

$$\frac{\Gamma, x{:}A \vdash_\Sigma B{:}K}{\Gamma \vdash_\Sigma \lambda\, x{:}A\,.B : \Lambda\, x{:}A\,.K}\ (A\lambda\Lambda\mathcal{I}) \qquad \frac{\Gamma \vdash_\Sigma B : \Lambda\, x{:}A\,.K \quad \Delta \vdash_\Sigma N{:}A \quad [\Xi';\Gamma;\Delta] \quad \Xi = \Xi' \backslash \kappa(\Gamma,\Delta)}{\Xi \vdash_\Sigma BN : K[N/x]}\ (A\Lambda\mathcal{E})$$

$$\frac{\Gamma, x!A \vdash_\Sigma B{:}K}{\Gamma \vdash_\Sigma \lambda\, x!A\,.B : \Lambda\, x!A\,.K}\ (A\lambda\Lambda!\mathcal{I}) \qquad \frac{\Gamma \vdash_\Sigma B : \Lambda\, x!A\,.K \quad !\Delta \vdash_\Sigma N{:}A \quad [\Xi;\Gamma;!\Delta]}{\Xi \vdash_\Sigma BN : K[N/x]}\ (A\Lambda!\mathcal{E})$$

$$\frac{\Gamma \vdash_\Sigma A{:}\mathsf{Type} \quad \Gamma \vdash_\Sigma B{:}\mathsf{Type}}{\Gamma \vdash_\Sigma A\&B{:}\mathsf{Type}}\ (A\&\mathcal{I})$$

$$\frac{\Gamma \vdash_\Sigma A{:}K \quad \Delta \vdash_\Sigma K' \text{ kind} \quad K \equiv K' \quad [\Xi;\Gamma;\Delta]}{\Xi \vdash_\Sigma A{:}K'}\ (A\equiv)$$

Valid Objects

$$\frac{\vdash_\Sigma !\Gamma \text{ context} \quad c!A \in \Sigma}{!\Gamma \vdash_\Sigma c{:}A}\ (Mc)$$

$$\frac{\Gamma \vdash_\Sigma A{:}\mathsf{Type}}{\Gamma, x{:}A \vdash_\Sigma x{:}A}\ (MVar) \qquad \frac{\Gamma \vdash_\Sigma A{:}\mathsf{Type}}{\Gamma, x!A \vdash_\Sigma x{:}A}\ (MVar!)$$

$$\frac{\Gamma, x{:}A \vdash_\Sigma M{:}B}{\Gamma \vdash_\Sigma \lambda\, x{:}A\,.M : \Lambda\, x{:}A\,.B}\ (M\lambda\Lambda\mathcal{I}) \qquad \frac{\Gamma \vdash_\Sigma M : \Lambda\, x{:}A\,.B \quad \Delta \vdash_\Sigma N{:}A \quad [\Xi';\Gamma;\Delta] \quad \Xi = \Xi' \backslash \kappa(\Gamma,\Delta)}{\Xi \vdash_\Sigma MN : B[N/x]}\ (M\Lambda\mathcal{E})$$

$$\frac{\Gamma, x!A \vdash_\Sigma M{:}B}{\Gamma \vdash_\Sigma \lambda\, x!A\,.M : \Lambda\, x!A\,.B}\ (M\lambda\Lambda!\mathcal{I}) \qquad \frac{\Gamma \vdash_\Sigma M : \Lambda\, x!A\,.B \quad !\Delta \vdash_\Sigma N{:}A \quad [\Xi;\Gamma;!\Delta]}{\Xi \vdash_\Sigma MN : B[N/x]}\ (M\Lambda!\mathcal{E})$$

$$\frac{\Gamma \vdash_\Sigma M{:}A \quad \Gamma \vdash_\Sigma N{:}B}{\Gamma \vdash_\Sigma \langle M,N\rangle : A\&B}\ (M\&\mathcal{I}) \qquad \frac{\Gamma \vdash_\Sigma M : A_0\&A_1}{\Gamma \vdash_\Sigma \pi_i M : A_i}\ (M\&\mathcal{E}_i)\ (i \in \{0,1\})$$

$$\frac{\Gamma \vdash_\Sigma M{:}A \quad \Delta \vdash_\Sigma A'{:}\mathsf{Type} \quad A \equiv A' \quad [\Xi;\Gamma;\Delta]}{\Xi \vdash_\Sigma M{:}A'}\ (M\equiv)$$

Table 15.2. $\lambda\Lambda$-calculus (continued)

the rules for abstraction, $\Lambda\mathcal{I}$ and $\Lambda!\mathcal{I}$, the type of extension determines the type of function formed, just as in **BI**.

$$Var\ \frac{\Gamma \vdash_\Sigma A{:}\mathsf{Type}}{\Gamma, x{:}A \vdash_\Sigma x{:}A} \qquad Var!\ \frac{\Gamma \vdash_\Sigma A{:}\mathsf{Type}}{\Gamma, x!A \vdash_\Sigma x{:}A}$$

$$\Lambda\mathcal{I}\ \frac{\Gamma, x{:}A \vdash_\Sigma M{:}B}{\Gamma \vdash_\Sigma \lambda x{:}A\,.M \ :\ \Lambda x{:}A\,.B} \qquad \Lambda!\mathcal{I}\ \frac{\Gamma, x!A \vdash_\Sigma M{:}B}{\Gamma \vdash_\Sigma \lambda x!A\,.M \ :\ \Lambda x!A\,.B}\ .$$

Before we discuss the rules for object-level application, we should like to motivate the notions of variable sharing and multiple occurrences. Consider the following example of a non-derivation:[4]

EXAMPLE 15.3 *Let* $A!\mathsf{Type},\ c!A \multimap \mathsf{Type} \in \Sigma$ *and note that the argument type,* cx*, is a dependent one; the linear* x *is free in it. The figure*

$$\dfrac{\dfrac{\dfrac{\vdots}{x{:}A \vdash_\Sigma cx{:}\mathsf{Type}}}{x{:}A, z{:}cx \vdash_\Sigma z{:}cx}}{x{:}A \vdash_\Sigma \lambda z{:}cx\,.z \ :\ \Lambda z{:}cx\,.cx} \quad \dfrac{\dfrac{\vdots}{x{:}A \vdash_\Sigma cx{:}\mathsf{Type}}}{x{:}A, y{:}cx \vdash_\Sigma y{:}cx}$$

$$\overline{x{:}A, x{:}A, y{:}cx \vdash_\Sigma (\lambda z{:}cx\,.z)y \ :\ cx}$$

is not a derivation. The problem is that an excess of linear xs now appear in the combined context after the application step. This is because of the fact that an x each is needed for the well-formedness of each premiss type but only one x is needed for the well-formedness of the conclusion type. Our solution is to recognize the two xs as two *distinct* occurrences of the *same* variable, the one occurring in the argument type cx, and to allow a notion of sharing of this variable. One implication of this solution is that repeated declarations of the same variable are allowed in contexts; there are side-conditions on the context formation rules which reflect this (but, for reasons of simplicity, were omitted before). It is now necessary to formally define a binding strategy for multiple occurrences; this we do in § 5 below. The sharing aspect is implemented via the κ function, defined in § 6 below.

We can now discuss the rules for function application:

[4]Note that the fact that $\Lambda z{:}cx\,.cx$ is just $cx \multimap cx$ is of no importance here, the dependency on x being the point of interest.

$$\Lambda\mathcal{E}\ \frac{\Gamma \vdash_\Sigma M : \Lambda x{:}A.B \quad \Delta \vdash_\Sigma N{:}A}{\Xi \vdash_\Sigma MN : B[N/x]} \quad \begin{array}{c}[\Xi';\Gamma;\Delta]\\ \Xi = \Xi' \backslash \kappa(\Gamma,\Delta)\end{array}$$

$$\Lambda!\mathcal{E}\ \frac{\Gamma \vdash_\Sigma M : \Lambda x!A.B \quad !\Delta \vdash_\Sigma N{:}A}{\Xi \vdash_\Sigma MN : B[N/x]}\ [\Xi;\Gamma;!\Delta].$$

The side-condition on these is as follows: firstly, join the premiss contexts; then, apply κ to maintain the linearity constraint. It may be seen that these side-conditions are type-theoretically and, via the propositions-as-types correspondence, logically natural.

An essential difference between linear and intuitionistic function application is that, for the latter, the context for the argument $N{:}A$ is an entirely intuitionistic one ($!\Delta$), which allows the function to use N as many times as it likes.

The definitional equality relation that we consider for the $\lambda\Lambda$-calculus is the $\beta\eta$-equality of terms at all three levels. The definitional equality relation, $\equiv$, between terms at each respective level is defined to be the symmetric and transitive closure of the parallel nested reduction relation. There is little difficulty (other than that for the $\lambda\Pi$-calculus [Coquand, 1991, Salvesen, 1990, van Daalen, 1980]) in taking the equality relation by the η-rule. Equality is discussed more fully in § 7.

4. Context Joining

The method of joining two contexts, Γ and Δ, to form a third context, Ξ, is a ternary relation $[\Xi;\Gamma;\Delta]$ defined as follows:

$$\frac{}{[\langle\rangle;\langle\rangle;\langle\rangle]}\ (\text{JOIN})$$

$$\frac{[\Xi;\Gamma;\Delta]}{[\Xi,x{:}A;\Gamma,x{:}A;\Delta]}\ (\text{JOIN-L}) \qquad \frac{[\Xi;\Gamma;\Delta]}{[\Xi,x{:}A;\Gamma;\Delta,x{:}A]}\ (\text{JOIN-R})$$

$$\frac{[\Xi;\Gamma;\Delta]}{[\Xi,x!A;\Gamma,x!A;\Delta,x!A]}\ (\text{JOIN-!}).$$

The join operation extends one context with the other, removing any duplicate intuitionistic variables.

5. Multiple Occurrences

The type theory allows multiple occurrences of variables. For example, if, given the evident typings for b and c,

$$a\,!\,\Lambda x{:}A\,.\,bx \multimap cx \multimap \mathsf{Type} \in \Sigma,$$

then $axyz$ is a valid type in the context

$$x{:}A, y{:}bx, x{:}A, z{:}cx.$$

The two occurrences of x may be seen as different "colourings" of x.[5] For the purposes of binding, we must be able to pick out the first occurrence of x from the second. We define the *leftmost free occurrence* of x in U and a corresponding binding strategy for it.

The leftmost linear occurrence of x in U is, basically, the first x, syntactically, in the body of U; *e.g.*, if $U = VM$, then the leftmost occurrence of x is defined as follows:

$$\begin{array}{lcll} lm_x(@M) & = & lm_x(M) & x, @ \text{ distinct} \\ lm_x(xM) & = & \{x\} & \end{array}$$

$$lm_x(VM) = \begin{cases} lm_x(V) & x \in LFV(V) \\ lm_x(M) & \text{otherwise,} \end{cases}$$

where @ ranges over atoms (constants or variables).

We define the leftmost occurrence of $x{:}A$ in a context Γ as the first declaration of $x{:}A$ in Γ. Similarly, the rightmost occurrence of $x{:}A$ in Γ is the last such declaration. The binding strategy now formalizes the concept of linearity constraint:

DEFINITION 15.4 *Assume $\Gamma, x{:}A, \Delta \vdash_\Sigma U{:}V$ and that $x{:}A$ is the rightmost occurrence of x in the context. Then x binds:*

1 The first leftmost occurrence of x in the co-domain of Δ, if there is such a declaration;

2 The unbound leftmost linear occurrences of x in $U{:}V$.

□

There is no linearity constraint for intuitionistic variables: the rightmost occurrence of $x!A$ in the context binds all the unbound xs used in the type of a declaration in Δ and all the occurrences of x in $U{:}V$.

[5]There is an erroneous claim in § 2.2 of [Ishtiaq and Pym, 1999], corrected in [Ishtiaq and Pym, 2000]. Although cxx is indeed a valid term, it is not an example requiring multiple occurrences of x: It follows that Example 3 (*ibid.*) is incorrect. A suitable replacement is given by Example 15.5, below.

The rules for deriving judgements are now read according to the strategy in place. For example, in the $\Lambda\mathcal{I}$ rule, the $\lambda(\Lambda)$ binds the leftmost occurrence of x in $M(B)$. Similarly, in the (admissible) cut rule, the term $N{:}A$ cuts with the leftmost occurrence of $x{:}A$ in the context $\Delta, x{:}A, \Delta'$. In the corresponding intuitionistic rules, the $\lambda!(\Lambda!)$ binds all occurrences of x in $M(B)$ and $N{:}A$ cuts all occurrences of $x!A$ in the context $\Delta, x!A, \Delta'$.

EXAMPLE 15.5 *If* $a\,!\,\Lambda x{:}A\,.\,bx \multimap cx \multimap \mathsf{Type} \in \Sigma$, *then* $axyz$ *is a valid type in the context*[6]

$$x{:}A, y{:}bx, x{:}A, z{:}cx.$$

Although we can construct terms such as $axyz$, which is well-typed in the context $x{:}A, y{:}bx, x{:}A, z{:}cx$, the main motivation for multiple occurrences may be seen as being to introduce the notion of sharing in this type-theoretic setting.

6. Variable Sharing

Sharing occurs when linear variables are needed for the well-formedness of the premiss types but not necessarily for the well-formedness of the conclusion type. This requirement is regulated by a function κ. We define κ by considering the situation when either of the two contexts Γ or Δ are of the form $\ldots, x{:}A$ or $\ldots, x{:}A, y{:}Bx$. The only case when the two declarations of $x{:}A$ are not identified with each other is when both Γ and Δ are of the form $\ldots, x{:}A, y{:}Bx$.

The function κ is defined for the binary, multiplicative rules as follows: For each $x{:}A$ occurring in the premiss contexts Γ and Δ, construct from right to left as follows:

$$\begin{array}{rcll}
\kappa(\Gamma,\Delta) & = & \{\} & \textit{if } \mathit{lin}(\Gamma)\cap \mathit{lin}(\Delta) = \emptyset \\[2ex]
\kappa(\Gamma,\Delta) & = & \{x{:}A \mid \text{either } (i) \textit{ there is no } y{:}B(x) & \\
 & & \quad \textit{to the right of } x{:}A \textit{ in } \Gamma & \\
 & & \quad \text{or } (ii) \textit{ there is no } y{:}B(x) & \\
 & & \quad \textit{to the right of } x{:}A \textit{ in } \Delta & \\
 & & \quad \text{or both } (i) \textit{ and } (ii)\} & \textit{otherwise.}
\end{array}$$

In the absence of sharing of variables, when the first clause only applies, we still obtain a useful linear dependent type theory, with a linear

[6]The proof of this judgement given in [Ishtiaq and Pym, 2000] is mis-typed but the correct proof should be clear.

dependent function space but without the dependency of the abstracting A_is on the previously abstracted variables.

Given the definition, we can now consider the following example, which corrects Example 15.3:

EXAMPLE 15.6 *Suppose $A!\mathsf{Type}, c!A \multimap \mathsf{Type} \in \Sigma$. Then we construct the following:*

$$\dfrac{\dfrac{\dfrac{\dfrac{\vdash_\Sigma c{:}A \multimap \mathsf{Type} \qquad \dfrac{\vdash_\Sigma A{:}\mathsf{Type}}{x{:}A \vdash_\Sigma x{:}A}}{x{:}A \vdash_\Sigma cx{:}\mathsf{Type}}\,\dagger}{x{:}A, z{:}cx \vdash_\Sigma z{:}cx}}{x{:}A \vdash_\Sigma \lambda\, z{:}cx\, .z \;:\; \Lambda\, z{:}cx\, .cx} \qquad \dfrac{\dfrac{\vdash_\Sigma c{:}A \multimap \mathsf{Type} \qquad \dfrac{\vdash_\Sigma A{:}\mathsf{Type}}{x{:}A \vdash_\Sigma x{:}A}}{x{:}A \vdash_\Sigma cx{:}\mathsf{Type}}\,\dagger}{x{:}A, y{:}cx \vdash_\Sigma y{:}cx}}{x{:}A, y{:}cx \vdash_\Sigma (\lambda\, z{:}cx\, .z)y \;:\; cx}\,\dagger\dagger.$$

The $\dagger$ denotes the context join to get $x{:}A$. The $\dagger\dagger$ side-condition is more interesting. Firstly, the premiss contexts are joined together to get $x{:}A, x{:}A, y{:}cx$. Then, κ removes the extra occurrence of $x{:}A$ and so restores the linearity constraint.

Note that we have been slightly economical in this definition in that it relies on the formation of a *set*, rather than a *multiset*, of variables. For example, in Example 15.6, $x{:}A$ occurs in $\kappa(\Gamma, \Delta)$ just once. It would seem that alternative formulations are possible.

The function κ is not required, *i.e.*, its use is vacuous, when certain restrictions of the $\lambda\Lambda$-calculus type theory are considered. For instance, if we restrict type-formation to be entirely intuitionistic so that type judgements are of the form $!\Gamma \vdash_\Sigma A{:}\mathsf{Type}$, then we recover the $\{\Pi, \multimap, \&\}$-fragment of Cervesato and Pfenning's $\lambda^{\Pi\multimap\&\top}$ type theory [Cervesato and Pfenning, 1996].

Before proceeding to summarize $\lambda\Lambda$'s theory of equality and its principal meta-theoretic properties, we conclude our discussion with a proof of the typing of the term $axyz$, introduced earlier, which requires both multiple occurrence are variable sharing.

EXAMPLE 15.7 *Let*

$$a\,!\,\Lambda x{:}A\,.\,bx \multimap cx \multimap \mathsf{Type} \in \Sigma,$$

then $axyz$ is a valid type in the context

$$x{:}A, y{:}bx, x{:}A, z{:}cx.$$

The typing proof goes as follows:

$$\dfrac{\dfrac{\dfrac{\vdash_\Sigma a : \Lambda x{:}A\,.bx \multimap cx \multimap \mathsf{Type} \quad x{:}A \vdash_\Sigma x{:}A}{x{:}A \vdash_\Sigma ax{:}bx \multimap cx \multimap \mathsf{Type}} \quad \dfrac{x{:}A \vdash_\Sigma bx{:}\mathsf{Type}}{x{:}A, y{:}bx \vdash_\Sigma y{:}bx}}{x{:}A, y{:}bx \vdash_\Sigma axy{:}cx \multimap \mathsf{Type}} \quad \dfrac{x{:}A \vdash_\Sigma cx{:}\mathsf{Type}}{x{:}A, z{:}cx \vdash_\Sigma z{:}cx}}{x{:}A, y{:}bx, x{:}A, z{:}cx \vdash_\Sigma axyz{:}\mathsf{Type}}.$$

The last two applications have a non-trivial κ action which forces one of the $x{:}A$s to be shared. It can be checked that all the constants used in the proof are well-typed. Again we should emphasize that κ relies on the formation of a set, rather than a multiset, of variables.

7. Equality

The definitional equality relation that we consider here is the $\beta\eta$-conversion of terms at all three levels, subject to the binding strategy. The parallel nested reduction form of $\beta\eta$-reduction is written as $\rightarrow$. The transitive closure of $\rightarrow$ is denoted by $\rightarrow^*$. The definitional equality relation, $\equiv$, between terms at each respective level is defined to be the symmetric and transitive closure of $\rightarrow$. The one-step reduction relation is written as $\rightarrow_1$.

The relation, subject to the binding strategy, is given by the rules in Table 15.3 below. We include just the rules for β-reduction; the rules for η-reduction follow the usual pattern [Harper et al., 1993, Coquand, 1991, Salvesen, 1990, van Daalen, 1980], *e.g.*,

$$\frac{\Gamma \vdash_\Sigma \lambda x{\in}A\,.Mx : B \quad \Gamma \vdash_\Sigma M : C \quad x \notin \mathrm{FV}(M)}{\Gamma \vdash_\Sigma \lambda x{\in}A\,.Mx \equiv M}.$$

This concludes our brief syntactic presentation of the type theory. We refer to this presentation as system **N**. We will write **N** proves $\Gamma \vdash_\Sigma M{:}A$, *etc.*, to mean that the assertion $\Gamma \vdash_\Sigma M{:}A$, *etc.* is derivable in the system **N**. Where appropriate, we will omit the "**N** proves". A term is said to be *well-typed* or *valid* in a signature and context if it can be shown to either be a kind, have a kind, or have a type in that signature and context. We speak similarly of valid contexts relative to a signature and of valid signatures.

8. Basic Properties

A summary of the major meta-theorems pertaining to the type theory and its reduction properties are given by the following:

$$\frac{}{U \to U}\ (\to refl) \qquad \frac{A \to A' \quad M \to M'}{\lambda\, x{\in}A\,.M \to \lambda\, x{\in}A'\,.M'}\ (\to M\lambda)$$

$$\frac{A \to A' \quad K \to K'}{\Lambda\, x{\in}A\,.K \to \Lambda\, x{\in}A'\,.K'}\ (\to K\Lambda) \qquad \frac{M \to M' \quad N \to N'}{MN \to M'N'}\ (\to Mapp)$$

$$\frac{A \to A' \quad B \to B'}{\Lambda\, x{\in}A\,.B \to \Lambda\, x{\in}A'\,.B'}\ (\to A\Lambda) \qquad \frac{M \to M' \quad N \to N'}{(\lambda\, x{\in}A\,.M)N \to M'[N'/x]}\ (\to M\beta)$$

$$\frac{A \to A' \quad B \to B'}{\lambda\, x{\in}A\,.B \to \lambda\, x{\in}A'\,.B'}\ (\to A\lambda) \qquad \frac{M \to M' \quad N \to N'}{\langle M, N\rangle \to \langle M', N'\rangle}\ (\to M\&)$$

$$\frac{A \to A' \quad M \to M'}{AM \to A'M'}\ (\to Aapp) \qquad \frac{M \to M'}{\pi_i M \to \pi_i M'}\ (\to M\pi)$$

$$\frac{B \to B' \quad N \to N'}{(\lambda\, x{\in}A\,.B)N \to B'[N'/x]}\ (\to A\beta) \qquad \frac{M \to M'}{\pi_0\langle M, N\rangle \to M'}\ (\to M\pi_0)$$

$$\frac{A \to A' \quad B \to B'}{A\&B \to A'\&B'}\ (\to A\&) \qquad \frac{N \to N'}{\pi_1\langle M, N\rangle \to N'}\ (\to M\pi_1)$$

Table 15.3. Parallel Nested Reduction

THEOREM 15.8 *1 All well-typed terms are Church-Rosser.*

2 Exchange, Weakening, Dereliction, Contraction and two forms of Cut (one each for $x{:}A$ and $x!A$, all occurrences of x replaced) are admissible.

3 If $\Gamma \vdash_\Sigma U{:}V$ and $\Gamma \vdash_\Sigma U{:}V'$, then $V \equiv V'$.

4 If $\lambda\, x{\in}A\,.U$ inhabits $\Lambda\, x{\in}B\,.V$, then $A \equiv B$.

5 If $\Gamma \vdash_\Sigma U{:}V$ and $U \to U'$, then $\Gamma \vdash_\Sigma U'{:}V$.

6 If $\Gamma \vdash_\Sigma U{:}V$, then U is strongly normalizing.

7 If $\Gamma \vdash_\Sigma M{:}A$, then $erase(M)$ (the type erasure of $M{:}A$) may be typed in the Curry type-assignment system.

8 All assertions of the $\lambda\Lambda$-calculus are decidable. □

The proof of this theorem, presented in [Ishtiaq and Pym, 1998], is obtained by adapting the techniques discussed in Harper, *et al.* and elsewhere [Harper et al., 1993, Coquand, 1991, Salvesen, 1990, van Daalen,

1980] to this setting (see [Ishtiaq and Pym, 1998, Ishtiaq, 1999, Ishtiaq and Pym, 2001] for more detail and discussion of the available choices). The proof of the Church-Rosser property is shown by proving confluence for the one-step reduction relation and then doing an induction on the number of reduction steps. The proof of strong normalization is by giving a "dependency- and linearity-less" translation of the $\lambda\Lambda$-calculus into the Curry-typable untyped λ-calculus. The translation is faithful and consistent and allows us to "reflect" the strong normalization property of the λ-calculus back to the $\lambda\Lambda$-calculus.

9. The Propositions-as-types Correspondence

The propositions-as-types correspondence [Barendregt, 1992] between intuitionistic logic and the lambda calculus associates proofs Φ of first-order sequents,

$$(X)\Delta \vdash \phi,$$

with typing judgements

$$\Gamma_X, \Gamma_\Delta \vdash M_\Phi : A_\phi,$$

in a first-order dependently-typed λ-calculus, where Γ_X corresponds to the first-order variables, X, Γ_Δ corresponds to the propositional context, Δ, M_Φ is a λ-term corresponding to the proof-object, Φ, and the type A_ϕ corresponds to ϕ.

As mentioned in Chapter 2, the $\alpha\lambda$-calculus stands in such a propositions-as-types correspondence with propositional **BI**.

The $\lambda\Lambda$-calculus [Ishtiaq, 1999, Ishtiaq and Pym, 1998] type theory is motivated by a consideration, *inter alia*, of linear logic. However, it is structurally also close to **BI**. In **BI**, we have the two kinds of function space, the linear one $\multimap$ and the intuitionistic one $\rightarrow$. Correspondingly, there are two kinds of quantifier, the linear one $\forall_{\mathbf{new}}$ and the intuitionistic one $\forall$. Proof-theoretically, these arise because of extra structure in the context: There are two distinct context-formation operators, the ";", which admits the structural rules of Weakening and Contraction, and the ",", which doesn't. We can add the Dereliction rule (see 15.1, below) to **BI**, with the result that consequences such as $\phi \mathrel{-\!\!*} \psi \vdash \phi \rightarrow \psi$ become provable. In $\lambda\Lambda$, we have two forms of context extension, the linear and the intuitionistic, connected by Dereliction. As with **BI**'s Dereliction rule (15.1), $\lambda\Lambda$'s Dereliction rule does not rely on the presence of a general modality, such as linear logic's !.

It follows that the $\lambda\Lambda$-calculus stands in a (slightly weak) propositions-as-types correspondence with the $(\wedge, \rightarrow, \mathrel{-\!\!*}, \forall, \forall_{\mathbf{new}})$-fragment of **BI** with the rule of *Dereliction*,

$$\frac{(X(Y,Y'))\Gamma(\Delta,\Delta') \vdash \phi}{(X(Y;!Y'))\Gamma(\Delta;\Delta') \vdash \phi} \; D, \tag{15.1}$$

where each $!Z$ denotes Z with each $x : A$ replaced by $x!A$ and each "," is replaced by ";", but without the unit operation taking a bunch $\Delta(\Gamma)$ to $\Delta(I,\Gamma)$. This operation changes the status of Γ within a derivation, so that a proposition which starts out in additive combination with its neighbours may be abstracted (or quantified) multiplicatively. No corresponding operation is possible in $\lambda\Lambda$, so that this propositions-as-types correspondence does not properly generalize that for $\alpha\lambda$ and propositional **BI**. (The converse situation is handled by Dereliction.)

The details of the correspondence, given in [Ishtiaq, 1999, Ishtiaq and Pym, 2000], which depend critically on the dereliction of certain "linear" judgements.[7] The addition of Dereliction represents the extent to which $\lambda\Lambda$ corresponds to linear logic; but note again that this form of dereliction does not rely on the presence of a full modality, such as linear logic's !.

The basic idea for the correspondence between the fragment of **BI**, without the unit rule given above but with Dereliction for **BI** as in (15.1), and the $\lambda\Lambda$-calculus is to consider ";" as intuitionistic extension and "," as linear extension. This is implemented by giving a translation of **BI** contexts which relies, to a certain extent, on the notion of dereliction. The idea of viewing the **BI** context joining connectives as context extension operators necessarily restricts the correspondence to a fragment, though a non-trivial one, of **BI**. The correspondence between the connectives is given by the following table:[8]

[7]The dereliction of "linear" judgements in the definition of the correspondence is a *weakness* of the result obtained and partially motivates the programme sketch in § 15.

[8]Note that [Ishtiaq, 1999] has some mis-statements, corrected in [Ishtiaq and Pym, 2000]: (i) the addition of Dereliction to the $(\wedge, \rightarrow, -\!\!*, \forall, \forall_{\mathbf{new}})$-fragment of **BI** was not mentioned, though it should be clear from the context that it is required; (ii) the translation of **BI**'s contexts to $\lambda\Lambda$'s contexts is mis-stated with the consequence that a condition required in the statement of soundness is missing; and (iii) the Contraction rule for multiplicatively combined variables $x : A, y : A$ for which we have Axiom(x, y), and the corresponding Contraction in the correspondence, were omitted. Similarly, the need for Dereliction should have been mentioned in [Ishtiaq and Pym, 1999], though no details are given therein. I am grateful to Peter O'Hearn for asking a question which drew attention to these mis-statements. It must be admitted that the propositions-as-types correspondence obtained in [Ishtiaq, 1999, Ishtiaq and Pym, 2000], which relies heavily on Dereliction, is rather weak; § 15 suggests a way forward.

BI	$\lambda\Lambda$
$\wedge$	$\&$
$\to$	$\to$
$-\!\!*$	$\multimap$
$\forall - ! -$	$\Lambda - ! -$
$\forall_{\mathbf{new}} - : -$	$\Lambda - : -$

So, one view of this correspondence is that the RLF meta-logic uses this fragment of **BI**, just as the LF meta-logic uses the $\{\to, \forall\}$-fragment of intuitionistic logic. The relationship between the fragment of **BI** and linear logic deserves some comment. Dereliction represents the extent to which the $\lambda\Lambda$-calculus also corresponds to a fragment of linear logic. Whilst the Dereliction

$$\frac{\Gamma, x : A \vdash_\Sigma M : B}{\Gamma, x!A \vdash_\Sigma M : B},$$

in which we assume that $x \notin \mathrm{FV}(B)$, renders $A \to B$ as $!A \multimap B$, linear logic has no quantifier corresponding to Λ, which follows the pattern one expects in **BI**:

$$\frac{\Gamma \vdash_\Sigma M : \Lambda x : A.B \quad \Delta \vdash_\Sigma N : A}{\Xi(\Gamma, \Delta) \vdash_\Sigma MN : B[N/x]}.$$

We conclude this section by remarking that to give a systematic analysis of the relationship between substructural logics and (dependent) type theories remains a challenging open problem. In particular, it would be interesting to formulate, if possible, a dependent type theory corresponding to a proper fragment, *i.e.*, without Dereliction, of **BI**. We return to this point in more detail in § 15.

10. Kripke Resource Semantics for $\lambda\Lambda$

As we have seen, the semantics of **BI**, a fragment of which corresponds to the internal logic of the $\lambda\Lambda$-calculus, may be understood, categorically, by a single category which carries two monoidal structures. It may also be understood, model-theoretically, by a unique combination of two familiar ideas: a Kripke-style possible worlds semantics and an Urquhart-style resource semantics. We will use the internal logic and its semantics to motivate an indexed categorical semantics for the type theory: indeed, we require that our models provide a semantics for both the $\lambda\Lambda$-calculus as a presentation of its internal logic and as a theory of functions. The mathematical structure we motivate will be quite modular; a sub-structure will model the intuitionistic $\{\to, \Pi\}$-fragment of the $\lambda\Lambda$-calculus (*i.e.*, the $\lambda\Pi$-calculus). The model theory of $\lambda\Lambda$ has

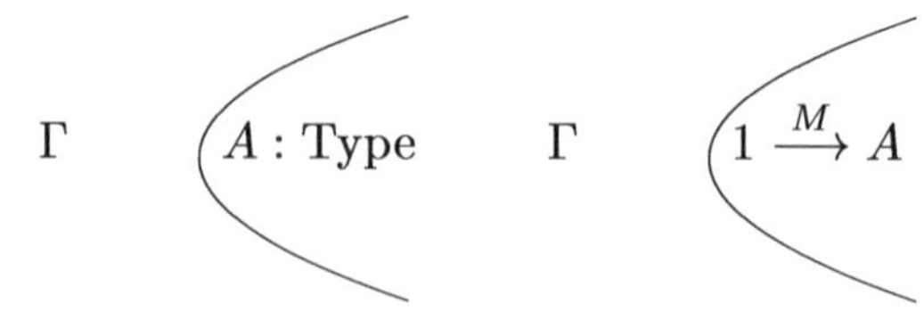

Figure 15.3. Fibred Kripke Models of Dependent Types

been studied in [Ishtiaq and Pym, 2001, Ishtiaq, 1999, Ishtiaq and Pym, 1999, Ishtiaq and Pym, 2000].

The semantics of $\lambda\Lambda$ is formulated in a fibred setting closely related to that used for predicate **BI**. The main difference between fibred models of **BI** and fibred models of $\lambda\Lambda$, which is the same as the main difference between fibred models of intuitionistic proofs and fibred models of intuitionistic dependent types, is that, in the case of dependent types, the base category and the fibres are mutually inductively defined. This indicated by Figure 15.3, in which it may be seen that the same fibred structured plays two rôles. On the one hand, it establishes which types are well-formed over a given context, Γ, while, on the other, it establishes which terms M inhabit such a type A.

We emphasize that, although we refer to resource semantics (structures, models) throughout our treatment of the semantics of $\lambda\Lambda$, motivated by our computational intuition, the mathematical structures we describe need not be interpreted in terms of resources.

11. Kripke Resource $\lambda\Lambda$-structure

The key issue in the syntax concerns the co-existing linear and intuitionistic function spaces and quantifiers. This distinction may be explained by reference to Kripke resource semantics.

A resource semantics elegantly explains the difference between the linear and intuitionistic connectives in that the action, or computation, of the linear connectives may be seen to consume resources. We consider this for the internal logic judgement $(X)\Delta \vdash \phi$. Let $\mathcal{M} = (M, \cdot, e, \sqsubseteq)$ be a preordered commutative monoid and recall that the forcing relation for the two implications is defined as follows:

1 $r \models \phi \multimap \psi$ if and only if, for all $s \in M$, if $s \models \phi$ then $r \cdot s \models \psi$;

2 $r \models \phi \to \psi$ if and only if, for all $s \in M$, if $s \sqsubseteq r$ and $s \models \phi$, then $s \models \psi$.

Recall furthermore that a similar pair of clauses defines the forcing relation for the two **BI** quantifiers. Here $D{:}\mathcal{M}^{op} \to \mathbf{Set}$ is a domain of individuals and $u \in [\![X]\!]r$ is an environment appropriate to the bunch of variables X at world r, where $[\![X]\!]$ is the interpretation of the bunch of variables X in $\mathbf{Set}^{\mathcal{M}^{op}}$:

1 $(X)u, r \models \forall x.\phi$ if and only if, for all $s \sqsubseteq r$ and all $d \in D(s)$,

$$(X;x)([\![X]\!](s \sqsubseteq r)u, d), s \models \phi;$$

2 $(X)u, r \models \forall_{\mathbf{new}} x.\phi$ if and only if, for all s and all $d \in D(s)$,

$$(X,x)[u,d], r \cdot s \models \phi.$$

Here, as usual, $(-,-)$ is cartesian pairing and $[-,-]$ is the pairing operation defined by Day's tensor product construction in $\mathbf{Set}^{\mathcal{M}^{op}}$.

Suppose we have a category $\mathcal{E}$, in which the propositions will be interpreted. Then we will index $\mathcal{E}$ in two ways for the purposes of interpreting the type theory. Firstly, we index it by a Kripke world structure $\mathcal{W}$. This is to let the functor category $[\mathcal{W}, \mathcal{E}]$ have enough strength to model the $\{\to, \forall\}$-fragment of the internal logic and so correspond to Kripke-style models for intuitionistic logic. Secondly, we index $[\mathcal{W}, \mathcal{E}]$ by a monoid $(R, +, 0)$ of resources, in the sense we have discussed in earlier chapters. Thus we obtain R-indexed sets of Kripke functors $\{\mathcal{J}_r{:}[\mathcal{W}, \mathcal{E}] \mid r \in R\}$. We remark that the separation of worlds from resources considered in this structure emphasizes a sort of "phase shift" [Girard, 1987, Hodas and Miller, 1994]. We briefly reconsider this choice in § 14.

We now consider how to model the propositions and so explicate the structure of $\mathcal{E}$. The basic judgement of the internal logic is $(X)\Delta \vdash \phi$, that ϕ is a proposition in the context Δ over the context X. One reading of this judgement, and perhaps the most natural, is to see X as an index for the propositional judgement $\Delta \vdash \phi$. This reading may be extended to the type theory, where, in the basic judgement $\Gamma \vdash_\Sigma M{:}A$, Γ may be seen as an index for $M{:}A$ or that $M{:}A$ depends on Γ for its meaning. Thus we are led to using the technology of indexed category theory. More specifically, in the case of the type theory, the judgement $\Gamma \vdash_\Sigma M{:}A$ is modelled as the arrow $1 \overset{[\![M]\!]}{\to} [\![A]\!]$ in the fibre over $[\![\Gamma]\!]$ in the strict indexed category $\mathcal{E}{:}\mathcal{C}^{op} \to \mathbf{Cat}$. Figure 15.4 suggests the set-up.

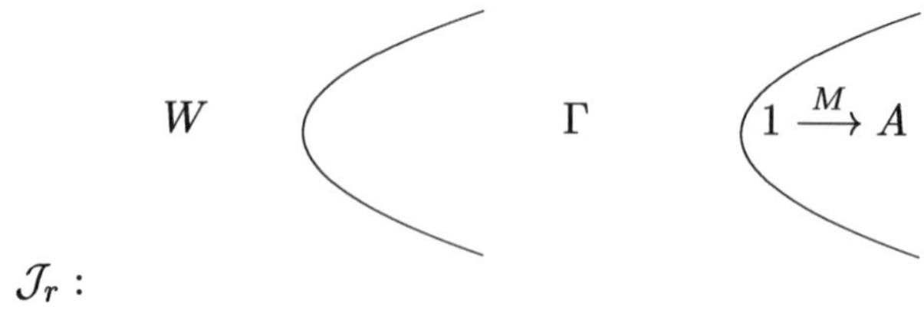

Figure 15.4. Fibred Models of $\lambda\Lambda$

We remark that this is not the only technique for modelling a typing judgement; Cartmell [Cartmell, 1994], Pitts [Pitts, 1992] and several other authors use a more "one-dimensional" structure which relies on the properties of certain classes of maps to model the intuitionistic fragment of the $\lambda\Lambda$-calculus. These are formally equivalent to the indexed approach but the latter is appealing for the main reason that it provides a technical separation of conceptually separate issues. For instance, at a logical level, the base and fibres deal, respectively, with terms and propositions.

We need the base category $\mathcal{C}$ to account for the structural features of the type theory and its internal logic, hence the following definition:

DEFINITION 15.9 *A* doubly monoidal category *is a category $\mathcal{C}$ equipped with two monoidal structures,* $(\otimes, I)$ *and* $(\times, 1)$*. $\mathcal{C}$ is called* cartesian doubly monoidal *if $\times$ is cartesian. We will use $\bullet$ to range over both multiplications.* □

There are a couple of comments we need to make about the monoidal structure on $\mathcal{C}$. Firstly, there is no requirement that the bifunctors $\otimes$ and $\times$ be symmetric, as the contexts which the objects are intended to model are (ordered) lists. Secondly, the use of the symbol $\times$ as one of the context extension operators suggests that $\times$ is a cartesian product. This is indeed the case when $\{\mathcal{J}_r \mid r \in R\}$ is a model of the internal logic, where there are no dependencies within the variable context X, but not when $\{\mathcal{J}_r \mid r \in R\}$ is a model of the type theory, where there are dependencies within Γ. In the latter case, we have the property that for each object D extended by $\times$, there is a first projection map $p_{D,A}{:}D \times A \to D$. There is no second projection map $q_{D,A}{:}D \times A \to A$ in $\mathcal{C}$, as A by itself may not correspond to a well-formed type. For modelling the judgement $\Gamma, x{\in}A \vdash_\Sigma x{:}A$, we do, however, require the existence of a map $1 \xrightarrow{q} [\![A]\!]$ in the fibre over $[\![\Gamma]\!] \bullet [\![A]\!]$.

A doubly monoidal category $\mathcal{C}$ with both exponentials or, alternatively, $\mathcal{C}$ equipped with two monoidal closed structures $(\times, \rightarrow, 1)$ and $(\otimes, \multimap, I)$, is a cartesian doubly closed category. Cartesian DCCs provide a class of models of propositional **BI** in which both function spaces are modelled within $\mathcal{C}$. We will work with the barer doubly monoidal category, requiring some extra structure on the fibres to model the function space. This may be seen as a natural extension to the semantics of bunches to account for dependency. It may be contrasted to the Barber–Plotkin model of **DILL** [Barber, 1996], which uses a pair of categories, a monoidal one and a cartesian one, together with a monoidal adjunction between them: this forces too much of a separation between the linear and intuitionistic parts of a context to be of use to us.

We now consider how the function spaces are modelled. In the intuitionistic case, the Weakening functor $p^*_{D,A}$ has a right adjoint $\Pi_{D,A}$ (which satisfies the Beck-Chevalley condition). In fact, this amounts to the existence of a natural isomorphism cur_W

$$\frac{Hom_{\mathcal{J}_r(W)(D\times A)}(p^*_{D,A}(C), B)}{Hom_{\mathcal{J}_r(W)(D)}(C, \Pi_{D,A}(B))} .$$

The absence of Weakening for the linear context extension operator means that we cannot model Λ in the same way. However, the structure displayed above suggests a way to proceed, just as discussed in § 1. It is sufficient to require the existence of a natural isomorphism $\Lambda_{D,A}$,

$$\frac{Hom_{\mathcal{J}_{r+r'}(W)(D\bullet A)}(1, B)}{Hom_{\mathcal{J}_r(W)(D)}(1, \Lambda_{D,A}\, x{\in}A\,.B)} ,$$

in the indexed category. Here, we use $\bullet$ to range over $\otimes$ and $\times$, and $\in$ to range over : and !. There are a couple of remarks that need to be made about the isomorphism. Firstly, it refers only to those homsets in the fibre whose source is 1. This restriction, which avoids the need to establish the well-foundedness of an arbitrary object over both D and $D \bullet A$, suffices to model the judgement $\Gamma \vdash_\Sigma M{:}A$ as an arrow $1 \stackrel{[\![M]\!]}{\rightarrow} [\![A]\!]$ in the fibre over $[\![\Gamma]\!]$: examples are provided by both the term and set-theoretic models that we will present later. Secondly, the extended context is defined in the $r + r'$-indexed structure. The reason for this may be seen by observing the form of the forcing clause for application in **BI**. Given these two remarks, the above isomorphism allows the formation of function spaces.

DEFINITION 15.10 (KRIPKE RESOURCE $\lambda\Lambda$-STRUCTURE) *Let* $(R, +, 0)$ *be a commutative monoid (of resources). A* Kripke resource $\lambda\Lambda$-structure *is an* R*-indexed set of functors*

$$\{\mathcal{J}_r{:}[\mathcal{W}, [\mathcal{C}^{op}, \mathbf{Cat}]] \mid r \in R\},$$

where $\langle \mathcal{W}, \leq \rangle$ *is a preorder,* $\mathcal{C}^{op} = \coprod_{W \in \mathcal{W}} \mathcal{C}_W^{op}$*, where* $W \in \mathcal{W}$ *and each* $\mathcal{C}_W$ *is a small doubly monoidal category, with* $1 \not\cong I$,[9] *and* **Cat** *is the category of small categories and functors such that*

1 *Each* $\mathcal{J}_r(W)(D)$ *has a terminal object,* $1_{\mathcal{J}_r(W)(D)}$*, preserved on the nose by each* $f^*(= \mathcal{J}_r(W)(f))$*, where* $f{:}E \longrightarrow D \in \mathcal{C}_W$*;*

2 *For each* $W \in \mathcal{W}$*,* $D \in \mathcal{C}_W$ *and each object* $A \in \mathcal{J}_r(W)(D)$*, there is a* $D \bullet A \in \mathcal{C}_W$.

For the cartesian extension, there are canonical first projections $D \times A \stackrel{p_{D,A}}{\longrightarrow} D$ *and canonical pullbacks*[10]

$$\begin{array}{ccc} E \times f^*(A) & \stackrel{f \times A}{\longrightarrow} & D \times A \\ {\scriptstyle p_{E,f^*A}}\downarrow & \lrcorner & \downarrow{\scriptstyle p_{D,A}} \\ E & \underset{f}{\longrightarrow} & D \end{array}$$

The pullback indicates, for the cartesian case, how to interpret realizations as tuples. In particular, for each $1 \stackrel{g}{\longrightarrow} A \in \mathcal{J}_r(W)(D)$ *there exists a unique arrow* $D \stackrel{\langle 1 \times g \rangle}{\longrightarrow} D \times A$*. It does not cover the case for the monoidal extension. For that, we require there to exist a unique* $D \stackrel{\langle 1 \otimes g \rangle}{\longrightarrow} D \otimes A$ *corresponding to each* $1 \stackrel{g}{\longrightarrow} A$ *over* D*, the tuples being given by the bifunctoriality of* $\otimes$, i.e., $(D =) D \otimes I \stackrel{\langle 1 \otimes f \rangle}{\longrightarrow} D \otimes A$.

For both extensions, there is a canonical second projection $1 \stackrel{q_{D,A}}{\longrightarrow} A$ *in the fibre over* $D \bullet A$.

These maps are required to satisfy the strictness conditions that

$$(1_D)^*(A) = A \text{ and } 1_D \bullet A = 1_{D \bullet A},$$

[9]Note that [Ishtiaq, 1999, Ishtiaq and Pym, 1999] each have a bad typographical error, corrected in [Ishtiaq and Pym, 2000]: They each have "$1 \cong I$" instead of $1 \not\cong I$.

[10]The projection $p_{D,A}$ being "canonical" means that $p^*_{D,A}$ is inclusion.

for each $A \in \mathcal{J}_r(W)(D)$,

$$g^*(f^*(A)) = (g;f)^*(A) \text{ and } (g \bullet f^*(A)); f \bullet A = (g;f) \bullet A$$

for each $F \xrightarrow{g} E$ *and* $E \xrightarrow{f} D$ *in* $\mathcal{C}_W$. *Moreover, for each* W *and* D, $D \bullet 1_{\mathcal{J}_r(W)(D)} = D$;

3 For each D, A, *there is a natural isomorphism* $\Lambda_{D,A}$,

$$\frac{Hom_{\mathcal{J}_{r+r'}(W)(D\bullet A)}(1, B)}{Hom_{\mathcal{J}_r(W)(D)}(1, \Lambda_{D,A}\, x{\in}A\,.B)}\ ,$$

in which the extended context is defined in the $r+r'$*-indexed functor. This natural isomorphism is required to satisfy the Beck-Chevalley condition: For each* $E \xrightarrow{f} D$ *in* $\mathcal{C}_W$ *and each* B *in* $\mathcal{J}_r(W)(D \bullet A)$

$$f^*(\Lambda_{D,A}B) = \Lambda_{E,f^*A}((f \bullet A)^*B)\ ;$$

4 Each category $\mathcal{J}_r(W)(D)$ *has cartesian products.* □

Our approach is modular enough to also provide a categorical semantics for the intuitionistic fragment of the $\lambda\Lambda$-calculus, the $\lambda\Pi$-calculus. Basically, we work with a single functor $\mathcal{J}_r{:}[\mathcal{W}, [\mathcal{D}^{op}, \mathbf{Cat}]]$, where $\mathcal{D}$ is a category with only the cartesian structure $(\times, 1)$ on it. The definition of Π as right adjoint to Weakening may be recovered from the natural isomorphism. We see this in the following lemma, which is motivated by the propositions-as-types correspondence which we discussed earlier:

LEMMA 15.11 *The natural isomorphism*

$$Hom_{\mathcal{J}(W)(D\bullet A)}(p^*_{D,A}(1), B) \cong Hom_{\mathcal{J}(W)(D)}(1, \Pi_{D,A}(B))$$

in the Kripke $\lambda\Pi$*-structure* $\mathcal{J}$ *is just the* $\Lambda_{D,A}$ *natural isomorphism in the* $D \times A$ *case in the Kripke resource* $\lambda\Lambda$*-structure.* □

For the proof, we provide a translation from a Kripke resource $\lambda\Lambda$-structure to a Kripke $\lambda\Pi$-structure which forgets the linear–intuitionistic distinction, translating both $\otimes$ and $\times$ in $\mathcal{C}$ to $\times$ in $\mathcal{D}$. The translation has some similarity with Girard's translation of $- \to -$ into $!- \multimap -$ [Girard, 1987]. Under the translation, the object $\Lambda\, x!A\,.B$ (in a particular $\mathcal{J}_r$) ends up as $\Pi\, x{:}A\,.B$ (in $\mathcal{J}$). If we uncurry this, we get the corresponding translation, as $p{:}D \times A \to D$ always exists in $\mathcal{C}$.

12. Kripke Resource Σ-$\lambda\Lambda$-model

We will restrict our discussion of semantics to the $M{:}A$-fragment. The treatment of the $A{:}K$-fragment is undertaken analogously — in a sense, the $A{:}K$-fragment has the same logical structure as the $M{:}A$-fragment but needs some extra structure. To interpret the kind Type, for instance, we must require the existence of a chosen object which obeys some equations regarding substitution and quantification. Treatments of the intuitionistic case may be found in, for example, [Cartmell, 1994, Seely, 1984, Streicher, 1988, Jacobs, 1998, Pitts, 1992, Pym, 2000a, Pym, 2000b, Pym, 2000c].

A Kripke resource model is a Kripke resource structure that has enough points to interpret not only the constants of Σ but also the $\lambda\Lambda$-calculus terms defined over Σ and a given context Γ. Formally, a Kripke resource model is made up of five components: a Kripke resource structure that has Σ-operations, an interpretation function, two $\mathcal{C}$-functors, and a satisfaction relation. Except for the structure, the components are defined, because of interdependences, simultaneously by induction on raw syntax.

Ignoring these interdependencies for a moment, we explain the purpose of each component of the model. Firstly, the Kripke resource structure provides the abstract domain in which the type theory is interpreted. The Σ-operations provide the points to interpret constants in the signature. Secondly, the interpretation $[\![-]\!]$ is a partial function which maps raw (that is, not necessary well-formed) contexts Γ to objects of $\mathcal{C}$, types over raw contexts A_Γ to objects in the category indexed by the interpretation of Γ, and terms over raw contexts M_Γ to arrows in the category indexed by the interpretation of Γ. Types and terms are interpreted, in the usual way, up to $\beta\eta$-equivalence. Thirdly, the $\mathcal{C}$-functors maintain the well-formedness of contexts with regard to joining and sharing. The model is also constrained so that multiple occurrences of variables in the context get the same interpretation. Fourthly, satisfaction is a relation on worlds and sequents axiomatizing the desired properties of the model. In stronger logics, such as intuitionistic logic, the abstract definition of the model is sufficient to derive the properties of the satisfaction relation. In our case, the definition is given more directly.

DEFINITION 15.12 (KRIPKE RESOURCE $\lambda\Lambda$-MODEL) *Let Σ be a $\lambda\Lambda$-calculus signature. A* Kripke resource Σ-$\lambda\Lambda$ model *is a 5-tuple*

$$\langle \{\mathcal{J}_r{:}[\mathcal{W}, [\mathcal{C}^{op}, \mathbf{Cat}]] \mid r \in R\}, [\![-]\!], \mathrm{join}, \mathrm{share}, \models_\Sigma \rangle\,,$$

where $\{\mathcal{J}_r{:}[\mathcal{W}, [\mathcal{C}^{op}, \mathbf{Cat}]] \mid r \in R\}$ *is a Kripke resource* $\lambda\Lambda$*-structure that has* Σ-operations, $[\![-]\!]$ *is an* interpretation *from the raw syntax of the* $\lambda\Lambda$*-calculus to components of* $\mathcal{J}_r{:}[\mathcal{W}, [\mathcal{C}^{op}, \mathbf{Cat}]]$, join *and* share *are* $\mathcal{C}$*-functors and* $\models_\Sigma$ *is a satisfaction relation on worlds and sequents, defined by simultaneous induction on the raw structure of the syntax as follows:*

1 *The Kripke resource* $\lambda\Lambda$*-structure has* Σ-operations *if, for all* W *in* $\mathcal{W}$,

(a) *Corresponding to each constant* $c!\Lambda\, x_1{\in}A_1 \ldots . \Lambda\, x_m{\in}A_m\, .\mathsf{Type} \in \Sigma$ *there is in each* $\mathcal{J}_r(W)([\![!\Phi]\!]^W_{\mathcal{J}_r})$ *an operation* op_c *such that*

$$op_c([\![M_{1\Gamma_1}]\!]^W_{\mathcal{J}_1}, \ldots, [\![M_{m\Gamma_m}]\!]^W_{\mathcal{J}_m})$$

is an object of $\mathcal{J}_r(W)([\![\Xi]\!]^W_{\mathcal{J}_r})$, *where*

$$[\![\Xi]\!]^W_{\mathcal{J}_r} = \mathrm{share\ join}([\![\Gamma_m]\!]^W_{\mathcal{J}_m}, \ldots, \mathrm{share\ join}([\![\Gamma_1]\!]^W_{\mathcal{J}_r}, [\![!\Phi]\!]^W_{\mathcal{J}_r}) \ldots)\ ;$$

(b) *Corresponding to each constant* $c!\Lambda\, x_1{\in}A_1 \ldots . \Lambda\, x_m{\in}A_m\, .A \in \Sigma$ *there is, in each*

$$\mathcal{J}_r(W)([\![!\Phi, x_1{\in}A_1, \ldots, x_m{\in}A_m]\!]^W_{\mathcal{J}_r}),$$

an arrow $1_{\mathcal{J}_r(W)(D)} \xrightarrow{op_c} [\![A]\!]^W_{\mathcal{J}_r}$, *where*

$$D = [\![!\Phi, x_1{\in}A_1, \ldots, x_m{\in}A_m]\!]^W_{\mathcal{J}_r};$$

2 *An* interpretation $[\![-]\!]^-_{\mathcal{J}_r}$, *in each such* $\mathcal{J}_r$, *satisfies, at each* W*:*[11]

(a) $[\![\langle\rangle]\!]^W_{\mathcal{J}_r} \simeq 1_{\mathcal{C}}$;

(b) $[\![\Gamma, x{:}A]\!]^W_{\mathcal{J}_{r+r'}} \simeq [\![\Gamma]\!]^W_{\mathcal{J}_r} \otimes [\![A_\Gamma]\!]^W_{\mathcal{J}_{r'}}$;

(c) $[\![\Gamma, x!A]\!]^W_{\mathcal{J}_r} \simeq [\![\Gamma]\!]^W_{\mathcal{J}_r} \times [\![A_\Gamma]\!]^W_{\mathcal{J}_r}$;

(d) $[\![\Gamma \xrightarrow{\langle M_1, \ldots, M_n\rangle} \Delta]\!]^W_{\mathcal{J}_r} \simeq [\![\Gamma]\!]^W_{\mathcal{J}_r} \xrightarrow{\langle [\![M_{1\Gamma_1}]\!]^W_{\mathcal{J}_1}, \ldots, [\![M_{n\Gamma_n}]\!]^W_{\mathcal{J}_n}\rangle} [\![\Delta]\!]^W_{\mathcal{J}_r}$,
where $[\![\Gamma]\!]^W_{\mathcal{J}_r} = \mathrm{share\ join}([\![\Gamma_n]\!]^W_{\mathcal{J}_n}, \ldots, \mathrm{share\ join}([\![\Gamma_2]\!]^W_{\mathcal{J}_2}, [\![\Gamma_1]\!]^W_{\mathcal{J}_1}) \ldots)$;

(e) $[\![\langle\rangle_\Gamma]\!]^W_{\mathcal{J}_r} \simeq 1_{[\![\Gamma]\!]^W_{\mathcal{J}_r}}$;

[11] Essentially, we build in the "environment model condition".

(f) $[\![(cM_1 \dots M_n)_\Gamma]\!]^W_{\mathcal{J}_r} \simeq op_c([\![M_{1\Gamma_1}]\!]^W_{\mathcal{J}_{r_1}}, \dots, [\![M_{n\Gamma_n}]\!]^W_{\mathcal{J}_{r_n}})$ *in* $\mathcal{J}_r(W)([\![\Gamma]\!]^W_{\mathcal{J}_r})$, *where* $[\![\Gamma]\!]^W_{\mathcal{J}_r} = \mathrm{share\,join}([\![\Gamma_n]\!]^W_{\mathcal{J}_n}, \dots, \mathrm{share}\, join([\![\Gamma_2]\!]^W_{\mathcal{J}_2}, [\![\Gamma_1]\!]^W_{\mathcal{J}_1}) \dots)$;

(g) $[\![\Lambda\, x{:}A\,.B_\Gamma]\!]^W_{\mathcal{J}_r} \simeq \Lambda_{[\![\Gamma]\!]^W_{\mathcal{J}_r}, [\![A_\Gamma]\!]^W_{\mathcal{J}_s}}([\![B_{\Gamma,x:A}]\!]^W_{\mathcal{J}_{r+s}})$, *where the extended context is defined in the* $r+s$*-indexed model;*

(h) $[\![\Lambda\, x!A\,.B_\Gamma]\!]^W_{\mathcal{J}_r} \simeq \Lambda_{[\![\Gamma]\!]^W_{\mathcal{J}_r}, [\![A_\Gamma]\!]^W_{\mathcal{J}_r}}([\![B_{\Gamma,x!A}]\!]^W_{\mathcal{J}_r})$;

(i) $[\![A\&B_\Gamma]\!]^W_{\mathcal{J}_r} \simeq [\![A_\Gamma]\!]^W_{\mathcal{J}_r} \times [\![B_\Gamma]\!]^W_{\mathcal{J}_r}$;

(j) $[\![\langle\rangle_\Gamma]\!]^W_{\mathcal{J}_r} \simeq 1_{[\![\Gamma]\!]^W_{\mathcal{J}_r}}$;

(k) $[\![c_{!\Gamma}]\!]^W_{\mathcal{J}_r} \simeq \Lambda^m(op_c)$;

(l) $[\![x_{\Gamma,x:A}]\!]^W_{\mathcal{J}_r} \simeq q_{[\![\Gamma,x:A]\!]^W_{\mathcal{J}_r}}$;

(m) $[\![\lambda\, x{:}A\,.M_\Gamma]\!]^W_{\mathcal{J}_r} \simeq \Lambda_{[\![\Gamma]\!]^W_{\mathcal{J}_r}, [\![A_\Gamma]\!]^W_{\mathcal{J}_r}}([\![M_{\Gamma,x:A}]\!]^W_{\mathcal{J}_r})$;

(n) $[\![\lambda\, x!A\,.M_\Gamma]\!]^W_{\mathcal{J}_r} \simeq \Lambda_{[\![\Gamma]\!]^W_{\mathcal{J}_r}, [\![!A_\Gamma]\!]^W_{\mathcal{J}_r}}([\![M_{\Gamma,x!A}]\!]^W_{\mathcal{J}_r})$;

(o) $[\![MN_\Xi]\!]^W_{\mathcal{J}_t} \simeq (\langle 1_{[\![\Gamma]\!]^W_{\mathcal{J}_r}}, [\![N_\Delta]\!]^W_{\mathcal{J}_s}\rangle^*(\Lambda^{-1}_{[\![\Gamma]\!]^W_{\mathcal{J}_r}, [\![A_\Gamma]\!]^W_{\mathcal{J}_s}}([\![M_\Gamma]\!]^W_{\mathcal{J}_r})))$, *where* $[\![\Xi']\!]^W_{\mathcal{J}_{r+s}} = \mathrm{join}\langle [\![\Gamma]\!]^W_{\mathcal{J}_r}, [\![\Delta]\!]^W_{\mathcal{J}_s}\rangle$ *and* $[\![\Xi]\!]^W_{\mathcal{J}_t} = \mathrm{share}([\![\Xi']\!]^W_{\mathcal{J}_{r+s}})$;

(p) $[\![\langle M, N\rangle_\Gamma]\!]^W_{\mathcal{J}_r} \simeq \langle [\![M_\Gamma]\!]^W_{\mathcal{J}_r}, [\![N_\Gamma]\!]^W_{\mathcal{J}_r}\rangle$;

(q) $[\![\pi_i(M)_\Gamma]\!]^W_{\mathcal{J}_r} \simeq \pi_i([\![M_\Gamma]\!]^W_{\mathcal{J}_r})$, *where* $i \in \{0, 1\}$.

Otherwise the interpretation is undefined;

3 *There exists a bifunctor* join *on* $\mathcal{C}$. *The purpose of* join, *on objects, is to extend the first object with the second, discarding any duplicate cartesian objects. The definition of* join *on objects is as follows:*

$$\begin{aligned}
\mathrm{join}\langle [\![\langle\rangle]\!]^W_{\mathcal{J}_r}, [\![\langle\rangle]\!]^W_{\mathcal{J}_r}\rangle &= [\![\langle\rangle]\!]^W_{\mathcal{J}_r} \\
\mathrm{join}\langle [\![\Gamma]\!]^W_{\mathcal{J}_r}, [\![\Delta, x{:}A]\!]^W_{\mathcal{J}_{s+t}}\rangle &= \mathrm{join}\langle [\![\Gamma]\!]^W_{\mathcal{J}_r}, [\![\Delta]\!]^W_{\mathcal{J}_s}\rangle \otimes [\![x{:}A]\!]^W_{\mathcal{J}_t} \\
\mathrm{join}\langle [\![\Gamma, x{:}A]\!]^W_{\mathcal{J}_{r+t}}, [\![\Delta]\!]^W_{\mathcal{J}_s}\rangle &= \mathrm{join}\langle [\![\Gamma]\!]^W_{\mathcal{J}_r}, [\![\Delta]\!]^W_{\mathcal{J}_s}\rangle \otimes [\![x{:}A]\!]^W_{\mathcal{J}_t} \\
\mathrm{join}\langle [\![\Gamma, x!A]\!]^W_{\mathcal{J}_r}, [\![\Delta, x!A]\!]^W_{\mathcal{J}_s}\rangle &= \mathrm{join}\langle [\![\Gamma]\!]^W_{\mathcal{J}_r}, [\![\Delta]\!]^W_{\mathcal{J}_s}\rangle \times [\![x!A]\!]^W_{\mathcal{J}_{r+s}}.
\end{aligned}$$

The definition of join *on morphisms is similar. It is easy to see that* $[\![\Xi]\!]^W_{\mathcal{J}_{r+s}} = \mathrm{join}\langle [\![\Gamma]\!]^W_{\mathcal{J}_r}, [\![\Delta]\!]^W_{\mathcal{J}_s}\rangle$.

There exists a functor share *on* $\mathcal{C}$. *The purpose of* share *is to regulate sharing of multiple occurrences of an object. The definition of* share *on objects is as follows:*

$$
\begin{array}{rcl}
\text{share}([\![\langle\rangle]\!]^{W}_{\mathcal{J}_r}) & = & [\![\langle\rangle]\!]^{W}_{\mathcal{J}_r} \\
\text{share}([\![\Xi]\!]^{W}_{\mathcal{J}_r}) & = & \text{join}\langle \text{share}([\![\Phi]\!]^{W}_{\mathcal{J}_{s+s'}}), [\![\Psi]\!]^{W}_{\mathcal{J}_{t+u+u'}}\rangle \\
 & & \quad \textit{if } [\![\Xi]\!]^{W}_{\mathcal{J}_t} = \text{join}\langle [\![\Gamma, x{:}A, \Gamma']\!]^{W}_{\mathcal{J}_{s+t+u}}, [\![\Delta, x{:}A, \Delta']\!]^{W}_{\mathcal{J}_{s'+t+u'}}\rangle, \\
 & & \quad \text{there is no } y{:}B(x) \in \Gamma', \Delta', \\
 & & \quad [\![\Phi]\!]^{W}_{\mathcal{J}_{s+s'}} = \text{join}\langle [\![\Gamma]\!]^{W}_{\mathcal{J}_s}, [\![\Delta]\!]^{W}_{\mathcal{J}_{s'}}\rangle \text{ and} \\
 & & \quad [\![\Psi]\!]^{W}_{\mathcal{J}_{t+u+u'}} = \text{join}\langle [\![x{:}A]\!]^{W}_{\mathcal{J}_t}, \text{join}\langle [\![\Gamma']\!]^{W}_{\mathcal{J}_u}, [\![\Delta']\!]^{W}_{\mathcal{J}_{u'}}\rangle\rangle \\
 & = & [\![\Xi]\!]^{W}_{\mathcal{J}_r} \ \textit{otherwise.}
\end{array}
$$

The definition of share *on morphisms is similar. The purpose of* share *is to ensure that the joined objects and morphisms are well-formed.*

Both join *and* share *"cut across interpretations" in that the result object is in a different R-indexed model from the argument object(s). This is necessary for defining the interpretation of function application;*

4 Satisfaction *in the model is a relation over worlds and sequents such that the following hold:*

(a) $\mathcal{J}_r, W \models_\Sigma (c{:}A)\ [!\Gamma]$ *if and only if* $c \in dom(\Sigma)$*;*

(b) $\mathcal{J}_r, W \models_\Sigma (x{:}A)\ [\Gamma, x{\in}A]$ *if and only if* $[\![\Gamma, x{\in}A]\!]^{W}_{\mathcal{J}_r}$ *is defined ;*

(c) $\mathcal{J}_r, W \models_\Sigma (M : \Lambda\, x{:}A.B)\ [\Gamma]$ *if and only if for all* $W \leq W'$ *and for all* $r' \in R$*, if* $\mathcal{J}_s, W' \models_\Sigma (N{:}A)\ [\Delta]$*, then* $\mathcal{J}_t, W' \models_\Sigma (MN{:}B[N/x])\ [\Xi]$*, where* $[\![\Xi]\!]^{W'}_{\mathcal{J}_{r+s}} = \text{join}\langle [\![\Gamma]\!]^{W'}_{\mathcal{J}_r}, [\![\Delta]\!]^{W'}_{\mathcal{J}_s}\rangle$ *and* $[\![\Xi]\!]^{W'}_{\mathcal{J}_t} = \text{share}([\![\Xi']\!]^{W'}_{\mathcal{J}_{r+s}})$*;*

(d) $\mathcal{J}_r, W \models_\Sigma (M{:}A\&B)\ [\Gamma]$ *if and only if* $\mathcal{J}_r, W \models_\Sigma (\pi_i(M){:}A)\ [\Gamma]$*, for* $i \in \{0, 1\}$*;*

(e) $\mathcal{J}_r, W \models_\Sigma (M : \Lambda\, x!A.B)\ [\Gamma]$ *if and only if for all* $W \leq W'$*, if* $\mathcal{J}_r, W \models_\Sigma (N{:}A)\ [!\Delta]$*, then* $\mathcal{J}_r, W \models_\Sigma (MN{:}B[N/x])\ [\Xi]$*, with* $[\![\Xi]\!]^{W}_{\mathcal{J}_r} = \text{join}\langle [\![\Gamma]\!]^{W}_{\mathcal{J}_r}, [\![!\Delta]\!]^{W}_{\mathcal{J}_r}\rangle$.

We require two further conditions on the model:

*1 (*Syntactic monotonicity*) If* $[\![X]\!]^{W}_{\mathcal{J}_r}$ *is defined, then* $[\![X']\!]^{W}_{\mathcal{J}_{r'}}$ *is defined, for all subterms* X' *of* X *and summands* r' *of* r*. This condition is needed for various inductive arguments. It is not automatic as the interpretation is defined over raw objects;*

*2 (*Accessibility*) The functor* $\mathcal{J}_r(W)$ *has domain* $\mathcal{C} = \amalg_{W \in \mathcal{W}} \mathcal{C}^{op}_{\mathcal{J}_r(W)}$*. So that* $[\![\Gamma]\!]^{W}_{\mathcal{J}_r} \in \mathcal{C}_W$ *and* $[\![\Gamma]\!]^{W'}_{\mathcal{J}_r} \in \mathcal{C}_{W'}$*. If there is an arrow* $W \leq W' \in \mathcal{W}$*, then*

(a) *there exists a functor* $\xi{:}\mathcal{C}_W \to \mathcal{C}_{W'}$ *such that* $\xi([\![X]\!]^{W}_{\mathcal{J}_r}) = [\![X]\!]^{W'}_{\mathcal{J}_r}$, *where* X *ranges over contexts, types and terms; and*

(b) *we have* $\mathcal{J}_r(W')([\![\Gamma]\!]^{W}_{\mathcal{J}_r}) = \mathcal{J}_r(W')([\![\Gamma]\!]^{W'}_{\mathcal{J}_r})$ *and* $\mathcal{J}_r(W)([\![\Gamma]\!]^{W}_{\mathcal{J}_r}) = \mathcal{J}_r(W)([\![\Gamma]\!]^{W'}_{\mathcal{J}_r})$, *for each context* Γ*; otherwise* $\mathcal{J}_r(W')([\![\Gamma]\!]^{W'}_{\mathcal{J}_r})$ *is undefined.*

□

A few remarks concerning Definition 15.12 are in order. The type theory has a structural freedom at the level of terms which, logically, allows the existence of multiple occurrences of the same proof. However, it may be that, in operating on the representation of two judgements, the same occurrence of an object in the base of the resulting representation is used to form the valid terms and types in both representations. This sharing requirement is regulated by the existence of the functor share on $\mathcal{C}$.

The second accessibility condition (b) on the model is the simplest one regarding the model-theoretic notion of *relativization*: that of interpreting constructs in one world and reasoning about them from the point of view of another. In the definition of model, and so in the sequel, the accessibility relation we take equates contexts, *etc.* over the worlds. A syntactic term may be seen, in a certain sense, as a "rigid designator", that is, one whose interpretation is the same over different worlds, for a semantic object. For example, suppose **N** proves $\Gamma \vdash_\Sigma M{:}A$. If $[\![M_\Gamma]\!]^{W}_{\mathcal{J}_r}$ is defined (given soundness this will be the case), then, for all $W \leq W' \in \mathcal{W}$, $[\![M_\Gamma]\!]^{W'}_{\mathcal{J}_r}$ is defined and equal to $[\![M_\Gamma]\!]^{W}_{\mathcal{J}_r}$. In a sense, the syntactic term M designates all objects $[\![M_\Gamma]\!]^{W'}_{\mathcal{J}_r}$.

We also remark that there are several notions of partiality in the model. Technically, the interpretation function is a partial one because it is defined for raw objects of the syntax. But partiality plays two other roles too. Firstly, there is dependent typing partiality to "bootstrap" the definition. Secondly, there is Kripke semantic partiality of information, in which the further up the world structure one goes, the more objects have defined interpretations. We refer to Streicher [Streicher, 1988], Pym [Pym, 1995a], and Mitchell and Moggi [Mitchell and Moggi, 1981] for some comments regarding these matters.

The following lemma follows easily from the definition:

LEMMA 15.13 join *and* share *are functors.*

PROOF We need to show that both join and share preserve identities and composition. We omit the details. □

We now consider various model-theoretic properties of the satisfaction relation.

LEMMA 15.14 (MONOTONICITY OF $\models_\Sigma$) *Let Σ be a signature and let*

$$\langle \{\mathcal{J}_r \mid r \in R\}, [\![-]\!], \text{join}, \text{share}, \models_\Sigma \rangle$$

be a Kripke resource model. If $\mathcal{J}_r, W \models_\Sigma (M{:}A)\ [\Gamma]$ and $W \leq W'$, then $\mathcal{J}_r, W' \models_\Sigma (M{:}A)\ [\Gamma]$.

PROOF By induction on the syntax of $M{:}A$. If $W \leq W'$, then, by accessibility, $[\![X]\!]^{W'}_{\mathcal{J}_r}$ is defined as $\xi[\![X]\!]^{W}_{\mathcal{J}_r}$, where X ranges over Γ, A and M. For each case of $M{:}A$, the conclusion is given by the definition of $\models_\Sigma$. □

LEMMA 15.15 ($\models_\Sigma$-FORCING VIA GLOBAL SECTIONS) *Let*

$$\langle \{\mathcal{J}_r \mid r \in R\}, [\![-]\!], \text{join}, \text{share}, \models_\Sigma \rangle$$

be a Kripke resource model. We have that $\mathcal{J}_r, W \models_\Sigma (M{:}A)\ [\Gamma]$ if and only if $[\![\Gamma]\!]^{W}_{\mathcal{J}_r}$, $[\![A_\Gamma]\!]^{W}_{\mathcal{J}_r}$ and $[\![M_\Gamma]\!]^{W}_{\mathcal{J}_r}$ are defined and $1_{\mathcal{J}_r(W)([\![\Gamma]\!]^{W}_{\mathcal{J}_r})} \overset{[\![M_\Gamma]\!]^{W}_{\mathcal{J}_r}}{\longrightarrow} [\![A_\Gamma]\!]^{W}_{\mathcal{J}_r}$ is an arrow in $\mathcal{J}_r(W)([\![\Gamma]\!]^{W}_{\mathcal{J}_r})$.

PROOF By induction on the structure of $M{:}A$.

($c{:}A$) For the $\Rightarrow$ direction, we require the model to have enough points, and so get such an arrow. The $\Leftarrow$ direction is immediate from the definition of $\models_\Sigma$;

($x{\in}A$) For the $\Rightarrow$ direction, the second projection map $1 \overset{q}{\to} A$ in the fibre over the context $\Gamma, x{\in}A$ gives us the required arrow. The $\Leftarrow$ direction is immediate from the definition of $\models_\Sigma$;

($\lambda x{:}A.M : \Lambda x{:}A.B$) For the $\Rightarrow$ direction, by induction hypothesis we have that

$$1_{\mathcal{J}_{r+s}(W)([\![\Gamma,x:A]\!]^{W}_{\mathcal{J}_{r+s}})} \overset{[\![M_{\Gamma,x:A}]\!]^{W}_{\mathcal{J}_{r+s}}}{\longrightarrow} [\![B_{\Gamma,x:A}]\!]^{W}_{\mathcal{J}_{r+s}}$$

is an arrow in $\mathcal{J}_{r+s}(W)([\![\Gamma, x{:}A]\!]^{W}_{\mathcal{J}_{r+s}})$. We then use the natural isomorphism Λ to get the arrow

$$1_{\mathcal{J}_r(W)([\![\Gamma]\!]^{W}_{\mathcal{J}_r})} \overset{[\![\lambda x:A.M_\Gamma]\!]^{W}_{\mathcal{J}_r}}{\longrightarrow} [\![\Lambda x{:}A.B_\Gamma]\!]^{W}_{\mathcal{J}_r}$$

in $\mathcal{J}_r(W)(\llbracket\Gamma\rrbracket^W_{\mathcal{J}_r})$. For the $\Leftarrow$ direction, suppose there exists an arrow

$$1_{\mathcal{J}_r(W)(\llbracket\Gamma\rrbracket^W_{\mathcal{J}_r})} \overset{\llbracket\lambda\, x{:}A\,.M_\Gamma\rrbracket^W_{\mathcal{J}_r}}{\longrightarrow} \llbracket\Lambda\, x{:}A\,.B_\Gamma\rrbracket^W_{\mathcal{J}_r}$$

in $\mathcal{J}_r(W)(\llbracket\Gamma\rrbracket^W_{\mathcal{J}_r})$. It follows immediately that the existence of an arrow

$$1_{\mathcal{J}_s(W)(\llbracket\Delta\rrbracket^W_{\mathcal{J}_s})} \overset{\llbracket N_\Delta\rrbracket^W_{\mathcal{J}_s}}{\longrightarrow} \llbracket A_\Delta\rrbracket^W_{\mathcal{J}_s}$$

implies the existence of an arrow

$$1_{\mathcal{J}_t(W)(\llbracket\Xi\rrbracket^W_{\mathcal{J}_t})} \overset{\llbracket MN_\Xi\rrbracket^W_{\mathcal{J}_t}}{\longrightarrow} \llbracket B[N/x]_\Xi\rrbracket^W_{\mathcal{J}_t},$$

where $\llbracket\Xi'\rrbracket^W_{\mathcal{J}_{r+s}} = \mathrm{join}\langle\llbracket\Gamma\rrbracket^W_{\mathcal{J}_r}, \llbracket\Delta\rrbracket^W_{\mathcal{J}_s}\rangle$ and $\llbracket\Xi\rrbracket^W_{\mathcal{J}_t} = \mathrm{share}\llbracket\Xi'\rrbracket^W_{\mathcal{J}_{s+r}}$. The definition of $\models_\Sigma$ then gives us $\mathcal{J}_r, W \models_\Sigma (\lambda\, x{:}A\,.M : \Lambda\, x{:}A\,.B)\ [\Gamma]$;

$(\lambda\, x!A\,.M : \Lambda\, x!A\,.B)$ For the $\Rightarrow$ direction, by the induction hypothesis we have that

$$1_{\mathcal{J}_r(W)(\llbracket\Gamma, x!A\rrbracket^W_{\mathcal{J}_r})} \overset{\llbracket M_{\Gamma,x!A}\rrbracket^W_{\mathcal{J}_r}}{\longrightarrow} \llbracket B_{\Gamma,x!A}\rrbracket^W_{\mathcal{J}_r}$$

is an arrow in $\mathcal{J}_r(W)(\llbracket\Gamma, x!A\rrbracket^W_{\mathcal{J}_r})$. We then use the natural isomorphism Λ to get the arrow

$$1_{\mathcal{J}_r(W)(\llbracket\Gamma\rrbracket^W_{\mathcal{J}_r})} \overset{\llbracket\lambda\, x!A\,.M_\Gamma\rrbracket^W_{\mathcal{J}_r}}{\longrightarrow} \llbracket\Lambda\, x!A\,.B_\Gamma\rrbracket^W_{\mathcal{J}_r}$$

in $\mathcal{J}_r(W)(\llbracket\Gamma\rrbracket^W_{\mathcal{J}_r})$. For the $\Leftarrow$ direction, suppose there exists an arrow

$$1_{\mathcal{J}_r(W)(\llbracket\Gamma\rrbracket^W_{\mathcal{J}_r})} \overset{\llbracket\lambda\, x!A\,.M_\Gamma\rrbracket^W_{\mathcal{J}_r}}{\longrightarrow} \llbracket\Lambda\, x!A\,.B_\Gamma\rrbracket^W_{\mathcal{J}_r}$$

in $\mathcal{J}_r(W)(\llbracket\Gamma\rrbracket^W_{\mathcal{J}_r})$. It follows immediately that the existence of an arrow

$$1_{\mathcal{J}_r(W)(\llbracket !\Delta \rrbracket^W_{\mathcal{J}_r})} \stackrel{\llbracket N_\Delta \rrbracket^W_{\mathcal{J}_r}}{\longrightarrow} \llbracket A_\Delta \rrbracket^W_{\mathcal{J}_r}$$

implies the existence of an arrow

$$1_{\mathcal{J}_r(W)(\llbracket \Xi \rrbracket^W_{\mathcal{J}_r})} \stackrel{\llbracket M N_\Xi \rrbracket^W_{\mathcal{J}_r}}{\longrightarrow} \llbracket B[N/x]_\Xi \rrbracket^W_{\mathcal{J}_r},$$

where $\llbracket \Xi \rrbracket^W_{\mathcal{J}_r} = \mathrm{join}\langle \llbracket \Gamma \rrbracket^W_{\mathcal{J}_r}, \llbracket !\Delta \rrbracket^W_{\mathcal{J}_r} \rangle$. The definition of $\models_\Sigma$ then gives us $\mathcal{J}_r, W \models_\Sigma (\lambda\, x!A\,.M : \Lambda\, x!A\,.B)\ [\Gamma]$;

(M:A&B) For the $\Rightarrow$ direction, by the induction hypothesis twice we have the arrows

$$1_{\mathcal{J}_r(W)(\llbracket \Gamma \rrbracket^W_{\mathcal{J}_r})} \stackrel{\llbracket \pi_0(M)_\Gamma \rrbracket^W_{\mathcal{J}_r}}{\longrightarrow} \llbracket A_\Gamma \rrbracket^W_{\mathcal{J}_r}$$

and

$$1_{\mathcal{J}_r(W)(\llbracket \Gamma \rrbracket^W_{\mathcal{J}_r})} \stackrel{\llbracket \pi_1(M)_\Gamma \rrbracket^W_{\mathcal{J}_r}}{\longrightarrow} \llbracket B_\Gamma \rrbracket^W_{\mathcal{J}_r}$$

in $\mathcal{J}_r(W)(\llbracket \Gamma \rrbracket^W_{\mathcal{J}_r})$. Recall that we can construct products in each $\mathcal{J}_r(W)(D)$. So we have the arrow

$$1_{\mathcal{J}_r(W)(\llbracket \Gamma \rrbracket^W_{\mathcal{J}_r})} \stackrel{\llbracket \langle \pi_0(M), \pi_1(M) \rangle_\Gamma \rrbracket^W_{\mathcal{J}_r}}{\longrightarrow} \llbracket A \& B_\Gamma \rrbracket^W_{\mathcal{J}_r}$$

in $\mathcal{J}_r(W)(\llbracket \Gamma \rrbracket^W_{\mathcal{J}_r})$. For the $\Leftarrow$ direction, by the induction hypothesis twice we have that $\mathcal{J}_r, W \models_\Sigma (\pi_0(M){:}A)\ [\Gamma]$ and $\mathcal{J}_r, W \models_\Sigma (\pi_1(M){:}B)\ [\Gamma]$. The definition of $\models_\Sigma$ then gives us

$$\mathcal{J}_r, W \models_\Sigma (M{:}A\&B)\ [\Gamma].$$

□

The substitution lemma for $\models_\Sigma$ has two cases, one for substituting a linear variable and one for substituting an intuitionistic one.

LEMMA 15.16 (SUBSTITUTIVITY OF $\models_\Sigma$) *Let Σ be a signature and let*

$$\langle\{\mathcal{J}_r \mid r \in R\}, [\![-]\!], \text{join}, \text{share}, \models_\Sigma\rangle$$

be a model.

1. *If $\mathcal{J}_r, W \models_\Sigma (U{:}V)\ [\Delta, x{:}A, \Delta']$, $\mathcal{J}_s, W \models_\Sigma (N{:}A)\ [\Gamma]$ and $[\![\Delta, \Delta'[N/x]]\!]^W_{\mathcal{J}_{r'}}$ is defined, then $\mathcal{J}_t, W \models_\Sigma (U{:}V[N/x])\ [\Xi]$, where $[\![\Xi']\!]^W_{\mathcal{J}_{s+r'}} = \text{join}\langle[\![\Gamma]\!]^W_{\mathcal{J}_s}, [\![\Delta, \Delta'[N/x]]\!]^W_{\mathcal{J}_{r'}}\rangle$ and $[\![\Xi]\!]^W_{\mathcal{J}_t} = \text{share}([\![\Xi']\!]^W_{\mathcal{J}_{s+r'}})$.*
2. *If $\mathcal{J}_r, W \models_\Sigma (U{:}V)\ [\Delta, x!A, \Delta']$, $\mathcal{J}_{r'}, W \models_\Sigma (N{:}A)\ [!\Gamma]$ and $[\![\Delta, \Delta'[N/x]]\!]^W_{\mathcal{J}_{r'}}$ is defined, then $\mathcal{J}_{r'}, W \models_\Sigma (U{:}V[N/x])\ [\Xi]$, where $[\![\Xi]\!]^W_{\mathcal{J}_{r'}} = \text{join}\langle[\![!\Gamma]\!]^W_{\mathcal{J}_{r'}}, [\![\Delta, \Delta'[N/x]]\!]^W_{\mathcal{J}_{r'}}\rangle$.*

PROOF By induction on the structure of the syntax and the functoriality of models.

1. The linear case is quite interesting as it shows an essential use of several of the model's components. In the following, we will omit the parameters on the interpretation for simplicity, though it may be seen, by induction, what these ought to be. Then, the basic argument is that, by the structure of the model, we can construct the following square in $\mathcal{C}_W$:

$$\begin{array}{ccc}
(\text{share join}\langle[\![\Delta]\!], [\![\Gamma]\!]\rangle) & \xrightarrow{\ e\ } & (\text{share join}\langle[\![\Delta]\!], [\![\Gamma]\!]\rangle) \otimes [\![A]\!] \\
\Big\downarrow f & & \Big\downarrow g \\
X & \xrightarrow[\ h\ ]{} & (\text{share join}\langle[\![\Delta]\!], [\![\Gamma]\!]\rangle) \otimes [\![A]\!] \bullet [\![\Delta']\!]
\end{array}$$

where

$$\begin{array}{rcl}
X & = & (\text{share join}\langle[\![\Delta]\!], [\![\Gamma]\!]\rangle) \bullet (\langle 1_{\text{share join}\langle[\![\Delta]\!],[\![\Gamma]\!]\rangle}, [\![N]\!]\rangle)^*[\![\Delta']\!] \\
e & = & \langle 1_{\text{share join}\langle[\![\Delta]\!],[\![\Gamma]\!]\rangle}, [\![N]\!]\rangle \\
f & = & \langle 1_{\text{share join}\langle[\![\Delta]\!],[\![\Gamma]\!]\rangle}, (\langle 1_{\text{share join}\langle[\![\Delta]\!],[\![\Gamma]\!]\rangle}, [\![N]\!]\rangle)^* d'\rangle \\
g & = & \langle\langle 1_{\text{share join}\langle[\![\Delta]\!],[\![\Gamma]\!]\rangle}, [\![N]\!]\rangle, d'\rangle \\
h & = & \langle\langle 1_{\text{share join}\langle[\![\Delta]\!],[\![\Gamma]\!]\rangle}, [\![N]\!]\rangle, 1_{[\![\Delta']\!]}\rangle
\end{array}$$

and $1 \xrightarrow{d'} [\![\Delta']\!]$ is, by induction, an arrow in the fibre over

$$(\text{share join}\langle[\![\Delta]\!], [\![\Gamma]\!]\rangle) \otimes [\![A]\!].$$

Then, by the functorial structure of the model, we have an arrow

$$1 \xrightarrow{(\langle\langle 1_{\text{share join}\langle[\![\Delta]\!],[\![\Gamma]\!]\rangle}, [\![N]\!]\rangle, 1_{[\![\Delta']\!]}\rangle)^*[\![U]\!]} (\langle\langle 1_{\text{share join}\langle[\![\Delta]\!],[\![\Gamma]\!]\rangle}, [\![N]\!]\rangle, 1_{[\![\Delta']\!]}\rangle)^*[\![V]\!]$$

in the fibre over the object

$$(\text{share join}\langle[\![\Delta]\!], [\![\Gamma]\!]\rangle) \bullet (\langle 1_{\text{share}([\![\Delta]\!]\otimes[\![\Gamma]\!])}, [\![N]\!]\rangle)^*[\![\Delta']\!].$$

2 The argument for the intuitionistic case is similar to the linear one, except that we use the pullback condition to extend the context with $[\![A]\!]$. □

13. Soundness and Completeness

In this section, we prove the (usual) soundness and completeness theorems for the $\lambda\Lambda$-calculus with respect to Kripke resource models. The proof of soundness uses the definition of interpretation and satisfaction in the model. The proof of completeness, via a term model construction, is more interesting, indicating a view of contexts as resources and worlds.

LEMMA 15.17 (CONTEXT AND TYPE INTERPRETATIONS) *Let Σ be a signature and let*

$$\langle\{\mathcal{J}_r \mid r \in R\}, [\![-]\!], \text{join}, \text{share}, \models_\Sigma\rangle$$

be a Kripke resource Σ-$\lambda\Lambda$ model.

1 If **N** *proves* $\vdash_\Sigma \Gamma$ context, *then, for those W where $[\![\Gamma]\!]^W_{\mathcal{J}_r}$ is defined, $[\![\Gamma]\!]^W_{\mathcal{J}_r} \in obj(\mathcal{C})$;*

2 If **N** *proves* $\Gamma \vdash_\Sigma A$:Type, *then, for those W where $[\![A_\Gamma]\!]^W_{\mathcal{J}_r}$ is defined, $[\![A_\Gamma]\!]^W_{\mathcal{J}_r} \in obj(\mathcal{J}_r(W)([\![\Gamma]\!]^W_{\mathcal{J}_r}))$.*

PROOF Follows from Definition 15.12. The proofs are done by induction on the structure of proofs of system **N** and, because of interdependencies, must be done simultaneously with the proof of Theorem 15.18. □

THEOREM 15.18 (SOUNDNESS) *Let Σ be a signature, let*

$$\langle\{\mathcal{J}_r \mid r \in R\}, [\![-]\!], \text{join}, \text{share}, \models_\Sigma\rangle$$

be a Kripke resource model and let W be any world in this model. If $\mathbf{N}$ *proves* $\Gamma \vdash_\Sigma M{:}A$ *and* $[\![\Gamma]\!]^W_{\mathcal{J}_r}$ *is defined,* $[\![A_\Gamma]\!]^W_{\mathcal{J}_r}$ *is defined and* $[\![M_\Gamma]\!]^W_{\mathcal{J}_r}$ *is defined, then* $\mathcal{J}_r, W \models_\Sigma (M{:}A)\ [\Gamma]$.

PROOF By induction on the structure of proofs of $\Gamma \vdash_\Sigma M{:}A$. The proof of soundness is done simultaneously with the proof of Lemma 15.17. We do some representative cases.

(Mc) Suppose $\mathbf{N}$ proves $!\Gamma \vdash_\Sigma c : \Lambda\, x_1{\in}A_1 \ldots . \Lambda\, x_m{\in}A_m\, .A$. By Definition 15.12, $\{\mathcal{J}_r \mid r \in R\}$ has enough points to interpret

$$c : \Lambda\, x_1{\in}A_1 \ldots . \Lambda\, x_m{\in}A_m\, .A$$

and $[\![c_{!\Gamma}]\!]^W_{\mathcal{J}_r} = \Lambda^m(op_c)$ (*i.e.*, m applications of the natural isomorphism on op_c) where

$$1_{\mathcal{J}_r(W)([\![!\Gamma, x_1\in A_1, \ldots, x_m\in A_m]\!]^W_{\mathcal{J}_r})} \xrightarrow{op_c} [\![A_{!\Gamma, x_1\in A_1,\ldots,x_m\in A_m}]\!]^W_{\mathcal{J}_r}.$$

It may be observed that $[\![c_{!\Gamma}]\!]^W_{\mathcal{J}_r}$ type-checks. By induction, we have that $[\![!\Gamma]\!]^W_{\mathcal{J}_r}$ is defined. So $\mathcal{J}_r, W \models_\Sigma (c : \Lambda\, x_1{\in}A_1 \ldots . \Lambda\, x_m{\in}A_m\, .A)\ [!\Gamma]$ follows.

($MVar$) Suppose $\mathbf{N}$ proves $\Gamma, x{:}A \vdash_\Sigma x{:}A$ because $\Gamma \vdash_\Sigma A{:}\mathsf{Type}$. By induction, we have that $[\![\Gamma, x{:}A]\!]^W_{\mathcal{J}_r}$ is defined. According to Definition 15.12, $[\![x_{\Gamma,x:A}]\!]^W_{\mathcal{J}_r} = q_{[\![\Gamma,x:A]\!]^W_{\mathcal{J}_r}}$ and the latter has the correct type. So we have shown

$$1_{\mathcal{J}_r(W)([\![\Gamma,x:A]\!]^W_{\mathcal{J}_r})} \xrightarrow{[\![x_{\Gamma,x:A}]\!]^W_{\mathcal{J}_r}} [\![A_\Gamma]\!]^W_{\mathcal{J}_r}$$

and $\mathcal{J}_r, W \models_\Sigma (x{:}A)\ [\Gamma, x{:}A]$ follows.

($M\lambda\Lambda\mathcal{I}$) Suppose $\mathbf{N}$ proves $\Gamma \vdash_\Sigma \lambda\, x{:}A\, .M : \Lambda\, x{:}A\, .B$ because $\Gamma, x{:}A \vdash_\Sigma M{:}B$. By induction, we have, for W such that $[\![B_{\Gamma,x:A}]\!]^W_{\mathcal{J}_r}$ is defined and that $\mathcal{J}_{r+s}, W \models_\Sigma (M{:}B)\ [\Gamma, x{:}A]$, *i.e.*, that

$$1_{\mathcal{J}_{r+s}(W)([\![\Gamma,x:A]\!]^W_{\mathcal{J}_{r+s}})} \xrightarrow{[\![M_{\Gamma,x:A}]\!]^W_{\mathcal{J}_{r+s}}} [\![B_{\Gamma,x:A}]\!]^W_{\mathcal{J}_{r+s}}.$$

We now use the natural isomorphism $\Lambda_{[\![\Gamma]\!]^W_{\mathcal{J}_r}, [\![A_\Gamma]\!]^W_{\mathcal{J}_s}}$ to get

$$1_{\mathcal{J}_r(W)([\![\Gamma]\!]^W_{\mathcal{J}_r})} \xrightarrow{[\![\lambda\, x:A\, .M_\Gamma]\!]^W_{\mathcal{J}_r}} [\![\Lambda\, x{:}A\, .B_\Gamma]\!]^W_{\mathcal{J}_r}.$$

So we obtain $\mathcal{J}_r, W \models_\Sigma (\lambda x{:}A\,.M : \Lambda x{:}A\,.B)\ [\Gamma]$.

$(M\Lambda\mathcal{E})$ Suppose **N** proves $\Xi \vdash_\Sigma MN{:}B[N/x]$ because $\Gamma \vdash_\Sigma M : \Lambda x{:}A\,.B$ and $\Delta \vdash_\Sigma N{:}A$ with $[\Xi';\Gamma;\Delta]$ and $\Xi = \Xi' \backslash \kappa\langle\Gamma,\Delta\rangle$. By the induction hypothesis twice we have that $\mathcal{J}_r, W \models_\Sigma (M : \Lambda x{:}A\,.B)\ [\Gamma]$, that is,

$$1_{\mathcal{J}_r(W)([\![\Gamma]\!]^W_{\mathcal{J}_r})} \stackrel{[\![M_\Gamma]\!]^W_{\mathcal{J}_r}}{\longrightarrow} [\![\Lambda x{:}A\,.B_\Gamma]\!]^W_{\mathcal{J}_r}$$

and $\mathcal{J}_s, W \models_\Sigma (N{:}A)\ [\Delta]$, that is,

$$1_{\mathcal{J}_s(W)([\![\Delta]\!]^W_{\mathcal{J}_{r'}})} \stackrel{[\![N_\Delta]\!]^W_{\mathcal{J}_s}}{\longrightarrow} [\![A_\Delta]\!]^W_{\mathcal{J}_s}.$$

Assume $W \leq W' \in \mathcal{W}$. By monotonicity and the definition of satisfaction, we have that $\mathcal{J}_t, W' \models_\Sigma (MN{:}B[N/x])\ [\Xi]$, where $[\![\Xi']\!]^{W'}_{\mathcal{J}_{r+s}} = \text{join}\langle[\![\Gamma]\!]^{W'}_{\mathcal{J}_r}, [\![\Delta]\!]^{W'}_{\mathcal{J}_s}\rangle$ and $[\![\Xi]\!]^{W'}_{\mathcal{J}_t} = \text{share}([\![\Xi']\!]^{W'}_{\mathcal{J}_{s+r}})$, that is

$$1_{\mathcal{J}_t(W')([\![\Xi]\!]^{W'}_{\mathcal{J}_t}} \stackrel{[\![MN_\Xi]\!]^{W'}_{\mathcal{J}_t}}{\longrightarrow} [\![B[N/x]_\Xi]\!]^{W'}_{\mathcal{J}_t}.$$

We check that this is the interpretation given by the model. According to Definition 15.12, $[\![MN_\Xi]\!]^W_{\mathcal{J}_t}$ is defined and is equal to, using monotonicity,

$$(\langle 1_{[\![\Gamma]\!]^{W'}_{\mathcal{J}_r}}, [\![N_\Delta]\!]^{W'}_{\mathcal{J}_s}\rangle^*(\Lambda_{[\![\Gamma]\!]^{W'}_{\mathcal{J}_r}, [\![A_\Gamma]\!]^{W'}_{\mathcal{J}_s}}([\![M_\Gamma]\!]^{W'}_{\mathcal{J}_r}))),$$

where Ξ is defined as above. We must check the types. Firstly, we already have

$$[\![M_\Gamma]\!]^{W'}_{\mathcal{J}_r} : 1_{\mathcal{J}_r(W')([\![\Gamma]\!]^{W'}_{\mathcal{J}_r})} \longrightarrow [\![\Lambda x{:}A\,.B_\Gamma]\!]^{W'}_{\mathcal{J}_r}.$$

Applying the natural isomorphism $\Lambda_{[\![\Gamma]\!]^{W'}_{\mathcal{J}_r}, [\![A_\Gamma]\!]^{W'}_{\mathcal{J}_s}}$ gives us

$$\Lambda_{[\![\Gamma]\!]^{W'}_{\mathcal{J}_{r+s}}, [\![A_\Gamma]\!]^{W'}_{\mathcal{J}_{r+s}}}([\![M_{\Gamma,x:A}]\!]^{W'}_{\mathcal{J}_{r+s}}) : 1_{\mathcal{J}_{r+s}(W')([\![\Gamma,x:A]\!]^{W'}_{\mathcal{J}_{r+s}})} \longrightarrow [\![B_{\Gamma,x:A}]\!]^{W'}_{\mathcal{J}_{r+s}}.$$

The functor $\langle 1_{[\![\Gamma]\!]^{W'}_{\mathcal{J}_r}}, [\![N_\Delta]\!]^{W'}_{\mathcal{J}_s}\rangle^*$ performs the required substitution. Finally the action of join and share gives us $[\![\Xi]\!]^{W'}_{\mathcal{J}_t}$.

$(M\&\mathcal{I})$ Suppose **N** proves $\Gamma \vdash_\Sigma \langle M, N\rangle{:}A\&B$ because **N** proves $\Gamma \vdash_\Sigma M{:}A$ and **N** proves $\Gamma \vdash_\Sigma N{:}B$. By induction hypothesis twice, we have that $\mathcal{J}_r, W \models_\Sigma (M{:}A)\ [\Gamma]$, that is

$$1_{\mathcal{J}_r(W)(\llbracket\Gamma\rrbracket^W_{\mathcal{J}_r})} \stackrel{\llbracket M_\Gamma\rrbracket^W_{\mathcal{J}_r}}{\longrightarrow} \llbracket A_\Gamma\rrbracket^W_{\mathcal{J}_r}$$

and that $\mathcal{J}_r, W \models_\Sigma (N{:}B)\ [\Gamma]$, that is

$$1_{\mathcal{J}_r(W)(\llbracket\Gamma\rrbracket^W_{\mathcal{J}_r})} \stackrel{\llbracket N_\Gamma\rrbracket^W_{\mathcal{J}_r}}{\longrightarrow} \llbracket B_\Gamma\rrbracket^W_{\mathcal{J}_r}.$$

Now, each category $\mathcal{J}_r(W)(D)$ in the model has products. We use this property in $\mathcal{J}_r(W)(\llbracket\Gamma\rrbracket^W_{\mathcal{J}_r})$ to construct

$$1_{\mathcal{J}_r(W)(\llbracket\Gamma\rrbracket^W_{\mathcal{J}_r})} \stackrel{\llbracket\langle M,N\rangle_\Gamma\rrbracket^W_{\mathcal{J}_r}}{\longrightarrow} \llbracket (A\&B)_\Gamma\rrbracket^W_{\mathcal{J}_r}$$

and $\mathcal{J}_r, W \models_\Sigma (\langle M, N\rangle{:}(A\&B))\ [\Gamma]$ follows.

$(M\&\mathcal{E}_i)$ Suppose **N** proves $\Gamma \vdash_\Sigma \pi_i(M){:}A_i$, for $i \in \{0,1\}$ because **N** proves $\Gamma \vdash_\Sigma M{:}A_0\&A_1$. By the induction hypothesis, we have that $\mathcal{J}_r, W \models_\Sigma (M{:}A_0\&A_1)\ [\Gamma]$, that is,

$$1_{\mathcal{J}_r(W)(\llbracket\Gamma\rrbracket^W_{\mathcal{J}_r})} \stackrel{\llbracket M_\Gamma\rrbracket^W_{\mathcal{J}_r}}{\longrightarrow} \llbracket (A_0\&A_1)_\Gamma\rrbracket^W_{\mathcal{J}_r}.$$

Then the definition of satisfaction allows us to construct, for $i \in \{0,1\}$,

$$1_{\mathcal{J}_r(W)(\llbracket\Gamma\rrbracket^W_{\mathcal{J}_r})} \stackrel{\llbracket \pi_i(M)_\Gamma\rrbracket^W_{\mathcal{J}_r}}{\longrightarrow} \llbracket A_{i\Gamma}\rrbracket^W_{\mathcal{J}_r}$$

and $\mathcal{J}_r, W \models_\Sigma (\pi_i(M){:}A_i)\ [\Gamma]$ follows.

$(M \equiv)$ It is convenient, as we are working in the $M{:}A$-fragment of the type theory, to observe that $\beta\eta$-equalities are generated by the rule

$$\frac{\Gamma, x{\in}A \vdash_\Sigma M{:}B \quad \Delta \vdash_\Sigma N{:}A}{\Xi \vdash_\Sigma (\lambda\, x{\in}A\,.M)N =_\beta M[N/x] \;:\; B[N/x]},$$

where $[\Xi'; \Gamma; \Delta]$ and $\Xi = \Xi' \backslash \kappa\langle \Gamma, \Delta \rangle$, and by the rule

$$\frac{\Gamma \vdash_\Sigma M : \Lambda x{\in}A.B}{\Gamma \vdash_\Sigma (\lambda y{\in}A.M)y =_\eta M : \Lambda x{\in}A.B},$$

where $y \notin FV(\Gamma, x{\in}A)$. Then, an application of the natural isomorphism and Lemma 15.16, allows us to show that if $M =_{\beta\eta} N$, then $[\![M]\!]^W_{\mathcal{J}_r} \simeq [\![N]\!]^W_{\mathcal{J}_r}$. Note that we make particular use of the definition of the interpretation of applications:

$$[\![MN_\Xi]\!]^W_{\mathcal{J}_t} \simeq (\langle 1_{[\![\Gamma]\!]^W_{\mathcal{J}_r}}, [\![N_\Delta]\!]^W_{\mathcal{J}_s}\rangle^* (\Lambda^{-1}_{[\![\Gamma]\!]^W_{\mathcal{J}_r}, [\![A_\Gamma]\!]^W_{\mathcal{J}_s}}([\![M_\Gamma]\!]^W_{\mathcal{J}_r}))),$$

where $[\![\Xi']\!]^W_{\mathcal{J}_{r+s}} = \text{join}\langle [\![\Gamma]\!]^W_{\mathcal{J}_r}, [\![\Delta]\!]^W_{\mathcal{J}_s}\rangle$ and $[\![\Xi]\!]^W_{\mathcal{J}_t} = \text{share}([\![\Xi']\!]^W_{\mathcal{J}_{r+s}})$.

□

We conclude our account of soundness with a remark abour Dereliction. The soundness of Dereliction follows from the interpretation of the two forms of axiom sequent,

$$\frac{\Gamma \vdash_\Sigma A : \mathsf{Type}}{\Gamma, x : A \vdash_\Sigma x : A} \quad \text{and} \quad \frac{\Gamma \vdash_\Sigma A : \mathsf{Type}}{\Gamma, x!A \vdash_\Sigma x : A}.$$

These are each interpreted by second projection maps in the fibres over $\Gamma, x : A \vdash_\Sigma x : A$ and $\Gamma, x!A \vdash_\Sigma x : A$, respectively. The latter projection exists whenever the former exists, and so, by induction on the structure of proofs, a term interpreted over $\Gamma, x : A$ may also be interpreted over $\Gamma, x!A$. Note, however, that the converse fails: the interpretation of a term over $\Gamma, x!A$ may rely upon the intuitionistic properties of extension with !.

We now turn to consider completeness. We begin with the appropriate definition of validity for $\models_\Sigma$.

DEFINITION 15.19 ($\models_\Sigma$-VALIDITY FOR $\lambda\Lambda$) $\Gamma \models_\Sigma M{:}A$, i.e., $M{:}A$ *is valid with respect to* Γ, *if and only if, for all models*

$$\langle \{\mathcal{J}_r \mid r \in R\}, [\![-]\!], \text{join}, \text{share}, \models_\Sigma \rangle$$

and all worlds W *such that* $[\![\Gamma]\!]^W_{\mathcal{J}_r}$, $[\![A_\Gamma]\!]^W_{\mathcal{J}_r}$ *and* $[\![M_\Gamma]\!]^W_{\mathcal{J}_r}$ *are defined,* $\mathcal{J}_r, W \models_\Sigma (M{:}A)\ [\Gamma]$. □

We now construct a term model. We work with $\beta\eta$-equivalence classes of contexts and realizations but suppress any explicit notational representation of this.

DEFINITION 15.20 *Let Σ be a signature. The base category $\mathcal{C}(\Sigma)$ of contexts and realizations is defined as follows:*

- *Objects: contexts Γ such that* $\mathbf{N}$ *proves* $\vdash_\Sigma \Gamma$ context;
- *Arrows: realizations* $\Gamma \stackrel{\langle M_1,\dots,M_n\rangle}{\longrightarrow} \Delta$ *such that*

$$\mathbf{N} \textit{ proves } \Gamma \vdash_\Sigma (M_i{:}A_i)[M_j/x_j]_{j=1}^{i-1},$$

where $\Delta = x_1{\in}A_1, \dots, x_n{\in}A_n$.

  - *Identities are* $x_1{\in}A_1, \dots, x_n{\in}A_n \stackrel{\langle x_1,\dots,x_n\rangle}{\longrightarrow} x_1{\in}A_1, \dots, x_n{\in}A_n$. *We will write the identity arrow on* Γ *as* 1_Γ;
  - *Composition is given by substitution. If* $f = \Gamma \stackrel{\langle M_1,\dots,M_n\rangle}{\longrightarrow} \Delta$ *and* $g = \Delta \stackrel{\langle N_1,\dots,N_p\rangle}{\longrightarrow} \Theta$, *then* $f;g = \Gamma \stackrel{\langle N_1[M_j/y_j]_{j=1}^n,\dots,N_p[M_j/y_j]_{j=1}^n\rangle}{\longrightarrow} \Theta$.

Throughout this definition, and in its use in the sequel, we have assumed that each Γ *and* M *denotes its* $\beta\eta$*-equivalence class.* □

$\mathcal{C}(\Sigma)$ is doubly monoidal by virtue of the two ways of extending the context.

PROPOSITION 15.21 *$\mathcal{C}(\Sigma)$ is a doubly monoidal category.*

PROOF The two context extension operators are taken to be an extension with A and an extension with $!A$. The units for each context extension operator require the following rules to be taken in the syntax of the type theory:

$$\frac{}{\vdash_\Sigma 1 \text{ context}} \qquad \frac{}{\vdash_\Sigma I \text{ context}}$$

together with the context equivalences which let I and 1 be, respectively, units of extension with A and extension with $!A$.

A little care is required in order to use the two context extensions to construct the two monoidal products. We define, inductively,

$$[\![\Gamma]\!] \otimes [\![\Delta]\!] = join([\![\Gamma]\!], [\![\Delta]\!])$$

$$\begin{aligned} 1 \times [\![\Gamma]\!](= [\![\Gamma]\!] \times 1) &= [\![\Gamma]\!] \\ [\![\Gamma]\!] \times [\![\Delta]\!] &= join([\![!\Gamma]\!], [\![!\Delta]\!]). \end{aligned}$$

Both $\otimes$ and $\times$ are associative. □

$\mathcal{C}(\Sigma)$ provides the base in the following definition of an indexed category:

DEFINITION 15.22 *We inductively define a strict indexed category*

$$\mathcal{E}(\Sigma){:}\mathcal{C}(\Sigma)^{op} \to \mathbf{Cat}$$

over the base category $\mathcal{C}(\Sigma)$ *as follows:*

- *For each* Γ *in* $\mathcal{C}(\Sigma)$*, the category* $\mathcal{E}(\Sigma)(\Gamma)$ *is defined as follows:*
  - *Objects: Types* A *such that* $\mathbf{N}$ *proves* $\Gamma \vdash_\Sigma A{:}\mathsf{Type}$*;*
  - *Morphisms:* $A \overset{M}{\to} B$ *where the object* M *is such that* $\Gamma, x{:}A \overset{M}{\to} y{:}B$ *in* $\mathcal{C}(\Sigma)$*. Composition is given by substitution;*
- *For each* $f{:}\Gamma \to \Delta$ *in* $\mathcal{C}(\Sigma)$*,* $\mathcal{E}(\Sigma)(f)$ *is a functor* $f^*{:}\mathcal{E}(\Sigma)(\Delta) \to \mathcal{E}(\Sigma)(\Gamma)$ *given by* $f^*(A) \overset{\text{def}}{=} A[f]$ *and* $f^*(M) \overset{\text{def}}{=} M[f]$*.*

□

We remark that each $\mathcal{E}(\Sigma)(\Gamma)$ is indeed a category. Note that the identity arrow $A \overset{1}{\to} A$ over Γ is given by the term $\lambda\, x{:}A\,.x$, corresponding to the definition of morphisms above. To see that this construction is correct, consider that the axiom sequent is of the form $\Gamma, x{:}A \vdash_\Sigma x{:}A$, with the side-condition that $\Gamma \vdash_\Sigma A{:}\mathsf{Type}$, thereby using the variables in Γ.

DEFINITION 15.23 *The category* $\mathcal{P}(\Sigma)$*, a full sub-category of* $\mathcal{C}(\Sigma)$*, is defined as follows:*

- *Objects:*
  - $\langle\rangle$ *is an object of* $\mathcal{P}(\Sigma)$*;*
  - *If* Γ *is an object of* $\mathcal{P}(\Sigma)$ *and there exists an arrow* $\Gamma \overset{\langle 1,M\rangle}{\to} \Gamma \times A$ *in* $\mathcal{C}(\Sigma)$*, then* $\Gamma \times A$ *is an object of* $\mathcal{P}(\Sigma)$*.*
- *Morphisms: The arrows just considered.*

LEMMA 15.24 *The tuple consisting of the set of objects in* $\mathcal{C}(\Sigma)$*, the context joining operation* $[-;-;-]$ *and the unit context* $\langle\rangle$ *defines a commutative monoid.*

PROOF For ease of argument, we will adopt the following notation: $\Gamma\Delta$ will denote the join of the contexts Γ and Δ.

We first show that $\langle\rangle$ behaves as a 2-sided identity. This is immediate because of the coherence equivalences between contexts.

Next, we show that the joining relation is associative: if Γ, Δ and Θ are valid contexts, then $\Gamma(\Delta\Theta) = (\Gamma\Delta)\Theta$, where $[\Gamma(\Delta\Theta);\Gamma;\Delta\Theta]$, $[\Delta\Theta;\Delta;\Theta]$, $[(\Gamma\Delta)\Theta;\Gamma\Delta;\Theta]$ and $[\Gamma\Delta;\Gamma;\Delta]$.

The proof of associativity is by induction on the length of the context $\Gamma(\Delta\Theta)$. The base case is when $\Gamma(\Delta\Theta) = \langle\rangle$. By the definition of the joining relation, this implies that $\Gamma = \langle\rangle$ and that $\Delta\Theta = \langle\rangle$. By the same argument, we know know that $\Delta = \langle\rangle$ and $\Theta = \langle\rangle$. We use the definition to construct $(\Gamma\Delta)\Theta$ which is also equal to $\langle\rangle$.

There are three inductive cases to consider, one for each of the (JOIN-L), (JOIN-R) and (JOIN-!) rules. For the first of these, we have $\Gamma(\Delta\Theta) = \Gamma(\Delta(\Theta', x{:}A))$ by (JOIN-L). By assumption, $\Gamma(\Delta(\Theta', x{:}A))$ splits into Γ and $\Delta(\Theta', x{:}A)$ and $\Delta(\Theta', x{:}A)$ splits into Δ and $\Theta', x{:}A$. By the induction hypothesis, Γ and Δ join to form $\Gamma\Delta$, and $\Gamma\Delta$ and Θ' join to form $(\Gamma\Delta)\Theta'$. By (JOIN-L), $(\Gamma\Delta)\Theta'$ and $x{:}A$ join to form $(\Gamma\Delta)\Theta', x{:}A = (\Gamma\Delta)\Theta$. The other two cases are argued similarly and we omit the details.

Lastly, we show the commutativity of the joining relation: if $[\Xi;\Gamma;\Delta]$, then $[\Xi;\Delta;\Gamma]$. The proof is by induction on the length of the context Ξ. For the base case, when $\Xi = \langle\rangle$, the proof is immediate. There are three inductive cases to consider, one for each of the (JOIN-L), (JOIN-R) and (JOIN-!) rules. For the first of these, suppose $[\Xi', x{:}A; \Gamma, x{:}A; \Delta]$. By the induction hypothesis, we have that if $[\Xi';\Gamma;\Delta]$, then $[\Xi;\Delta;\Gamma]$. Then an application of (JOIN-R) gives us $[\Xi', x{:}A; \Delta; \Gamma, x{:}A]$. The other two cases are argued similarly and we omit the details. □

As joining is associative, we can informally say "Γ, Δ and Θ join to form $\Gamma\Delta\Theta$". That is, we can talk about n-way joining and there need be no confusion.

We remark that in logics, such as intuitionistic logic or **BI**, which include conjunctions and disjunctions, one must develop the notion of prime theory. Prime theories have exactly the structure required by the semantic clauses for the connectives and are used to prove completeness. The construction of prime theories is not necessary in the minimal cases of both the $\lambda\Pi$- and $\lambda\Lambda$-calculi, where function spaces are the only connectives. (The $\lambda\Lambda$-calculus does have the additive conjunction, but the term model inherits enough structure from the syntax to push the definitions through.)

LEMMA 15.25 (MODEL EXISTENCE) *There is a Kripke Σ-$\lambda\Lambda$ model*

$$\langle\{\mathcal{T}(\Sigma)_\Delta\}, [\![-]\!]^{-}_{\mathcal{T}(\Sigma)_\Delta}, \mathrm{join}, \mathrm{share}, \models_\Sigma\rangle$$

with a world W_0 such that if $\Gamma \not\vdash_\Sigma M{:}A$, then $\mathcal{T}(\Sigma)_\Delta, W_0 \not\models_\Sigma (M{:}A)[\Gamma]$.

PROOF We construct such a model out of the syntax of the $\lambda\Lambda$-calculus.

The Kripke Σ-$\lambda\Lambda$ structure $\mathcal{T}(\Sigma)_\Delta$ is defined as follows. The category of worlds is taken to be $\mathcal{P}(\Sigma)$. The base category is the co-product of $\mathcal{C}(\Sigma)_\Delta$, where each $\Delta \in ob(\mathcal{P}(\Sigma))$. The indexing monoid is given by the context joining relation $[-;-;-]$, as defined by Lemma 15.24. The functor $\mathcal{T}(\Sigma)_\Delta$, indexed by an element $\Delta \in obj(\mathcal{C}(\Sigma))$, is defined as follows:

$$\mathcal{T}(\Sigma)_\Delta(\Theta)(\Gamma) \quad = \quad \begin{cases} \textit{Objects}: & \text{Types } A \text{ such that } \mathbf{N} \text{ proves } \Psi \vdash_\Sigma A{:}\mathsf{Type} \\ \textit{Arrows}: & \mathcal{E}(\Sigma)(\Psi) \text{ arrows} \end{cases}$$

where $[\Phi';\Theta;\Delta]$, $\Phi = \Phi' \backslash \kappa\langle\Theta,\Delta\rangle$ and $[\Psi';\Gamma;\Phi]$, $\Psi = \Psi' \backslash \kappa\langle\Gamma,\Phi\rangle$.

From the algebraic presentation of the type theory given by Definition 15.20, Proposition 15.21, Definition 15.22 and Definition 15.23, we can see that $\mathcal{T}(\Sigma)_\Delta(\Theta)(\Gamma)$ is a category. We also need to check that $\mathcal{T}(\Sigma)_\Delta$ is a functor.

We next check that $\mathcal{T}(\Sigma)_\Delta$ is a Kripke structure. Each of the following points refers to those of Definition 15.10.

1 The terminal object in each $\mathcal{T}(\Sigma)_\Delta(\Theta)(\Gamma)$ is taken to be the unit additive context 1. We choose this because the proof theory has the judgement $\mathbf{N}$ proves $\Gamma \vdash_\Sigma 1$ so that 1 always exists in each fibre. 1 contains no free variables and so is always preserved on the nose by any f^*;

2 The map $q_{\langle\Gamma,A\rangle}$ is given by the term x where $\Gamma, x{\in}A \vdash_\Sigma x{:}A$. The first projection map for an intuitionistically extended context

$$\Gamma = x_1{\in}A_1, \ldots, x_n{\in}A_n, x_{n+1}!A_{n+1}$$

is defined by $p(\Gamma) = x_1{\in}A_1, \ldots, x_n{\in}A_n$. This is well-defined because Weakening is admissible in the syntax. We need to check that the appropriate square is a pullback. This may be done using the properties of substitution and we omit the details.

The two extensions, $D \to D \otimes A$ and $D \to D \times A$, are given by the context extension rules of the type theory.

We need to check the strictness conditions. This too may be done using the properties of substitution and we omit the details.

3 The natural isomophism is given by the abstraction and application rules of the type theory:

$$\frac{\Gamma, x{\in}A \vdash_\Sigma M{:}B}{\Gamma \vdash_\Sigma \lambda x{\in}A\,.M \,:\, \Lambda x{\in}A\,.B}\,,$$

where $x{\in}A$, recall, ranges over both linear $x{:}A$ and intuitionistic $x!A$ declarations. We need to check that these meet the Beck-Chevalley condition. This may be done using the properties of substitution and we omit the details.

4 The products in each $\mathcal{T}(\Sigma)_\Theta(\Delta)(\Gamma)$ are given by the $(M\&\mathcal{I})$ and $(M\&\mathcal{E}_i)$ rules.

We sketch the construction of the model. $\mathcal{T}(\Sigma)_\Delta$ is the Kripke $\lambda\Lambda$-structure defined above. The Σ-operations of the model are given by the constants declared in the signature Σ. The interpretation $[\![-]\!]^-_{\mathcal{T}(\Sigma)_-}$ is the obvious one in which a term (type) is interpreted by the class of terms definitionally equivalent to the term (type) in the appropriate component of $\mathcal{T}(\Sigma)$. The functors join and share are defined by the joining relation $[-;-;-]$ and κ, respectively.

The satisfaction relation $\models_\Sigma$ in $\mathcal{T}(\Sigma)$ is given by provability in the type theory. That is, $\mathcal{T}(\Sigma)_\Theta, \Delta \models_\Sigma (M{:}A)\ [\Gamma]$ is defined to be $\Xi \vdash_\Sigma M{:}A$, where Ξ is the sharing-sensitive join of Θ, Δ and Γ. We must check that this relation satisfies the inductive clauses of the satisfaction relation:

1 $!\Xi \vdash_\Sigma c{:}A$ if and only if $c{:}A \in \Sigma$ is immediate as the Σ-operations are the $c{:}A$s;

2 $\Xi, x{:}A \vdash_\Sigma x{:}A$ if and only if $\vdash_\Sigma \Xi, x{:}A$ context by induction on the structure of proofs of both hypotheses;

3 $\Xi \vdash_\Sigma M{:}\Lambda\, x{:}A\,.B$ if and only if $\Phi \vdash_\Sigma N{:}A$ implies $\Psi \vdash_\Sigma MN{:}B[N/x]$, where $[\Psi';\Xi;\Phi]$ and $\Psi = \Psi' \backslash \kappa\langle\Xi,\Phi\rangle$, holds, in one direction, by $(M\Lambda\mathcal{E})$ and, in the other, by an application of Cut. The intuitionistic case is similar;

4 $\Xi \vdash_\Sigma M{:}A\&B$ if and only if $\Xi \vdash_\Sigma \pi_0(M){:}A$ and $\Xi \vdash_\Sigma \pi_1(M){:}B$ is immediate by the & rules.

The conditions on the models are met as follows:

1 Monotonicity is met by the fact that all terms are well-defined. *i.e.*, constructed in accordance with the proof rules. so a valid term will only ever be constructed from valid sub-terms; and

2 Accessibility is provided by the posetal nature of $\mathcal{P}(\Sigma)$.

It follows routinely from the definitions of $\mathcal{C}(\Sigma)$ and $\mathcal{E}(\Sigma)$ that

$$\mathcal{T}(\Sigma)_\Theta, \Delta \models_\Sigma (M{:}A)\ [\Gamma] \text{ if and only if } \Xi \vdash_\Sigma M{:}A,$$

where Ξ is the sharing-sensitive join of Θ, Δ and Γ.

We can now finish the proof of model existence. We assume the premiss $\Gamma \not\vdash_\Sigma M{:}A$. Then, at the initial node ($W_0 = \langle\rangle$), the model constructed from the syntax has the required property; that $\mathcal{T}(\Sigma)_\Theta, W_0 \not\models_\Sigma (M{:}A)\ [\Phi]$, where $[\Gamma;\Theta;\Phi]$. □

THEOREM 15.26 (COMPLETENESS)

$$\Gamma \vdash_\Sigma M{:}A \quad \textit{if and only if} \quad \Gamma \models_\Sigma M{:}A.$$

PROOF Theorem 15.18 (Soundness) shows the forward direction. For the other, we assume $\Gamma \not\vdash_\Sigma M{:}A$ and apply Lemma 15.25. □

In fact, we might hope for a more abstract completeness theorem. Specifically, we might hope for a more abstract definition of a class of models of $\lambda\Lambda$, together a result to the effect that Kripke resource models are complete for that class — perhaps via a covering theorem.

Turning briefly back to logical frameworks, and recalling that

$$\textit{Framework} = \textit{Language} + \textit{Representation},$$

we remark that our semantic analysis has had little to say about RLF: we have considered only the 'language' part.

Both of these last two points are deferred to another occasion, partly because we consider $\lambda\Lambda$ and its semantics to be at best a partial analysis of substructural dependent types. We return to this point in § 15.

14. A Class of Set-theoretic Models

We describe a set-theoretic class of concrete Kripke resource models, in which the Kripke resource $\lambda\Lambda$-structure

$$\{\mathcal{J}_r{:}[\mathcal{W}, [\mathcal{C}^{op}, \mathbf{Cat}]] \mid r \in R\}$$

is given by $\mathbf{BIFam}{:}[\mathcal{C}, [Ctx^{op}, \mathbf{Set}]]$, where $\mathcal{C}$ is a small monoidal category and Ctx is a small set-theoretic category of "contexts". The model is a construction on the category of families of sets and exploits Day's construction to define the linear dependent function space.

We begin with a description of the indexed category of families of sets, $\mathbf{Fam}{:}[Ctx^{op}, \mathbf{Cat}]$. The base, Ctx, is a small set-theoretic category defined inductively as follows: The objects of Ctx, called "contexts", are (*i.e.*, their denotations are) sets and the arrows of Ctx, called "realizations", are set-theoretic functions. For each $D \in obj(Ctx)$,

$\mathbf{Fam}(D) = \{y \in B(x) \mid x \in D\}$. The fibre may be described as a discrete category whose objects are the ys and whose arrows are the maps $1_y{:}y \to y$ corresponding to the identity functions $id{:}\{y\} \to \{y\}$ on y considered as a singleton set. If $E \xrightarrow{f} D$ is an arrow in Ctx, then $\mathbf{Fam}(f) = f^*{:}\mathbf{Fam}(D) \to \mathbf{Fam}(E)$ re-indexes the set $\{y \in B(x) \mid x \in D\}$ over D to the set $\{f(z) \in B(f(z)) \mid z \in E\}$ over E. We are viewing **Set** within **Cat**; each object of **Set** is seen as an object, a discrete category, in **Cat**. Because of this, the category of families of sets may just be considered as a presheaf $\mathbf{Fam}{:}[Ctx^{op}, \mathbf{Set}]$, rather than as an indexed category; we will adopt this view in the sequel.

We can explicate the structure of Ctx by describing **Fam** as a contextual category [Cartmell, 1994]. The following definition is from Streicher [Streicher, 1988]:

DEFINITION 15.27 *The contextual category* **Fam**, *together with its length and denotation* DEN:**Fam** $\to$ **Set**, *is described as follows:*

1. 1 *is the unique context of length* 0 *and*

$$\mathrm{DEN}(1) = \{\emptyset\};$$

2. If D is a context of length n and $A{:}\mathrm{DEN}(D) \to$ **Set** *is a family of sets indexed by elements of $DEN(D)$, then $D \times A$ is a context of length $n+1$ and*

$$\mathrm{DEN}(D \times A) = \{\langle x, y\rangle \mid x \in \mathrm{DEN}(D),\ y \in A(x)\}.$$

If D and E are objects of the contextual category **Fam**, *then the morphisms between them are simply the functions between* $\mathrm{DEN}(D)$ *and* $\mathrm{DEN}(E)$. □

The co-domain of the denotation, *i.e.*, **Set**, allows the definition of an extensional context extension $\times$. But **Set** does not have enough structure to define an intensional context extension $\otimes$. In order to be able to define both $\times$ and $\otimes$, we denote **Fam** not in **Set** but in a presheaf $\mathbf{Set}^{\mathcal{C}^{op}}$, where $\mathcal{C}$ is a monoidal category. We emphasize that, in general, $\mathcal{C}$ may be any monoidal category and, therefore, we are actually going to describe a class of set-theoretic models. For simplicity, we take $\mathcal{C}^{op}$ to be a partially-ordered commutative monoid $\mathcal{M} = (M, \cdot, e, \sqsubseteq)$. The cartesian structure on the presheaf gives us the $\times$ context extension and a restriction of Day's tensor product [Day, 1970] gives us the $\otimes$ context extension.

We remark that the restriction of Day's tensor product we consider is merely this: consider the set-theoretic characterization of Day's tensor product as tuples $\langle x, y, f\rangle$ and, of all such tuples, consider only those where the y is an element of the family of sets in x. This is quite concrete, in the spirit of the Cartmell–Streicher models, and is not a general construction for a fibred Day product.

Within the contextual setting, we then have the following definition:

DEFINITION 15.28 *The contextual category* **BIFam**, *together with its length and denotation* DEN:**BIFam** $\to$ $\mathbf{Set}^{\mathcal{M}}$, *is described as follows:*

1. 1 *is a context of length* 0 *and*

$$\mathrm{DEN}(1)(Z) = \{\emptyset\}\,;$$

2. I *is a context of length* 0 *and*

$$\mathrm{DEN}(I)(-) = \mathcal{M}[-, I]\,;$$

3. If D *is a context of length* n *and* A:DEN$(D)(X) \to \mathbf{Set}^{\mathcal{M}}$ *is a family of* $\mathcal{M}$*-sets indexed by elements of* $DEN(D)(X)$, *then*

(a) $D \times A$ *is a context of length* $n+1$ *and*

$$\mathrm{DEN}(D \times A)(X) = \\ \{\langle x, y\rangle \mid x \in \mathrm{DEN}(D)(X),\ y \in (A(x))(X)\}\,;$$

(b) $D \otimes A$ *is a context of length* $n+1$ *and*

$$\mathrm{DEN}(D \otimes A)(Z) = \{\langle x, y, f\rangle \in \\ \textstyle\int^{X,Y} \mathrm{DEN}(D)(X) \times (A(x))(Y) \times \mathcal{M}[Z, X \otimes Y]\}\,.$$

Here we have used the characterization of Day's tensor product as tuples, with the restriction, to account for dependency, of the triples $\langle x, y, f\rangle$ *to those in which* $y \in A(x)(Y)$.

If D *and* E *are objects of* **BIFam**, *then the morphisms between them are the functions between* DEN$(D)(X)$ *and* DEN$(E)(Y)$. **BIFam** *is* **Fam** *parametrized by* $\mathcal{M}$*; objects that were interpreted in* **Set** *are now interpreted in* $\mathbf{Set}^{\mathcal{M}}$. □

Now consider **BIFam** in an indexed setting. By our earlier argument relating indexed and contextual presentations of families of sets, **BIFam** may be seen as a functor category **BIFam**:$[Ctx^{op}, \mathbf{Set}^{\mathcal{M}}]$. This is not quite the presheaf setting we require. However, if we calculate

$$\begin{aligned} [Ctx^{op}, [\mathbf{Set}^{\mathcal{M}}]] &\cong [Ctx^{op} \times \mathcal{M}, \mathbf{Set}] \\ &\cong [\mathcal{M} \times Ctx^{op}, \mathbf{Set}] \\ &\cong [\mathcal{M}, [Ctx^{op}, \mathbf{Set}]], \end{aligned}$$

then this restores the indexed setting and also reiterates the idea that $\mathcal{M}$ parametrizes **Fam**. The right adjoint to $\otimes$, given by Day's construction, provides the isomorphism required to define the linear dependent function space, Λ.

Lastly, we say what the R and $\mathcal{W}$ components of the concrete model are. Define $(R, +, 0) = (M, \cdot, e)$ and define $(\mathcal{W}, \leq) = (M/\sim, \sqsubseteq)$, where the quotient of M by the relation $w \sim w \cdot w$ is necessary because of the separation of worlds from resources (*cf.* **BI**'s semantics [O'Hearn and Pym, 1999, Pym, 1999, Urquhart, 1972]). This allows us to define $\mathcal{J}_r(w) = \mathbf{BIFam}(r \cdot w)$. The quotiented $\mathcal{M}$ maintains the required properties of monotonicity and bifunctoriality of the internal logic forcing relation. We then check that $\mathbf{BIFam}(r \cdot w)$ does simulate $\mathcal{J}_r(w)$, and that **BIFam** is a Kripke resource $\lambda\Lambda$-structure.

THEOREM 15.29 **BIFam**:$[\mathcal{M}, [Ctx^{op}, \mathbf{Set}]]$ *is a Kripke resource* $\lambda\Lambda$*-structure and may be extended to a Kripke resource model.* □

Definition 15.28 above comprises the main part of the proof that **BIFam** is a Kripke resource structure. It describes how Ctx may have two kinds of extension. These extensions are then used to describe two kinds of function space in **BIFam**. For the linear case, for instance, $\Lambda x{:}A.B$ is defined as the following set:

$$\begin{array}{l}\{f : \mathbf{BIFam}(Y)(A(x)) \to \\ \cup_y\{\mathbf{BIFam}(X \otimes Y)(B(x,y)) \mid y \in \mathbf{BIFam}(Y)(A(x))\} \mid \\ \forall a \in \mathbf{BIFam}(Y)(A(x))\ f(a) \in \mathbf{BIFam}(X \otimes Y)(B(x,a))\},\end{array}$$

where $x \in \mathbf{BIFam}(X)(D)$. The intuitionistic function space is defined analogously, with the "resource" X over which the sets are defined staying the same. The natural isomorphism is given by abstraction and application in this setting.

In order to extend **BIFam** to a model, the structure must have enough points to interpret the constants of the signature. We can work with an arbitrary signature and interpret constants and variables as the functors $Const{:}\mathcal{M} \to \mathbf{Set}$ and $D{:}\mathcal{M} \to \mathbf{Set}$ respectively. The interpretation function $[\![-]\!]^X_{\mathbf{BIFam}}$ is parametrized over worlds/resources X. The interpretation of contexts is defined using the same idea as the construction of the category Ctx:

1 $[\![\Gamma, x{:}A]\!]^{X \otimes Y}_{\mathbf{BIFam}} \simeq [\![\Gamma]\!]^X_{\mathbf{BIFam}} \otimes [\![A_\Gamma]\!]^Y_{\mathbf{BIFam}}$;

2 $[\![\Gamma, x!A]\!]^X_{\mathbf{BIFam}} \simeq [\![\Gamma]\!]^X_{\mathbf{BIFam}} \times [\![A_\Gamma]\!]^X_{\mathbf{BIFam}}$.

The interpretation of functions is defined using the abstraction and application. We must also define instances of the functors *join* and *share* for this setting; these are defined along the same lines as those for the term model. Finally, satisfaction is a relation over M and $[Ctx^{op}, \mathbf{Set}]$ with the clauses reflecting the properties — in particular, those of application — of the example model:

1 $X \models_\Sigma f{:}\Lambda\, x{:}A\,.B\ [D]$ if and only if $Y \models a{:}A\ [E]$ implies

$$X \otimes Y \models_\Sigma f(a){:}B[a/x]\ [D \otimes E];$$

2 $X \models_\Sigma f{:}\Pi\, x{:}A\,.B\ [D]$ if and only if $X \models_\Sigma a{:}A\ [E]$ implies

$$X \models_\Sigma f(a){:}B[a/x]\ [D \times E].$$

15. Towards Systematic Substructural Type Theory

We have already explained the extent to which the propositions-as-types **BI** and $\lambda\Lambda$ is weaker than is normally considered appropriate. The deficiencies are consequences of the presence of Dereliction in $\lambda\Lambda$ and the focus on extension, rather than monoidal products, in the combinatorics of $\lambda\Lambda$'s contexts.

So it is natural to ask whether there is a (dependent) type theory which is fully in correspondence with (predicate) **BI**. This is a conceptually and technically challenging question. We sketch here a possible, though somewhat speculative, development of such a type theory, highlighting some possible difficulties.

The structure of **BI**'s contexts, or bunches, *i.e.*, doubly monoidal, must be reflected in the contextual structure of the type theory. However, at first sight, the nature of a type theory relies on the formation of types over a context, so permitting the extension of the context by the type. So, if the extension of the context is to be by a monoidal product, then it would seem natural to form not merely types but bunches over a given bunch. To see this, consider Figure 15.5 which represents the bunch $x_1 : A_1\,, ((x_2 : A_2\,, x_3 : A_3)\,; x_4 : A_4)$. The possible dependencies of types on variables are determined by the bunched version of the left-to-right ordering that is usual in dependent type theory: x_i is below x_j just in case x_i is on the left-hand branch of the tree rooted at the least upper bound (a "," or ";") of x_i and x_j.

We must ask, however, how we might formally form such a context. Consider the top-level combinator, a comma. The right-hand branch of

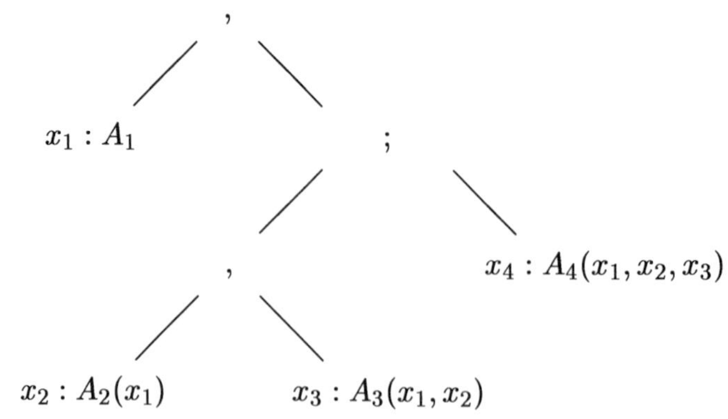

Figure 15.5. Dependent Bunches

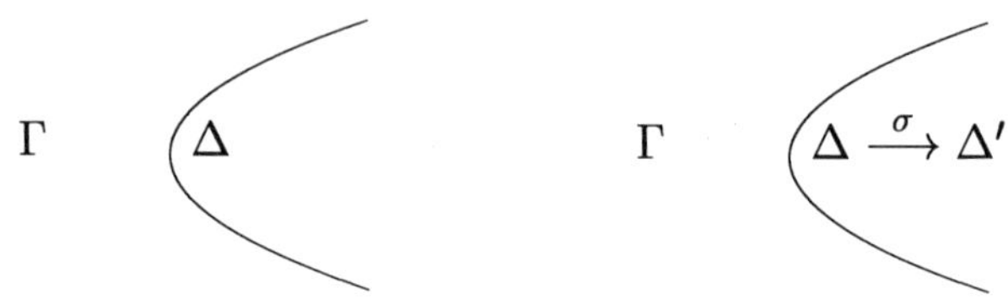

Figure 15.6. Fibred Models of Bunched Types

the comma is only a well-formed bunch in the presence of the left-hand branch, $x_1 : A_1$. This suggests that we need a judgment of the form

$$\frac{\Gamma \vdash_\Sigma \Delta \text{ bunch}}{\vdash_\Sigma \Gamma \circ \Delta \text{ bunch}},$$

where $\circ$ stands for (appropriate versions of) the monoidal operations, "," or ";".

This rule suggests that the general semantic structure required to capture the proof-theoretic structure of the type theory should be fibred, with sufficient structure in the fibres to interpret not merely types (and terms) over a context, or here bunch, but bunches themselves, dependent on the base bunch.

The essentially fibred structure of models raises another issue, however. If we follow the pattern of our Kripke resource models of $\lambda\Lambda$, then we interpret axiom sequents via second projection maps into fibres, so that a sequent of the form $\Gamma\,, x : A \vdash x : A$ is interpreted by a second projection, "$q_\otimes$" from $[\![\Gamma]\!] \otimes [\![A_\Gamma]\!]$ into the fibre over it and, similarly, a sequent of the form $\Gamma\,; x : A \vdash x : A$ is interpreted by a second projection, "$q_\times$" from $[\![\Gamma]\!] \times [\![A_\Gamma]\!]$ into the fibre over it. A consequence of this form of interpretation is the semantic admissibility of dereliction,

$$\frac{\Gamma\,, x : A \vdash M : B}{\Gamma\,; x : A \vdash M : B},$$

since the conditions for the existence of the second projection maps in the fibres are the same for both "," and ";". with the consequence that we can convert "," into ";" (though not *vice versa* since a proof starting with ";" may not satisfy the various conditions required for one starting with ",").[12]

It seems that the solution to this difficulty lies, essentially, in the form of judgement in this setting, namely

$$\Gamma \vdash_\Sigma \Delta : \mathsf{bunch},$$

with the corresponding typing judgement, $\Gamma \vdash_\Sigma \sigma : \Delta$, or more generally, perhaps

$$\Gamma \vdash_\Sigma \Delta \xrightarrow{\sigma} \Delta',$$

as sketched in Figure 15.6. The axiom judgement might then be taken to be something like

$$\Gamma \vdash_\Sigma 1_\Gamma : \Gamma,$$

where 1_Γ is the base identity $\langle x_1 \circ \ldots \circ x_m \rangle$ on $\Gamma = x_1 : A_1 \circ \ldots \circ x_m : A_m$, where $\circ$ is used to stand for (the appropriate notions of) "," or ";". Semantically, we require the interpretation of the identity on Γ to be an arrow in the fibre over Γ. Thus we should be able to avoid the need for the second projection maps in the fibres and the consequent admissibility of Dereliction.

The corresponding, multiplicative and additive, abstraction and application rules might then be something like

$$\frac{\Gamma, \Gamma' \vdash_\Sigma \sigma : \Delta}{\Gamma \vdash_\Sigma \lambda\Gamma'.\sigma : \Lambda\Gamma'.\Delta} \qquad \frac{\Gamma; \Gamma' \vdash_\Sigma \sigma : \Delta}{\Gamma \vdash_\Sigma \pi\Gamma'.\sigma : \Pi\Gamma'.\Delta}$$

and

$$\frac{\Gamma \vdash_\Sigma \sigma : \Lambda\Gamma'.\Delta \quad \Theta \vdash_\Sigma \rho : \Gamma'}{\Gamma\,, \Theta \vdash_\Sigma \mathbf{app}_\Lambda(\sigma, \rho) : \Delta[\rho/\Gamma']} \qquad \frac{\Gamma \vdash_\Sigma \sigma : \Pi\Gamma'.\Delta \quad \Theta \vdash_\Sigma \rho : \Gamma'}{\Gamma\,; \Theta \vdash_\Sigma \mathbf{app}_\Pi(\sigma, \rho) : \Delta[\rho/\Gamma']}$$

[12]The soundness of Dereliction is proved by induction on the structure of derivations: The projection maps give the base case.

Semantically, we require the existence of function spaces in the usual way (*q.v.* § 11).

Substitution (or Cut) would then be something like

$$\frac{\Gamma(\Gamma') \vdash_\Sigma \sigma : \Delta \quad \Theta \vdash_\Sigma \rho : \Gamma'}{\Gamma(\Theta) \vdash_\Sigma \textbf{let } x'_1 \circ \ldots \circ x'_m \textbf{ be } \rho \textbf{ in } \sigma : \Delta},$$

where $\Gamma' = x'_1 : A'_1 \circ \ldots \circ x'_m : A'_m$ and where the **let** – **be** – **in** – construct has reduction rules of the usual form,

$$\textbf{let } x_1 \circ \ldots \circ x_m \textbf{ be } \langle M_1 \circ \ldots \circ M_m \rangle \textbf{ in } \sigma(x_1 \circ \ldots \circ x_m) \succcurlyeq \sigma(M_1 \circ \ldots \circ M_m)$$

and

$$\mathbf{app}_\Pi(\lambda\Gamma.\sigma, \rho) \succcurlyeq \textbf{let } x_1 \circ \ldots \circ x_m \textbf{ be } \rho \textbf{ in } \sigma,$$

where $\Gamma = x_1 : A_1 \circ \ldots \circ x_m : A_m$ and $\rho = \langle M_1 \circ \ldots \circ M_m \rangle$.

A full development of a type theory along the line sketched herein is beyond out present scope.

However, we must surely be able to combine directly an arbitrary pair of bunches, say Γ and Δ, to form a third bunch Ξ. Such combinations are not a simple matter: Consider, recalling the discussion of § 3, how to define the rule of application for the multiplicative dependent function space,

$$\frac{\Gamma \vdash_\Sigma M : \Lambda x : A.B \quad \Delta \vdash_\Sigma N : A}{\Xi \vdash_\Sigma MN : B[N/x]},$$

where Ξ is somehow constructed by combining Γ and Δ. The combination is not simple because Γ and Δ must, in general, have variables in common, so that the typehood of A may be established on both the left- and right-hand branches of the proof. For example, recall Example 15.6 and suppose, in $\lambda\Lambda$, $A!\mathsf{Type}, c!A \multimap \mathsf{Type} \in \Sigma$. Then we construct the following instance of dependent function space application:

$$\frac{x{:}A \vdash_\Sigma \lambda z{:}cx.z : \Lambda z{:}cx.cx \quad x{:}A, y{:}cx \vdash_\Sigma y{:}cx}{x{:}A, y{:}cx \vdash_\Sigma (\lambda z{:}cx.z)y : cx}.$$

Here the key point is that the $x{:}A$ is required, in order to establish the type-hood of cx on both the left- and right-hand branches of the proof. In $\lambda\Lambda$, we regulate the multiplicative combination of variables in order to eliminate the duplication of $x{:}A$ in the conclusion and, as we have seen, in $\lambda\Lambda$ multiple occurrences of variables, standing for distinct occurrences of the *same* proof, are permitted; recall Example 15.5:

$$x{:}A, y{:}bx, x{:}A, z{:}cx \vdash_\Sigma dxyz : \mathsf{Type}.$$

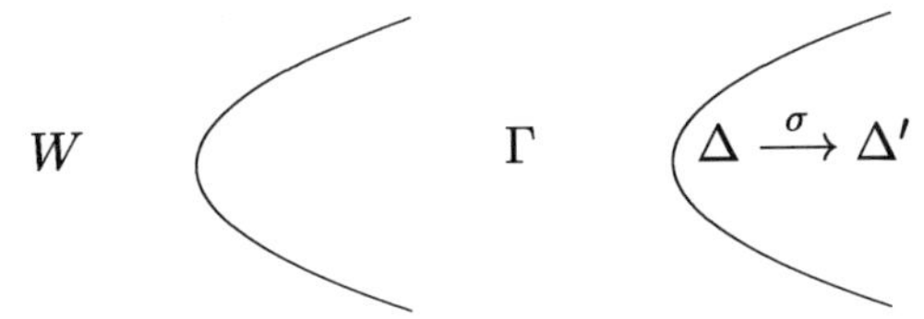

Figure 15.7. Kripke Models of Bunched Types

However, it remains open to fully understand the range of possibilities in this area.

We expect the character of the intended semantics, from the point of view of model-theoretic meaning rather than merely the translation of syntax, to be driven by the choice of structure to be carried by the collection (category) of worlds. Initially, following our existing treatments for **BI** and $\lambda\Lambda$, we intend a semantics in the style of Kripke's semantics for intuitionistic logic. To this end, we expect that an appropriate semantics for bunched dependent types would parametrize the fibred structure corresponding to the type system over a category of worlds such as a preordered monoid.

The meaning of a bunch varies, in general, from world to world. Accordingly, we suggest that an appropriate semantics structure will involve an indexing, in the sense of Kripke (resource) semantics of the fibred structure of typing over a category of worlds, as sketched in Figure 15.7.

We conclude this section by remarking that the Kripke-style models for dependent type theory that we have considered here are perhaps rather too closely tied to the syntaxes of the languages. Clearly, we should like to understand more abstract treatments of the semantics of susbstructural dependent type theories, understanding the Kripke-style models as leading examples and, ideally, obtaining suitable covering properties.

Chapter 16

THE SHARING INTERPRETATION, II

1. Logic Programming in Predicate BI

We have already seen, in Chapter 9, how proof-search in propositional **BI** gives rise to logic programming with a "sharing interpretation". However, a more substantial view of logic programming is based on predicate logic and, in **BI**, this too receives a semantics based on the sharing and non-sharing of resources. This section also represents joint work with Pablo Armelín.

Just as in the propositional case, our notion of logic programming is based on a goal-directed operational semantics realized by the class of uniform proofs, which is complete for hereditary Harrop sequents. In the predicate case, we have the following additional structure:

- The class of hereditary Harrop formulæ is extended as follows (again, we simplify for brevity):

$$\begin{array}{rrcl} \textit{Program clauses} & P & ::= & I \mid A \mid P \wedge P \mid G \to A \mid \\ & & & P * P \mid G \mathrel{-\!\!*} A \\ & & & \forall x.P \mid \forall_{\mathbf{new}} x.P \\ \\ \textit{Goals} & G & ::= & \top \mid I \mid A \mid G \wedge G \mid G \vee G \mid \\ & & & P \to G \mid G * G \mid P \mathrel{-\!\!*} G \\ & & & \exists x.P \mid \exists_{\mathbf{new}} x.P \end{array}$$

A sequent is said to be hereditary Harrop if it is of the form

$$(X)P \vdash G,$$

where P is a finite set of program clauses and G is a goal. As usual, we require only that X be sufficient to establish the well-formedness of $P \vdash G$.

- The identity group rules, *i.e.*, resolution and axiom, are now driven by substitutions, calculated by unification. Given an atomic goal, A, we invoke the program, using a resolution step. Consider first the intuitionistic setting. Suppose the program includes a proposition of the form $\forall x.G \to B$, in which B is atomic, such that there is a substitution σ for x such that $B\sigma = A$. Then we can immediately proceed to the *sub-goal* $G\sigma$:

  $$\frac{P \text{ ?- } G\sigma}{P \text{ ?- } A} \quad (\forall x.G \to B \text{ in } P,\ B\sigma = A\sigma),$$

  in which, as previously, we write ?- to denote putative consequence. The *answer substitution* for a computation is the composition of the substitutions calculated at identity group rules.

  In **BI**, the resolution rule is adapted to handle each kind (*i.e.*, additive or multiplicative combinations) of clauses. For example, consider a program which includes a clause of the form

  $$\forall_{\mathbf{new}} x.G \mathrel{-\!\!*} A.$$

  The resolution rule would then go as follows (again, we simplify, for brevity):

  $$\frac{\dfrac{\vdots}{(X)P \text{ ?- } G} \qquad \dfrac{}{(Y)A\sigma \text{ ?- } B\sigma}\,(A\sigma = B\sigma)\mathit{Axiom}}{(X,Y)P\,,\ \forall_{\mathbf{new}} y.G \mathrel{-\!\!*} A \text{ ?- } B}\,\mathit{Resolution}.$$

  It should be clear that the variables in the bunch Y required to form the substitution σ are *local* to the given resolution step. Were the program of the form $(X;Y)P\,;\ \forall y.G \mathrel{-\!\!*} A$, say, then the bunch used to form σ would be *global* (up to isomorphism) and could persist in the remaining computation, *i.e.*, up the left-hand branch.

- In intuitionistic logic, the "logical variables" in a goal arise from existential quantifications. In **BI**, we have two kinds of logical variable, each of which gives rise to a form of logical variable:

  - $(X)P$?- $\exists x.G$: in this case the answer substitution for x *cannot* share variables with the program, $\mathcal{P}$;

– $(X)P$?- $\exists_{\mathbf{new}} x.G$: in this case the answer substitution for x *may* share variables with the program, P.

We conjecture that the main use of free variables, together with the distinction between the multiplicative and additive quantifiers, on logic programming based on **BI** will be in the theory and implementation of) a system of modules (*cf.* Miller's notion of module in [Miller, 1981]), for which multiplicative signatures might also be useful.

EXAMPLE 16.1 ([ARMELÍN AND PYM, 2001]) *A logic programming language based on predicate* **BI** *is very good for modelling the interactions between opposing groups of entities, such as political parties or warring states. We give a simple example due to Pablo Armelín, whose expected Ph.D. thesis*[1] *should contain more details.*[2]

To indicate that two individuals, x and y, stand in a relation in which they may fight, we write[3]

$$\forall x, y.\mathrm{p}(x) * \mathrm{p}(y) \multimap\!\!* \,\mathrm{fight}(x, y)$$

We give a small program to calculate whether given individuals may fight, using the syntax of Armelín's implemented logic programming language, which we call here BLP, *based on the hereditary Harrop fragment of predicate* **BI**. *Here* p(x) *means that* x *is a person. The initial bunch structure says that* a1 *and* a2 *belong to the same group and that, for example,* a1 *and* a3 *belong to belligerent groups.* [–] *is used to denote additive (universal) quantification (*BLP *uses* [[–]] *to denote multiplicative (universal) quantification, not used here but see our remarks on modules).*

```
(p(a1);p(a2)),
(p(a3);p(a4)),
[x,y]fight(x,y) *- p(x) * p(y)
```

If more than two groups are involved, then the program must be modified. One way is to (additively) decorate each group (of individuals) with the multiplicative unit, thereby ensuring that it may be ignored:

[1]Department of Computer Science, Queen Mary, University of London.
[2]This example, like that of the logic for pointers discussed in Chapter 9, makes use only of additive quantifiers. Nevertheless, we include it in this chapter, rather than in Chapter 9, since the rôle of quantifiers seems more central. Moreover, the possible uses of the multiplicative quantifiers in this context are intriguing.
[3]We emphasize that in the present implementation, without free variables, we could use the multiplicative $\forall_{\mathbf{new}}$ here. Morally, however, it is $\forall_{\mathbf{new}}$ and would be essentially so in a system with free variables.

```
(p(a1);p(a2);I),
(p(a3);p(a4);I),
(p(a5);p(a6);I)
```

Another way is to add garbage collection to the main clause, adding $\top$*, written as* T *in* BLP*, so that once* x *and* y *have been found, and leftovers are disregarded:*

```
[x,y]fight(x,y) *- p(x) * p(y) * T
```

This latter solution is preferable because it avoids redundant solutions.

In contrast, the Prolog [Clocksin and Mellish, 1994, Clocksin, 1997] code for the problem of interacting groups would be something like

```
p(a1,t1).
p(a2,t2).
p(a3,t2).
p(a4,t2).
fight(X,Y) :- p(X,T) , p(Y,U) , T \= U.
```

Notice that it uses the tags T *and* U *which are not naturally part of the logical description of the problem, to distinguish the groups and must perform calculations with the tags.*

*Consider now that political parties sometimes split into rival factions and that each faction may want to keep its former allies (*i.e., *individuals with whom they do not fight). The following modification should be self-explanatory:*

```
(p(a1);p(a2)),
(p(a3);(p(a41);p(a42)),
((p(a43);p(a44))),
[x,y]fight(x,y) *- p(x) * p(y) * T
```

Notice that the defining clause required no modification. In contrast, the modification of the Prolog program would require something like the addition of an extra tag to reflect the extra structure:

```
p(a1,t1,_).
p(a2,t2,_).
p(a3,t2,_).
p(a4,t2,_).
fight(X,Y) :- p(X,T,_) , p(Y,U,_) , T \= U.
fight(X,Y) :- p(X,T,V) , p(Y,U,W) , T = U , V \= W.
```

Notice that we have had to modify the whole program to account for the extra, essentially non-logical, tag. In fact, one can do a little better in Prolog, by using lists of tags as a second argument:

```
p(a1,t1).
p(a2,t1).
p(a3,t2).
p(a41,[t2,t1]).
p(a42,[t2,t1]).
p(a43,[t2,t2]).
p(a44,[t2,t2]).

fight(X,Y) :- p(X,U) , p(Y,S) , mismatch(U,S).
mismatch([H1|_],[H2|_]) :- H1 \= H2.
mismatch([H1|T1],[H2|T2]) H1 = H2 , mismatch(T1,T2).
```

Nevertheless, the complexity of the solution in Prolog is somewhat greater than the one in BLP.

The bunched structure also helps to give fine control over the scope of predicates, specially implications. We can think of a variety of ways in which constants can be predicated. For example, a2 *might be a special kind of person. It would be possible to modify the program in the following way:*

```
(p(a1);q(a2);[x]p(x) <- q(x)),
(p(b1);(p(b21);q(b22)),(p(b23);p(b24))),
[x,y]fight(x,y)*- p(x)*p(y)*T
```

Now this program says that a2 *is a* q *but also that all* qs *are* ps. *However, this relation between* ps *and* qs *holds only for the group formed by* a1 *and* a2, i.e., *is* local *to that world. Other* qs *appearing in other places in the program, for example* b22, *will not be picked up by the* local *implication, or* $\rightarrow$, *(which matches the ";" combining* p(a1) *and* q(a2)*).*

The formal operational semantics of BLP is described in [Armelín and Pym, 2001].

The significance of predicate **BI** for modules in logic programming, in which we conjecture a substantial rôle for free variables, remains to be investigated.

2. ML with References in RLF

We present an encoding in the RLF logical of the the programming language ML [Milner et al., 1997] extended with references (MLR), a reworking of an example in Cervesato and Pfenning [Cervesato, 1996, Cervesato and Pfenning, 1996]. In our reworking, we exploit the use of the Λ which is not available to Cervesato and Pfenning. Consequently, we are in the full $\lambda\Lambda$-calculus type theory, in which κ's action is non-trivial.

The basic MLR logic judgement is of the form $S \triangleright K \vdash_{MLR} i \longrightarrow a$ which means: the program i is evaluated with the store S and continuation K and leaves an answer (a store–expression pair) a. The signature Σ_{MLR} begins with the declarations $store!\mathsf{Type}$, $cont!\mathsf{Type}$, $instr!\mathsf{Type}$ and $ans!\mathsf{Type}$ to represent the syntactic categories of store, continuations, expressions and answers. Evaluation is represented by the following declaration:

$$ev\,!\,cont \multimap instr \multimap answer \multimap \mathsf{Type}\ .$$

We are really only interested in the rule for evaluating re-assignment. This may be stated as follows:

$$\frac{S, c = v', S' \ \triangleright\ K \vdash_{\mathrm{MLR}} \bullet \longrightarrow A}{S, c = v, S' \ \triangleright\ K \vdash_{\mathrm{MLR}} \mathtt{ref}\ c := v' \longrightarrow A},$$

where $\bullet$ is the MLR unit expression.

The ML memory is modelled by a set of (cell,expression)-pairs. Each such pair is represented by a linear hypothesis of type *contains* which holds a lvalue (the cell) and its rvalue (the expression).

$$cell\,!\,\mathsf{Type} \qquad exp\,!\,\mathsf{Type} \qquad contains\,!\,cell \multimap exp \multimap \mathsf{Type}.$$

The rule for re-assignment evaluation is encoded as follows:

$$\begin{array}{lll} \text{EV-REASS} & ! & \Lambda\, c!cell\,.\,\Lambda\, v, v'!exp\,. \\ & & ((contains\ c\ v') \multimap (ev\ K\ \bullet\ A)) \multimap \\ & & (\Lambda\, v{:}exp\,.\,(contains\ c\ v)) \multimap (ev\ K\ (c := v')\ A), \end{array}$$

where the assignment instruction $c := v$ is shown in the usual (infix) form for reasons of readability. The rule may also be encoded in such a fashion that the linear property of the memory is formalized via the Λ quantifier. We will illustrate this idea soon. For now, based on our re-working of the MLR example, we can state the following by referring to [Cervesato and Pfenning, 1996].

THEOREM 16.2 (REPRESENTATION FOR MLR) *The encoding functions are compositional bijections: for all stores S of shape $\langle c_1, v_1\rangle, \ldots, \langle c_n, v_n\rangle$, continuations K, instructions i and answers A (which are closed except for possible occurrences of free cells),*

$$S \triangleright K \vdash_{\mathrm{MLR}} \Pi{:}i \longrightarrow a \quad \textit{if and only if}$$
$$\left[\begin{array}{l} c_1!\mathit{cell}, \ldots, c_n!\mathit{cell}, p_1{:}(\mathit{contains}\ c_1\ \mathcal{E}(v_1)), \\ \qquad \ldots, p_m{:}(\mathit{contains}\ c_m\ \mathcal{E}(v_m)) \end{array}\right] \vdash_{\Sigma_{\mathrm{MLR}}} M_\Pi{:}(\mathit{ev}\ \mathcal{E}(K)\ \mathcal{E}(i)\ \mathcal{E}(a)),$$

where Π is a proof object of MLR and M_Π is a canonical object of the $\lambda\Lambda$-calculus.

One property that it is desirable to show for the MLR logic is type preservation; in the context of a store Ω, if $S \triangleright K \vdash_{\mathrm{MLR}} i \longrightarrow a$, i is a valid instruction of type τ, K is a valid continuation of type $\tau \to \tau'$ and S is a valid store, then a is a valid answer of type τ. The main difference in our reworking of this example is how the proof of type preservation for the EV-REASS rule, prEV-REASS, is encoded.

$$\begin{array}{lll} \text{prEV-REASS} & ! & \Lambda c!\mathit{cell}\,.\,\Lambda v, v'!\mathit{exp}\,.\,\Lambda p{:}(\mathit{contains}\ c\ v)\,. \\ & & (\Lambda p'{:}(\mathit{contains}\ c\ v')\,.\,(\mathit{prCell}\ p'\ c\ v') \multimap (\mathit{ev}\ K\ \bullet\ A)) \multimap \\ & & (\mathit{prCell}\ p\ c\ v) \multimap (\mathit{prEv}\ K\ (x := v')\ A) \end{array}$$

In the above type, *prCell* and *prEv* are the proofs of type preservation over cells and for evaluations, respectively. We note that the types of p and p' have no linear free variables in them. That is, the type theory we have employed in the encoding does not involve the notion of sharing.

Now, the cells could have been quantified intuitionistically (as they are in [Cervesato and Pfenning, 1996]) instead of linearly. In that case, a sub-proof of $\Gamma \vdash_\Sigma$ prEV-REASS:U, where U is the above type of prEV-REASS, would consist of an instance of Π-introduction. However, this would allow us to admit garbage: (cell,expression)-pairs which are occupying memory space but not being used. The linear quantification gives us a better representation of memory management, *i.e.*, of garbage collection. The encoding above realizes the intuition that we are making general statements about linear variables, so the Λ and not the Π quantifier should be used.

The encoded version of MLR type preservation may be stated and shown as in [Cervesato and Pfenning, 1996]. We omit the details.

We conclude by remarking that this provides another example of an explicitly "spatial" interpretation of bunched logic.

References

[OED, 1976] (1976). *Concise Oxford Dictionary (Second Edition)*. Oxford University Press.

[Abramsky et al., 1992] Abramsky, S., Gabbay, D. M., and Maibaum, T. S. E., editors (1992). *Background: Computational Structures*, volume 2 of *Handbook of Logic in Computer Science*. Oxford University Press, Oxford, England.

[Amadio and Curien, 1998] Amadio, R. and Curien, P.-L. (1998). *Domains and Lambda-Calculi*. Cambridge University Press.

[Ambler, 1992] Ambler, S. (1992). First order linear logic in symmetric monoidal closed categories. PhD thesis, University of Edinburgh.

[Anderson and Belnap, 1975] Anderson, A. and Belnap, N. (1975). *Entailment: the Logic of Relevance and Necessity, volume I.* Princeton University Press.

[Anderson et al., 1992] Anderson, A., Dunn, J., and Belnap, N. (1992). *Entailment: the Logic of Relevance and Necessity, volume II.* Princeton University Press.

[Apt, 1989] Apt, K. (1989). Ten years of Hoare's logic: A survey. *ACM Transactions on Programming Languages and Systems*, 3(4):79–108.

[Armelín and Pym, 2001] Armelín, P. and Pym, D. (2001). Bunched logic programming (extended abstract). In *Proc. IJCAR 2001*, number 2083 in LNAI, pages 289–304. Springer.

[Avron, 1991] Avron, A. (1991). Simple consequence relations. *Information and Computation*, 91(1):105–139.

[Avron et al., 1992] Avron, A., Honsell, F., Mason, I., and Pollack, R. (1992). Using typed lambda calculus to implement formal systems on a machine. *Journal of Automated Reasoning*, 9:309–354.

[Avron et al., 1998] Avron, A., Honsell, F., Miculan, M., and Paravano, C. (1998). Encoding modal logics in a logical framework. *Studia Logica*, 60(1).

[Barber, 1996] Barber, A. (1996). Dual intuitionistic linear logic. Technical Report ECS-LFCS-96-347, University of Edinburgh.

[Barber and Plotkin, 1997] Barber, A. and Plotkin, G. (1997). Dual intuitionistic linear logic. Draft.

[Barendregt, 1992] Barendregt, H. (1992). Lambda calculi with types. In [Abramsky et al., 1992], pages 117–309.

[Barendregt, 1984] Barendregt, H. P. ((revised edition) 1984). *The Lambda Calculus: Its Syntax and Semantics*, volume 103 of *Studies in Logic and the Foundations of Mathematics*. North-Holland, Amsterdam.

[Barr, 1979] Barr, M. (1979). *$*$-autonomous categories*, volume 752 of *LNM*. Springer.

[Barr and Wells, 1995] Barr, M. and Wells, C. (1995). *Category Theory for Computing Science (second edition)*. Prentice-Hall International, London.

[Barwise, 1989] Barwise, J. (1989). Situations, facts, and true propositions. In *The Situation in Logic*, number 17 in CSLI Lecture Notes. CSLI Publications.

[Barwise and Perry, 1983] Barwise, J. and Perry, J. (1983). *Situations and attitudes*. MIT Press.

[Belnap, 1982] Belnap, N. (1982). Display logic. *Journal of Philosophical Logic*, 11:375–414.

[Bénabou, 1985] Bénabou, J. (1985). Fibered categories and the foundations of naive category theory. *J. Symbolic Logic*, 50:10–37.

[Benton et al., 1993] Benton, N., Bierman, G., de Paiva, V., and Hyland, M. (1993). A term calculus for intuitionistic linear logic. In Bezen, M. and Groote, J. F., editors, *Typed Lambda Calculi and Applications*, volume 664 of *Lecture Notes in Computer Science*, pages 75–90, Utrecht, The Netherlands. Springer-Verlag, Berlin.

[Benton et al., 1992] Benton, P., Bierman, G., de Paiva, V., and Hyland, J. (1992). Term assignment for intuitionistic linear logic (preliminary report). Technical report, University of Cambridge, Computer Laboratory. Report 262.

[Benton, 1994] Benton, P. N. (1994). A mixed linear and non-linear logic: proofs, terms and models (preliminary report). Technical Report 352, University of Cambridge Computer Laboratory.

[Bierman, 1995] Bierman, G. (1995). What is a categorical model of intuitionistic linear logic? In *Proceedings of Second International Conference on Typed λ-calculi and Applications*, volume 902 of *Lecture Notes in Computer Science*, pages 78–93. Springer-Verlag, Berlin.

[Boolos, 1998] Boolos, G. (1998). Don't eliminate cut. In Jeffrey, R., editor, *Logic, Logic, and Logic*, pages 365–369. Harvard University Press.

[Brookes et al., 1995] Brookes, S., Main, M., Melton, A., and Mislove, M., editors (1995). *Mathematical Foundations of Programming Semantics, Eleventh Annual Conference*, volume 1 of *Electronic Notes in Theoretical Computer Science*, Tulane University, New Orleans, Louisiana. Elsevier Science.

[Cardelli and Gordon, 2000] Cardelli, L. and Gordon, A. (Boston, Massachusetts, 2000). Anytime, anywhere: modal logics for mobile processes. In *Conference Record of the 27th. Annual ACM SIGPLAN-SIGACT Symposium on Pronciples of Programming Languages.* ACM, New York.

[Cartmell, 1994] Cartmell, J. (1994). Generalised algebraic theories and contextual categories. *Annals of Pure and Applied Logic*, 32:209–243.

[Cervesato, 1996] Cervesato, I. (1996). *A Linear Logical Framework.* Ph.D. thesis, Università di Torino.

[Cervesato and Pfenning, 1996] Cervesato, I. and Pfenning, F. (1996). A linear logical framework. In Clarke, E., editor, *Proc. 11th LICS, New Brunswick, NJ*, pages 264–275. IEEE Computer Society Press.

[Chellas, 1980] Chellas, B. (1980). *Modal Logic: an introduction.* Cambridge University Press.

[Clocksin, 1997] Clocksin, W. (1997). *Clause and effect.* Springer-Verlag.

[Clocksin and Mellish, 1994] Clocksin, W. and Mellish, C. (1994). *Programming in Prolog.* Springer-Verlag.

[Coquand, 1991] Coquand, T. (1991). An algorithm for testing conversion in type theory. In Huet, G. and Plotkin, G., editors, *Logical Frameworks*, pages 255–279. Cambridge University Press.

[Dam, 1990] Dam, M. F. (1990). *Relevance logic and concurrent computation.* Ph.D. thesis, University of Edinburgh.

[Day, 1970] Day, B. J. (1970). On closed categories of functors. In Mac Lane, S., editor, *Reports of the Midwest Category Seminar*, volume 137 of *Lecture Notes in Mathematics*, pages 1–38. Springer-Verlag, Berlin-New York.

[Day, 1973] Day, B. J. (1973). An embedding theorem for closed categories. In Dold, A. and Eckmann, B., editors, *Proceedings of the Sydney Category Seminar 1972/73*, volume 420 of *Lecture Notes in Mathematics*, pages 55–65. Springer-Verlag, Berlin.

[Devlin, 1990] Devlin, K. (1990). Infons and types in an information-based logic. In *Situation Theory and Its Applications (Volume 1)*, number 22 in CSLI Lecture Notes. CSLI Publications.

[Dummett, 1977] Dummett, M. (1977). *Elements of Intuitionism.* Oxford University Press.

[Dunn, 1975] Dunn, J. (1975). Conseqution formulation of positive R with cotenability and t. In [Anderson and Belnap, 1975], pp381–391.

[Dunn, 1986] Dunn, J. M. (1986). Relevant logic and entailment. In [Gabbay and Guenthner, 1986], pages 117–224.

[Eilenberg and Kelly, 1965] Eilenberg, S. and Kelly, G. M. (1965). Closed categories. In Eilenberg, S. et al., editors, *Proceedings of the Conference on Categorical Algebra*, pages 421–562, La Jolla, California. Springer-Verlag, New York, 1966.

[Engberg and Winskel, 1993] Engberg, U. and Winskel, G. (Gdansk, Poland, 1993). Completeness results for linear logic on Petri nets. In *Proceedings of the Conference on Mathematical Foundations of Computer Science*, volume 711 of *LNCS*, pages 442–452. Springer-Verlag.

[Fitting, 1983] Fitting, M. (1983). *Proof Methods for Modal and Intuitionistic Logics.* D. Reidel.

[Foltz et al., 1980] Foltz, F., Lair, C., and Kelly, G. M. (1980). Algebraic categories with few monoidal biclosed structures or none. *J. Pure and Applied Algebra*, 17:171–177.

[Gabbay, 1996] Gabbay, D. (1996). *Labelled Deductive Systems; principles and applications. Vol 1: Basic Principles.* Oxford University Press.

[Gabbay and Guenthner, 1986] Gabbay, D. and Guenthner, F., editors (1986). *Handbook of Philosophical Logic, vol. III: Alternatives to Classical Logic.* Number 166 in Synthese Library. D. Reidel, Dordrecht, Holland.

[Galmiche and Larchey-Wendling, 1998] Galmiche, D. and Larchey-Wendling, D. (1998). Provability in intuitionistic linear logic from a new interpretation on petri nets - extended abstract. *Electronic Notes in Theoretical Computer Science*, 17. 18 pages.

[Galmiche and Méry, 2001a] Galmiche, D. and Méry, D. (2001a). Proof-search and countermodel generation in propositional BI logic - extended abstract -. In *Proc. International Symposium on Theoretical Aspects of Computer Software TACS 2001, Sendai, Japan*, LNCS. Springer.

[Galmiche and Méry, 2001b] Galmiche, D. and Méry, D. (2001b). Semantic Tableaux for Propositional **BI**, I. Submitted. Title is provisional.

[Galmiche et al., 2001] Galmiche, D., Méry, D., and Pym, D. (2001). Semantic Tableaux for Propositional **BI**, II. Draft. Title is provisional.

[Galmiche et al., 2002] Galmiche, D., Méry, D., and Pym, D. (2002). Resource Tableaux. Manuscript, available at http://www.bath.ac.uk/~cssdjp.

[Galmiche and Pym, 2000] Galmiche, D. and Pym, D. (2000). Proof-search in type-theoretic languages: an introduction. *Theoretical Computer Science*, 232:5–53.

[Gentzen, 1934] Gentzen, G. (1934). Untersuchungen über das logische Schliessen. *Mathematische Zeitschrift*, 39:176–210, 405–431.

[Ghani, 1995] Ghani, N. (1995). $\beta\eta$-equality for coproducts. In *Typed Lambda-Calculi and Applications*, volume 902 of *LNCS*, pages 171–185. Springer Verlag.

[Gillies, 1996] Gillies, D. (1996). *Artificial intelligence and scientific method.* Oxford University Press.

[Girard, 1989] Girard, J. (1989). Towards a geometry of interaction. *Contemporary Mathematics 92: Categories in Computer Science and Logic*, 69–108.

[Girard, 1972] Girard, J.-Y. (1972). *Interprétation Fonctionnelle et Elimination des Coupures de l'Arithmétique d'Ordre Supérieur.* Thèse de doctorat d'état, Université Paris VII.

[Girard, 1987] Girard, J.-Y. (1987). Linear logic. *Theoretical Computer Science*, pages 1–102.

[Girard, 1993] Girard, J.-Y. (1993). On the unity of logic. *Annals of Pure and Applied Logic*, 59:201–217.

[Girard et al., 1989] Girard, J.-Y., Lafont, Y., and Taylor, P. (1989). *Proofs and Types*. Cambridge University Press.

[Goossens et al., 1994] Goossens, M., Mittelbach, F., and Samarin, A. (1994). *The LaTeX Companion*. Addison Wesley.

[Gray, 1974] Gray, J. W. (1974). *Formal Category Theory — Adjointness for 2-Categories*, volume 391 of *Lecture Notes in Math.* Springer.

[Hansen, 1973] Hansen, P. B. (1973). *Operating System Principles*. Prentice Hall.

[Harland and Pym, 1997] Harland, J. and Pym, D. (1997). Resource-distribution via Boolean constraints. In *Proc. CADE-14*, number 1249 in LNAI, pages 222–236. Springer.

[Harland et al., 1996] Harland, J., Pym, D., and Winikoff, M. (1996). Programming in Lygon: an overview. In Wirsing, M. and Nivat, M., editors, *Proc. AMAST '96*, volume 1101 of *LNCS*, pages 391–405. Springer.

[Harper et al., 1994] Harper, R., Sannella, D., and Tarlecki, A. (1994). Structured theory presentations and logic representations. *Ann. Pure Appl. Logic*, 67:113–160.

[Harper et al., 1987] Harper, R. W., Honsell, F., and Plotkin, G. D. (1987). A framework for defining logics (extended abstract). In *Proc. LICS 87*. IEEE Computer Society Press.

[Harper et al., 1993] Harper, R. W., Honsell, F., and Plotkin, G. D. (1993). A framework for defining logics. *Journal of the ACM*, 40(1):143–184.

[Hermida, 1993] Hermida, C. A. (1993). *Fibrations, Logical Predicates and Indeterminates*. PhD thesis, University of Edinburgh. Report CST-103-93, Department of Computer Science.

[Heyting, 1989] Heyting, A. (1989). *Intuitionism: An Introduction*. Cambridge University Press, Cambridge.

[Hoare, 1985] Hoare, C. A. R. (1985). *Communicating Sequential Processes*. Prentice-Hall International, London.

[Hodas and Miller, 1994] Hodas, J. and Miller, D. (1994). Logic programming in a fragment of intuitionistic linear logic. *Information and Computation*, 110(2):327–365.

[Hodges, 1993] Hodges, W. (1993). *Logic*. Penguin.

[Howard, 1980] Howard, W. (1980). The formulæ-as-types notion of construction. In [Seldin and Hindley, 1980], pages 479–490.

[Im and Kelly, 1986] Im, G. B. and Kelly, G. M. (1986). A universal property of the convolution monoidal structure. *J. Pure and Applied Algebra*, 43:75–88.

[Ishtiaq, 1999] Ishtiaq, S. (1999). A relevant analysis of natural deduction. PhD thesis, Queen Mary and Westfield College, University of London.

[Ishtiaq and O'Hearn, 2001] Ishtiaq, S. and O'Hearn, P. (2001). **BI** as an assertion language for mutable data structures. In *28th ACM-SIGPLAN Symposium on Principles of Programming Languages, London*, pages 14–26. Association for Computing Machinery.

[Ishtiaq and Pym, 1998] Ishtiaq, S. and Pym, D. (1998). A relevant analysis of natural deduction. *Journal of Logic and Computation*, 8(6):809–838.

[Ishtiaq and Pym, 1999] Ishtiaq, S. and Pym, D. (1999). Kripke resource models of a dependently-typed, bunched λ-calculus (extended abstract). In Flum, J. and Rodríguez-Artalejo, M., editors, *Computer Science Logic*, volume 1683 of *LNCS*, pages 235–249. Springer.

[Ishtiaq and Pym, 2000] Ishtiaq, S. and Pym, D. (2000). Corrections and remarks. Research Report RR-00-04, Department of Computer Science, Queen Mary and Westfield College, University of London, London. ISSN 1470-5559.

[Ishtiaq and Pym, 2001] Ishtiaq, S. and Pym, D. (2001). Kripke resource models of a dependently-typed, bunched λ-calculus. To appear: Journal of Logic and Computation. Manuscript available at http://www.bath.ac.uk/~cssdjp.

[Jacobs, 1998] Jacobs, B. (1998). *Categorical Logic and Type Theory.* Elsevier.

[Jay, 1989a] Jay, C. (1989a). Languages for monoidal categories. *Journal of Pure and Applied Algebra*, 59(1):61–85.

[Jay, 1989b] Jay, C. (1989b). A note on natural numbers objects in monoidal categories. *Studia Logica*, XLVIII(3).

[Jay, 1990] Jay, C. (1990). The structure of free closed categories. *Journal of Pure and Applied Algebra*, 66:271–285.

[Johnstone, 1980] Johnstone, P. (1980). Open maps of toposes. *Manuscripta Mathematica*, 31:217–247.

[Kant, 1800] Kant, I. (1800). *Immanuel Kants Logik (Edited by G.B. Jäsch).* Friedrich Nicolovius, Königsberg. In translation: R.S. Hartman and W. Schwarz, Dover Publications, Inc., 1988.

[Kelly, 1982] Kelly, G. (1982). *Basic Concepts of Enriched Category Theory.* Cambridge University Press.

[Kleene, 1968] Kleene, S. (1968). *Mathematical Logic.* Wiley and Sons.

[Kowalski, 1979] Kowalski, R. (1979). *Logic for Problem-solving.* North-Holland, Elsevier.

[Kripke, 1965] Kripke, S. A. (1965). Semantical analysis of intuitionistic logic I. In Crossley, J. N. and Dummett, M. A. E., editors, *Formal Systems and Recursive Functions*, pages 92–130. North-Holland, Amsterdam.

[Lambek, 1958] Lambek, J. (1958). The mathematics of sentence structure. *American Mathematical Monthly*, 65:154–170.

[Lambek, 1968] Lambek, J. (1968). Deductive Systems and Categories I. *J. Math. Systems Theory*, 2:278–318.

[Lambek, 1969] Lambek, J. (1969). Deductive Systems and Categories II. *Springer LNM*, 86:76–122.

[Lambek, 1972] Lambek, J. (1972). Deductive Systems and Categories III. *Springer LNM*, 274:57–82.

[Lambek, 1993] Lambek, J. (1993). From categorial grammar to bilinear logic. In Schroeder-Heister, P. and Došen, K., editors, *Substructural Logic*, pages 207–238. Oxford University Press.

[Lambek and Scott, 1986] Lambek, J. and Scott, P. (1986). *Introduction to Higher-Order Categorical Logic*. Cambridge University Press.

[Larchey-Wendling and Galmiche, 2000] Larchey-Wendling, D. and Galmiche, D. (2000). Quantales as completions of ordered monoids: Revised semantics for intuitionistic linear logic. *Electronic Notes in Theoretical Computer Science*, 35. 15 pages.

[Lawvere, 1969] Lawvere, F. W. (1969). Adjointness in foundations. *Dialectica*, 23:281–296.

[Mac Lane, 1971] Mac Lane, S. (1971). *Categories for the Working Mathematician*. Springer-Verlag, New York.

[Mac Lane and Moerdijk, 1992] Mac Lane, S. and Moerdijk, I. (1992). *Sheaves in Geometry and Logic*. Springer-Verlag, New York.

[Martin-Löf, 1996] Martin-Löf, P. (1996). On the meanings of the logical constants and the justifications of the logical laws. Also Technical Report 2, Scuola di Specializziazione in Logica Matematica, Università di Siena, 1982.

[Mason, 1986] Mason, I. (1986). Hoare's Logic in LF. Technical Report ECS-LFCS-87-32, Laboratory for Foundations of Computer Science, Department of Computer Science, University of Edinburgh, The King's Buildings, Edinburgh EH9 3JZ, Scotland, U.K.

[Mendelson, 1987] Mendelson, E. (1987). *Introduction to Mathematical Logic*. Van Nostrand, Princeton.

[Meyer, 1982] Meyer, A. (1982). What is a model of the lambda calculus ? *Information and Control*, 52:87–122.

[Miller, 1981] Miller, D. (1981). A logical analysis of modules in logic programming. *J. Logic. Programming*, 6(1& 2):431–483.

[Miller et al., 1991] Miller, D., Nadathur, G., Pfenning, F., and Ščedrov, A. (1991). Uniform proofs as a foundation for logic programming. *Annals of Pure and Applied Logic*, 51:125–157.

[Milner, 1975] Milner, R. (1975). Processes: a mathematical model of computing agents. In Rose, H. E. and Shepherdson, J. C., editors, *Logic Colloquium'73*, pages 157–174. North-Holland, Amsterdam.

[Milner, 1989] Milner, R. (1989). *Communication and Concurrency.* Prentice Hall, New York.

[Milner, 1999] Milner, R. (1999). *Communicating and mobile systems: The pi-calculus.* Cambridge University Press.

[Milner et al., 1997] Milner, R., Tofte, M., Harper, R., and MacQueen, D. (1997). *The Definition of Standard ML (Revised).* MIT Press.

[Mitchell and Moggi, 1981] Mitchell, J. and Moggi, E. (1981). Kripke-style models for typed lambda calculus. *Annals of Pure and Applied Logic*, 51:99–124.

[O'Hearn, 2000] O'Hearn, P. (2000). On Bunched Typing. Manuscript.

[O'Hearn, 1999] O'Hearn, P. (LNCS 1581, 1999). Resource interpretations, bunched implications and the $\alpha\lambda$-calculus (preliminary version). In Girard, J.-Y., editor, *Proc TLCA '99.* Springer-Verlag.

[O'Hearn and Pym, 1999] O'Hearn, P. and Pym, D. (June 1999). The logic of bunched implications. *Bulletin of Symbolic Logic*, 5(2):215–244.

[O'Hearn et al., 1995] O'Hearn, P. W., Power, A. J., Takeyama, M., and Tennent, R. D. (1995). Syntactic control of interference revisited. In [Brookes et al., 1995]. Also in [O'Hearn and Tennent, 1997a], pages 189–226.

[O'Hearn et al., 1999] O'Hearn, P. W., Power, A. J., Takeyama, M., and Tennent, R. D. (1999). Syntactic control of interference revisited. *Theoretical Computer Science*, 228(1-2):211–252. Preliminary version in [Brookes et al., 1995] and [O'Hearn and Tennent, 1997a], vol 2.

[O'Hearn and Tennent, 1997a] O'Hearn, P. W. and Tennent, R. D., editors (1997a). *Algol-like Languages*, volume 2. Birkhäuser, Boston.

[O'Hearn and Tennent, 1997b] O'Hearn, P. W. and Tennent, R. D., editors (1997b). *Algol-like Languages*, volume 1. Birkhäuser, Boston.

[Pinto and Dyckhoff, 1985] Pinto, L. and Dyckhoff, R. (1985). Loop-free construction of counter-models for intuitionistic propositional logic. In Behara/Fritsch/Lintz, E., editor, *Symposia Gaussiana, Conf. A.*, pages 225–232. Walter de Gruyter and Co., Berlin-New York.

[Pitts, 1992] Pitts, A. (1992). Categorical logic. In Abramsky, S., Gabbay, D., and Maibaum, T., editors, *Handbook of Logic in Computer Science, Volume 6*, pages 264–275. Oxford University Press.

[Plotkin, 1978] Plotkin, G. D. (1978). *The category of complete partial orders: a tool for making meanings.* Lecture notes for the Summer School on Foundations of Artificial Intelligence and Computer Science, Pisa.

[Plotkin, 1980] Plotkin, G. D. (1980). Lambda definability in the full type hierarchy. In [Seldin and Hindley, 1980], pages 363–373.

[Polakow and Pfenning, 1999] Polakow, J. and Pfenning, F. (1999). Natural deduction for intuitionistic non-commutative linear logic. In Girard, J.-Y., editor, *Proceedings of the Fourth International Conference on Typed Lambda-Calculi and Applications, LNCS 1581*, pages 295–309. Springer-Verlag.

[Prawitz, 1965] Prawitz, D. (1965). *Natural Deduction: A Proof-Theoretical Study.* Almquist and Wiksell, Stockholm.

[Prawitz, 1971] Prawitz, D. (1971). Ideas and results in proof theory. In *Proceedings of the Second Scandinavian Logic Symposium.* North Holland.

[Prawitz, 1978] Prawitz, D. (1978). Proofs and the meaning and completeness of the logical constants. In J. Hintikka, J. N. and Saarinen, E., editors, *Essays on mathematical anmd philosophical logic*, pages 25–40. D. Rediel, Dordrecht.

[Pym, 1990] Pym, D. (1990). *Proofs, Search and Computation in General Logic.* Ph.D. thesis, Univ of Edinburgh.

[Pym, 1992] Pym, D. (1992). A relevant analysis of natural deduction. Lecture at EU Types Workshop, Baastad, Sweden.

[Pym, 1995a] Pym, D. (1995a). Functorial Kripke models of the $\lambda\Pi$-calculus. Invited Lecture, Newton Institute (Cambridge), Semantics of Computation Programme, Workshop on Category Theory and Logic Programming.

[Pym, 1995b] Pym, D. (1995b). A note on the proof theory [of] the $\lambda\Pi$-calculus. *Studia Logica*, 54:199–230.

[Pym, 1996] Pym, D. (1996). A note on representation and semantics in logical frameworks. In *Proc. CADE-13 Workshop, Proof-search in type-theoretic languages.*

[Pym, 1999] Pym, D. (1999). On bunched predicate logic. In *Proc. LICS'99*, pages 183–192. IEEE Computer Society Press.

[Pym, 2000a] Pym, D. (2000a). Functorial Kripke-Beth-Joyal Models of the $\lambda\Pi$-calculus I: Type Theory and Internal Logic. Manuscript.

[Pym, 2000b] Pym, D. (2000b). Functorial Kripke-Beth-Joyal Models of the $\lambda\Pi$-calculus II: The LF Logical Framework. Manuscript.

[Pym, 2000c] Pym, D. (2000c). Functorial Kripke-Beth-Joyal Models of the $\lambda\Pi$-calculus III: Logic Programming and Its Semantics. Manuscript.

[Pym, 2001] Pym, D. (2001). Notes towards a semantics for proof-search. *Electronic Notes in Theoretical Computer Science*, 37:18 pages. http://www.elsevier.nl/locate/entcs/volume37.html.

[Pym and Harland, 1994] Pym, D. and Harland, J. (1994). A uniform proof-theoretic investigation of linear logic programming. *J. Logic. Computat.*, 4:175–207.

[Pym et al., 2000] Pym, D., O'Hearn, P., and Yang, H. (2000). Possible worlds and resources: The semantics of **BI**. Manuscript. Available at http://www.bath.ac.uk/~cssdjp.

[Pym and Ritter, 2001] Pym, D. and Ritter, E. (2001). On the semantics of classical disjunction. *Journal of Pure and Applied Algebra*, 159:315–338.

[Pym and Wallen, 1991] Pym, D. and Wallen, L. (1991). Proof-search in the $\lambda\Pi$-calculus. In Huet, G. and Plotkin, G., editors, *Logical Frameworks*, pages 309–340. Cambridge University Press.

[Pym and Wallen, 1992] Pym, D. and Wallen, L. (1992). Logic programming via proof-valued computations. In Broda, K., editor, *ALPUK92, Proc. 4th U.K. Conference on Logic Programming*, pages 253–262. Springer Verlag.

[Read, 1988] Read, S. (1988). *Relevant Logic: A Philosophical Examination of Inference.* Basil Blackwell.

[Read, 2000] Read, S. (2000). Truthmakers, disjunction and necessity. In Wansing, H., editor, *Essays in Non-classical Logic.*

[Reisig, 1998] Reisig, W. (1998). *Distributed Algorithms: Modelling and Analyis with Petri Nets.* Springer.

[Restall, 1999] Restall, G. (1999). *An Introduction to Substructural Logics.* Routledge.

[Retoré, 1998] Retoré, C. (1998). Pomset logic: a non-commutaive extension of classical linear logic. In *Computer Science Logic, Paderborn, 1995*, LNCS. Springer.

[Reynolds, 2000] Reynolds, J. (2000). Lectures on reasoning about shared mutable data structure. Tandil, Argentina.

[Reynolds, 1978] Reynolds, J. C. (1978). Syntactic control of interference. In *Conference Record of the Fifth Annual ACM Symposium on Principles of Programming Languages*, pages 39–46, Tucson, Arizona. ACM, New York. Also in [O'Hearn and Tennent, 1997b], pages 273–286.

[Reynolds, 1981] Reynolds, J. C. (1981). The essence of Algol. In de Bakker, J. W. and van Vliet, J. C., editors, *Algorithmic Languages*, pages 345–372, Amsterdam. North-Holland, Amsterdam. Also in [O'Hearn and Tennent, 1997b], pages 67-88.

[Ritter et al., 2000] Ritter, E., Pym, D., and Wallen, L. (2000). On the intuitionistic force of classical search. *Theoretical Computer Science*, 232:299–333.

[Ruet and Fages, 1998] Ruet, P. and Fages, F. (1998). Concurrent constraint programming and non-commutative logic. In *Computer Science Logic '97*, LNCS. Springer.

[Salvesen, 1990] Salvesen, A. (1990). A proof of the Church-Rosser property for the Edinburgh LF with η-conversion. Lecture given at the First Workshop on Logical Frameworks, Sophia-Antipolis, France, May 1990.

[Schroeder-Heister, 1983] Schroeder-Heister, P. (1983). Generalised rules for quantifiers and the completness of the intuitionistic operators &, ∨, ⊃, ⅄, ∀, ∃. In *et al.*, M. R., editor, *Computation and Proof Theory, Logic Coloquium Aachen*, volume 1104 of *LNM*, pages 399–426. Springer-Verlag.

[Scott, 1974] Scott, D. (1974). Rules and derived rules. In Stenlund, S., editor, *Logical theory and semantical analysis*, pages 147–161. Reidel: Dordrecht.

[Seely, 1983] Seely, R. A. G. (1983). Hyperdoctrines, natural deduction and the Beck condition. *Zeitschr. für Math. Logik und Grundlagen der Math.*, 29:505–542.

[Seely, 1984] Seely, R. A. G. (1984). Locally-cartesian closed categories and type theory. *Math. Proc. Camb. Philos. Soc.*, 95:33–48.

[Seldin and Hindley, 1980] Seldin, J. P. and Hindley, J. R., editors (1980). *To H. B. Curry: Essays in Combinatory Logic, Lambda Calculus and Formalism.* Academic Press.

[Statman, 1982] Statman, R. (1982). Completeness, invariance and lambda-definability. *J. Symbolic Logic*, 47:17–26.

[Statman, 1985a] Statman, R. (1985a). Equality between functionals. In *Harvey Friedman's Research on the Foundations of Mathematics.* North Holland.

[Statman, 1985b] Statman, R. (1985b). Logical relations and the typed λ-calculus. *Information and Computation*, 65:85–97.

[Streicher, 1988] Streicher, T. (1988). Correctness and completeness of a categorical semantics of the calculus of constructions. PhD thesis, Universität Passau, 1988.

[Sundholm, 1986] Sundholm, G. (1986). Proof theory and meaning. In [Gabbay and Guenthner, 1986], pages 471–506.

[Szabo, 1978] Szabo, M. (1978). *Algebra of Proofs.* North Holland, Amsterdam.

[Tait, 1967] Tait, W. (1967). The intensional interpretation of functionals of finite type. *J. Symbolic Logic*, 32.

[Tarski, 1956] Tarski, A. (1956). *Logic, Semantics, Metamathematics.* Oxford University Press, Oxford.

[Taylor, 2002] Taylor, P. (2002). "diagrams" and "prooftree" packages for LaTeX. Available from www.ctan.org, occasionally revised.

[Troelstra and Schwichtenberg, 1996] Troelstra, A. and Schwichtenberg, H. (1996). *Basic Proof Theory.* Cambridge University Press, Cambridge.

[Trolestra, 1992] Trolestra, A. (1992). *Lectures on Linear Logic.* Number 29 in Lecture Notes. CSLI.

[Urquhart, 1972] Urquhart, A. (1972). Semantics for relevant logics. *Journal of Symbolic Logic*, pages 1059–1073.

[van Daalen, 1980] van Daalen, D. T. (1980). *The Language Theory of AUTOMATH.* PhD thesis, Technical University of Eindhoven, The Netherlands.

[van Dalen, 1983] van Dalen, D. (1983). *Logic and Structure.* Springer, Berlin, second edition.

[van Dalen, 1986] van Dalen, D. (1986). Intuitionistic logic. In [Gabbay and Guenthner, 1986], pages 225–339.

[Wainer and Wallen, 1992] Wainer, S. and Wallen, L. (1992). Basic proof theory. In P. Aczel, H. S. and Wainer, S., editors, *Proof Theory*, pages 1–26. Cambridge University Press.

[Warner, 1983] Warner, F. W. (1983). *Foundations of Differentiable Manifolds and Lie Groups.* Springer-Verlag.

[Winikoff and Harland, 1995] Winikoff, M. and Harland, J. (1995). Implementing the logic programming language Lygon. In Lloyd, J., editor, *Proc. ILPS '95.* MIT Press.

[Winskel, 1993] Winskel, G. (1993). *The Formal Semantics of Programming Languages: An Introduction.* The MIT Press, Cambridge, Mass., and London, England.

[Yetter, 1990] Yetter, D. (1990). Quantales and (non-commutative) linear logic. *J. Symb. Logic*, 55(1):41–64.

Index

APPLIED LOGIC SERIES

1. D. Walton: *Fallacies Arising from Ambiguity.* 1996 ISBN 0-7923-4100-7
2. H. Wansing (ed.): *Proof Theory of Modal Logic.* 1996 ISBN 0-7923-4120-1
3. F. Baader and K.U. Schulz (eds.): *Frontiers of Combining Systems.* First International Workshop, Munich, March 1996. 1996 ISBN 0-7923-4271-2
4. M. Marx and Y. Venema: *Multi-Dimensional Modal Logic.* 1996 ISBN 0-7923-4345-X
5. S. Akama (ed.): *Logic, Language and Computation.* 1997 ISBN 0-7923-4376-X
6. J. Goubault-Larrecq and I. Mackie: *Proof Theory and Automated Deduction.* 1997 ISBN 0-7923-4593-2
7. M. de Rijke (ed.): *Advances in Intensional Logic.* 1997 ISBN 0-7923-4711-0
8. W. Bibel and P.H. Schmitt (eds.): *Automated Deduction - A Basis for Applications.* Volume I. Foundations - Calculi and Methods. 1998 ISBN 0-7923-5129-0
9. W. Bibel and P.H. Schmitt (eds.): *Automated Deduction - A Basis for Applications.* Volume II. Systems and Implementation Techniques. 1998 ISBN 0-7923-5130-4
10. W. Bibel and P.H. Schmitt (eds.): *Automated Deduction - A Basis for Applications.* Volume III. Applications. 1998 ISBN 0-7923-5131-2
    (Set vols. I-III: ISBN 0-7923-5132-0)
11. S.O. Hansson: *A Textbook of Belief Dynamics.* Theory Change and Database Updating. 1999 Hb: ISBN 0-7923-5324-2; Pb: ISBN 0-7923-5327-7
    Solutions to exercises. 1999. Pb: ISBN 0-7923-5328-5
    Set: (Hb): ISBN 0-7923-5326-9; (Pb): ISBN 0-7923-5329-3
12. R. Pareschi and B. Fronhöfer (eds.): *Dynamic Worlds from the Frame Problem to Knowledge Management.* 1999 ISBN 0-7923-5535-0
13. D.M. Gabbay and H. Wansing (eds.): *What is Negation?* 1999 ISBN 0-7923-5569-5
14. M. Wooldridge and A. Rao (eds.): *Foundations of Rational Agency.* 1999 ISBN 0-7923-5601-2
15. D. Dubois, H. Prade and E.P. Klement (eds.): *Fuzzy Sets, Logics and Reasoning about Knowledge.* 1999 ISBN 0-7923-5911-1
16. H. Barringer, M. Fisher, D. Gabbay and G. Gough (eds.): *Advances in Temporal Logic.* 2000 ISBN 0-7923-6149-0
17. D. Basin, M.D. Agostino, D.M. Gabbay, S. Matthews and L. Viganò (eds.): *Labelled Deduction.* 2000 ISBN 0-7923-6237-3
18. P.A. Flach and A.C. Kakas (eds.): *Abduction and Induction.* Essays on their Relation and Integration. 2000 ISBN 0-7923-6250-0
19. S. Hölldobler (ed.): *Intellectics and Computational Logic.* Papers in Honor of Wolfgang Bibel. 2000 ISBN 0-7923-6261-6

APPLIED LOGIC SERIES

20. P. Bonzon, M. Cavalcanti and Rolf Nossum (eds.): *Formal Aspects of Context.* 2000 ISBN 0-7923-6350-7
21. D.M. Gabbay and N. Olivetti: *Goal-Directed Proof Theory*. 2000 ISBN 0-7923-6473-2
22. M.-A. Williams and H. Rott (eds.): *Frontiers in Belief Revision*. 2001 ISBN 0-7923-7021-X
23. E. Morscher and A. Hieke (eds.): *New Essays in Free Logic*. In Honour of Karel Lambert. 2001 ISBN 1-4020-0216-5
24. D. Corfield and J. Williamson (eds.): *Foundations of Bayesianism*. 2001 ISBN 1-4020-0223-8
25. L. Magnani, N.J. Nersessian and C. Pizzi (eds.): *Logical and Computational Aspects of Model-Based Reasoning*. 2002 Hb: ISBN 1-4020-0712-4; Pb: ISBN 1-4020-0791-4
26. D.J. Pym: *The Semantics and Proof Theory of the Logic of Bunched Implications*. 2002 ISBN 1-4020-0745-0
27. P.B. Andrews: *An Introduction to Mathematical Logic and Type Theory: To Truth Through Proof.* Second edition. 2002 ISBN 1-4020-0763-9

KLUWER ACADEMIC PUBLISHERS – DORDRECHT / BOSTON / LONDON